Novel Results in Particle Physics

(Vanderbilt, 1982)

AIP Conference Proceedings
Series Editor: Hugh C. Wolfe
Number 93
Particles and Fields Subseries No. 27

Novel Results in Particle Physics

(Vanderbilt, 1982)

Edited by
R. S. Panvini, M. S. Alam, S. E. Csorna
Vanderbilt University

American Institute of Physics
New York 1982

L.C. Catalog Card No. 82-73954
ISBN 0-88318-192-4
DOE CONF- 8205116

FOREWORD

This was the fifth international conference on particle physics at Vanderbilt, held May 22-24, 1982. The first conference in 1973 occurred shortly after the CERN ISR and Fermilab (then NAL) began yielding data and many of the talks at that time were from experiments with the new machines. In 1980 we had a topical conference on e^+e^- interactions; this year the title and theme of the conference was Novel Results in Particle Physics. Results presented this time included such diverse topics as quark searches (and discoveries?), evidence for a monopole, proton decay experiments, and the latest data from the new CERN proton-antiproton collider. Significant results from more conventional experiments were also presented. The theory talks stressed the physics of 100 GeV and beyond.

This conference was organized with the expert advice from many people. E. Berger organized the theory talks and C. Baltay was our expert on neutrino experiments; both also helped in planning other aspects of the conference. D. Coyne and B. Wiik advised us on e^+e^- experiments and a variety of people including D. Berley, J. Orear, and R. Slansky made suggestions and helped organize the quark search talks. K. Moffeit and N. Reay advised us about charmed particle lifetime experiments.

Many aspects of the local organization of this conference were handled by my colleagues M. S. Alam and S. E. Csorna. Finally, and most importantly, credit goes again to my wife Doria who really made the conference possible by contributing her considerable organizational skills and personal commitment towards getting the job done.

With all of the energy and enthusiasm provided by the people mentioned above, the bills could not have been paid without the assistance of grants from the National Science Foundation, the Department of Energy, and the Vanderbilt University Research Council.

Robert S. Panvini
Conference Chairman
September 1982

TABLE OF CONTENTS

The Search for Fractionally Charged Particles

Klaus S. Lackner[*] *and George Zweig*[**]
Los Alamos National Laboratory
Los Alamos, NM 87545

1. Do Fractionally Charged Particles Exist in Isolation?

Quarks, the constituents of hadrons and the fermion fields of quantum chromodynamics, have fractional charges $-\frac{1}{3}e$ and $\frac{2}{3}e$. All charges are integral multiples of $\frac{1}{3}e$ and not e, as was previously believed. Therefore it is natural to ask if isolated particles of fractional charge exist, either as an intrinsic part of matter, or as particles that can be produced at high energy accelerators. This question can only be answered by experiment, and remains interesting even if quantum chromodynamics turns out to be an absolutely confining theory of quarks. For example, small deviations from the standard version of quantum chromodynamics [1], or the incorporation of quantum chromodynamics into a more comprehensive theory [2–12], could require the existence of free fractionally charged particles [1)].

---

[*] Presently at the Stanford Linear Accelerator
[**] On leave from the California Institute of Technology

[1)] "There are more things in heaven and earth, Horatio,
Than are dreamt of in your philosophy."
[Shakespeare: *Hamlet* I. v.]

2. The Search for Fractionally Charged Particles.

Quark production in high energy reactions is expected to be prohibitively small. Even if unconfined quarks exist, their production cross section has been estimated [13, 14] to be

$$\sigma \propto \exp\left\{-\frac{M_q^2}{M_0^2}\right\},$$

where M_q is the mass of the isolated quark and M_0 is a symmetry breaking mass somewhere in the interval 0.3 to 3 GeV/c^2. The basic assumption underlying this estimate is that a quark-antiquark pair is connected by a string that has a certain probability, per unit length per unit time, of breaking into two mesons. This string of gluonic flux fades out at distances larger than the symmetry breaking length $^1/_{M_0}$. Free quarks are produced if the quark-antiquark pair is separated further than this length, without breaking the string that binds them.

In any case, sufficiently heavy fractionally charged particles cannot be produced at accelerators since the threshold energy for their production cannot be reached.

On the other hand, the "big bang" provides a potential source of fractionally charged particles, for example free quarks [14−24]. Presumably there was a time very early in the history of the universe when the mean separation between quarks was much less than their mean separation in hadrons. As the universe expanded and cooled, quarks eventually clustered in groups of three, forming protons and neutrons. In a universe with free quarks, not all quarks found partners. These isolated quarks (or their fractionally charged decay products) interacted with nucleons and the helium and lithium nuclei present, possibly forming complex systems of relatively small fractional charge. Subsequently a substantial fraction of the matter from the early universe entered stars, enabling quarks or other fractionally charged particles to interact with heavier nuclei.

The "big bang" could, in fact, be a copious source for the production of fractionally charged particles. Fractionally charged particles that do not interact strongly have small annihilation cross sections. Therefore they survive intact with high probability [25]. Strongly interacting particles may also be copiously produced through color fluctuations [26].

Presently, the search for fractionally charged particles appears to be more promising in matter than at accelerators.

3. The Search in Matter.

In a series of publications [27 – 30] now spanning many years, the Fairbank group at Stanford has insisted that isolated fractionally charged particles exist as an intrinsic part of matter. The evidence presented is impressive. Although older experiments seemingly contradict their results, careful analysis of the experimental conditions has shown that this is definitely not the case [31]. Furthermore, reviewers [32 – 35] have over-simplified the results of these experiments, making them appear to be more negative than their authors originally claimed. Reviewers have also perpetuated inaccurate views of the chemistry of fractionally charged atoms, encouraging poorly designed experiments. Finally, older experiments [36 – 39], performed for other reasons, have been inappropriately interpreted by reviewers as establishing stringent upper limits on the concentration of fractionally charged particles.

A striking example of misinterpretation is provided by the 1959 Hillas-Cranshaw experiment [37] which was designed to check the cancellation of electron and proton charges. At the time of this experiment there were speculations that a small residual charge ($\approx 10^{-18} e$) on the hydrogen atom would drive the expansion of the universe. However, Hillas and Cranshaw showed this residual charge to be less than $(1\pm3)\times10^{-20}e$. Subsequently this near equality of charges was interpreted to give an upper limit of 10^{-22} fractionally charged particles per nucleon [32–35]. This was reputedly one of the best upper limits on the concentration of fractionally charged particles in matter, in apparent contradiction with the Fairbank group's lower limit of 1 per 10^{20}.

However, there is an error in this reinterpretation of the experiment of Hillas and Cranshaw. They essentially measured the bulk charge of a gas streaming out of a metal container by determining the change of charge in the container with an electrometer. This change in charge sets an upper limit on the average charge carried away by each gas atom or molecule. This was later thought to set a limit on the number of fractionally charged particles leaving the container [32–35].

However, any fractionally charged particle inside the container would, either through convection or diffusion, have found its way to the metallic walls where it would have been bound by its image charge. Therefore, it would not have been free to escape from the container with the pressurized gas. Other objections to the reinterpretation of the Hillas-Cranshaw experiment have also been discussed by Lyons [40]. This experiment should never have found its way into lists of searches for fractionally charged particles.

Even less comprehensible is the fact that another experiment [39], which measures the cancellation of electron and proton charge by deflecting a beam of molecules in an electric field strong enough to detect charges smaller than $10^{-18}e$ has been quoted [33] as yielding a limit on the concentration of fractionally charged particles.

Although many experimenters emphasize that their experiments are sensitive only to certain kinds of fractionally charged particles, these limitations are often lost when the results of these experiments are quoted by other authors. For example, an experiment looking for fractionally charged particles in a wide variety of natural materials was only sensitive to particles with mass less than $10\,GeV/c^2$ [41], a point that was sometimes overlooked in the literature [34, 35]. Similarly, some experiments are sensitive only to certain values of fractional charge, but these restrictions are often forgotten. The negative evidence for the existence of fractionally charged particles is much weaker than is commonly believed.

Although the chemistry of fractionally charged atoms had already been addressed by Zel'dovich et al. [15], this work was overlooked by all reviewers. Instead, inaccurate remarks like the following appear in the literature:

> "Such an atom [with fractional charge] will have chemical properties somewhat different from the ordinary element with the same atomic number, but with the number of electrons being the same, the properties will tend to be qualitatively similar." [42];

or,

> "In general, quarked-atoms would have the same number of electrons as their normal (un-quarked) counterparts and their chemical properties would be qualitatively similar. However, the quarked-atoms would in general be more reactive because of their net charge." [33].

An accurate view of the chemistry of fractionally charged atoms has recently been given elsewhere [43 − 49].

A neglect of the chemistry of fractionally charged atoms has either weakened or invalidated a number of experimental results. The fact that fractionally charged atoms are always ions causes several problems. As was noted in our discussion of the Hillas-Cranshaw experiment, fractionally charged atoms are "sticky". They are attracted to their image charges on metallic surfaces, and induced dipoles or surfaces charges on dielectrics. Essentially all searches in gaseous media or non-polar liquids are suspect [50, 37, 51, 42, 52].

Conversely, fractionally charged atoms in polar liquids will tend to remain there [15, 53]. Thus, enrichment schemes involving the evaporation of fractionally charged atoms from sea water and their collection on electrified plates are doomed to failure [42, 51].

The search for fractionally charged atoms in the atmosphere is also complicated by their ionic nature. They will either attach themselves to dust particles, or be swept by the ambient electric field of the atmosphere [53]. The existence of this electric field was not mentioned in an experiment [42] which used an "electrified fence" to collect fractionally charged particles from the air. This technique uses the same properties for their concentration which makes their residence in the air about the fence so unlikely.

It is important to note that most experiments looked for fractionally charged particles produced by cosmic rays. Indeed, their negative results were converted into "equivalent" upper limits on the cosmic ray flux [34]. They were, therefore, not optimized to look for fractionally charged atoms left over from the "big bang", a source which we now believe to be more promising.

Experiments that look for fractionally charged particles assume, either explicitly or implicitly, certain production mechanisms and chemical behavior for these objects. Therefore a negative result is difficult to interpret. In particular, it is usually impossible to give an accurate upper limit to the cosmic abundance of fractionally charged particles. Nevertheless experimenters, and more so reviewers, summarize or compare the results of these experiments in units of so many fractionally charged particles per nucleon. For example, Morpurgo, Gallinaro and Palmieri [54] quote an upper limit of 5×10^{-19} fractionally charged particles per nucleon. In order to intelligently interpret the negative result of this experiment, the chemical composition of the samples must be known. Unfortunately the only information given is that they were made of pyrolytic graphite (pyrolytic at 2700°C). These samples were picked for their "exceedingly high diamagnetic susceptivity", in order to levitate large samples. Pyrolytic graphite is known to be an extremely pure form of carbon. The suppression of any impurity, including possible fractionally charged atoms, below their cosmic abundance is very strong. Since no fractionally charged atom looks chemically like carbon with respect to electron structure, electronegativity and crystal radius, fractionally charged atoms will not be easily incorporated into the pyrolytic graphite structure. In fact, fractionally charged atoms are not likely to be present in the precursor gas from which the pyrolytic graphite was made. The cosmic abundance of fractionally charged particles could be many orders of magnitude larger than the limit quoted by this experiment.

Braginskii, Zel'dovich, Martynov, and Migulin recognized that working with highly refined materials is dangerous [55]. They coated their graphite particles with meteoritic substance and the residue from evaporated water. The methodology of this experiment is excellent, although the exact nature of the materials sampled is not given [2)].

The Fairbank group looks for fractionally charged particles in ultrapure niobium spheres. Although the same objections applied to the experiment of Morpurgo et al. apply here, "nothing succeeds like success." Note, however, that particles of charge $^1/_3 e$ and $^2/_3 e$ may be extremely soluble in niobium, making a search in this metal not entirely hopeless [3)]. The solubility of hydrogen in

---

[2)] The meteorite is described only as a "rock meteor", while the water, which is apparently not sea water since it gives only 0.5g residue from 100 liters, is of uncertain origin.

[3)] The extent to which particles of charge $^1/_3 e$ or $^2/_3 e$ are soluble in metals deserves further investigation.

niobium, tantalum, palladium and other selected metals is well-known.

Ordinary elements are highly concentrated in certain rocks by geochemical and geophysical forces. The same would be true for fractionally charged atoms. Therefore, an effective search for fractionally charged atoms in matter requires an understanding of their chemistry.

4. The Chemistry of Fractionally Charged Atoms.

The chemistry of fractionally charged atoms has been considered in a series of papers [45–49], where their electronegativities, oxidation numbers and crystal radii are established. These quantities are conceptually simple and are helpful in making predictions.

The easiest way to understand the chemical behavior of fractionally charged atoms is to draw analogies with the chemistry of ordinary atoms. These analogies are based on similarities in chemical properties like those discussed in the papers quoted above.

A particle of positive charge $^4/_3 e$ provides a good example. It will form the center of an atom with one electron. It may therefore be viewed as a hydrogen atom with an additional charge of $^1/_3 e$ in its center, and may be referred to as $H(^1/_3)$. The electronegativity [4)] of $H(^1/_3)$ is much larger than that of any ordinary element, including that of fluorine [46]. Therefore, in its interactions with ordinary matter $H(^1/_3)$ acquires an additional electron, thereby filling its $1s$ shell and forming the negative ion $H(^1/_3)^-$. Both $H(^1/_3)$ and fluorine come only in the oxidation state -1. Like F^-, $H(^1/_3)^-$ has a spherically symmetric rare gas configuration.

[4)] The electronegativity of an atom is defined as the average of its ionization potential and electron affinity.

The crystal radii of $H(^1/_3)^-$ and F^- are approximately equal [47]. Therefore $H(^1/_3)$ can substitute for fluorine in ionic crystals. If fractional charges occur in the form of $H(^1/_3)$, then they would be found, for example, substituting for fluorine in fluorapatite, $Ca_5(PO_4)_3F$. The F^- in fluorapatite is often replaced by Cl^- or OH^-. These ions, which differ from F^- in crystal radius, may be used as controls. Their concentration indicates the past promiscuity of the fluorine site in the apatite crystal.

5. A Search Strategy.

In future searches for fractionally charged particles the following materials should be examined:

a) The same niobium used in the experiments of the Fairbank group. The impurities present in this niobium depend strongly on the manufacturing history of the sample. Therefore a negative result using niobium of different origin would not necessarily contradict the results of the Fairbank group.

b) Carbonaceous chondrites. This material seems to contain all non-volatile elements in approximately their cosmic abundance. The virtue of this material is that its composition has not undergone unknown geochemical differentiation.

c) Iron meteorites. Perhaps their composition approximates that of the center of the earth. Particles which are depleted on the surface of the earth might be found in these meteorites.

d) Fluorapatite and other special minerals that are expected to host certain fractionally charged atoms. Knowledge of the chemistry of fractionally charged atoms makes it possible to identify minerals that are most likely to hold them. For each type of fractionally charged atom, a small number of preferred mineral sites can be found. Minerals, where fractionally charged atoms competed for sites ordinarily occupied by rare elements, are particularly interesting, since competition for these sites was less extreme.

e) Pegmatites and aplites. Rare elements, whose chemical properties differ substantially from the most abundant mineral forming elements of the crust, tend to concentrate there.

6. Outlook.

In summary, the experimental results of the Fairbank group are not in contradiction with any other measurements; the search for fractionally charged atoms in matter was complicated by their largely unknown chemistry.

Today, the chemistry of fractionally charged atoms is being actively studied, and several new techniques for the detection of fractionally charged atoms are being developed [56 – 59]. These techniques permit the examination of a wide variety of materials and will make a systematic search for fractionally charged particles finally possible.

References

[1] R. Slansky, T. Goldman, and G. L. Shaw, Phys. Rev. Lett. **47** (1981) 887.

[2] F. Wilczek and A. Zee, Phys. Rev. **D16** (1977) 860.

[3] A. De Rujula, R. C. Giles, and R. L. Jaffe, Phys. Rev. **D17** (1978) 285.

[4] A. Zee, Phys. Lett. **84B** (1979) 91.

[5] H. Harari, Phys. Lett. **86B** (1979) 83.

[6] M. A. Shupe, Phys. Lett. **86B** (1979) 87.

[7] D. Horn, Observation of Quarks, Tel Aviv University, Ramat Aviv, Israel (1980), Preprint TAUP 894-80.

[8] L. F. Li and F. Wilczek, Phys. Lett. **107B** (1981) 64.

[9] E. W. Kolb, G. Steigman, and M. S. Turner, Phys. Rev. Lett. **47** (1981) 1357.

[10] H. Goldberg, T. W. Kephart, and M. T. Vaughn, Phys. Rev. Lett. **47** (1981) 1429.

[11] M. I. Strikman, Phys. Lett. **105B** (1981) 230.

[12] V. Gupta and P. Kabir, Phys. Rev. **D25** (1982) 867.

[13] J. D. Bjorken, Stanford Linear Accelerator Center, Stanford, California (1979), SLAC-PUB-2366.

[14] R. V. Wagoner, Free and Bound Quarks - Accelerators and the Early Universe, Stanford Univ., Stanford, California (1980), Preprint ITP-673.

[15] Ya. B. Zel'dovich, L. B. Okun', and S. B. Pikel'ner, Soviet Physics Uspekhi **8** (1966) 702.

[16] Ya. B. Zel'dovich, Comm. Astrophys. Space Phys. **2** (1970) 12.

[17] Ya. B. Zel'dovich, L. B. Okun', and S. B. Pikel'ner, Phys. Lett. **17** (1965) 164.

[18] Ya. B. Zel'dovich, Adv. Astron. Astrophys. **3** (1965) 241.

[19] R. V. Wagoner and G. Steigman, Phys. Rev. **D20** (1979) 825.

[20] R. V. Wagoner, "The Early Universe," **Physical Cosmology**, (Les Houches, Session XXXII, North Holland Publisher, 1979).

[21] G. Steigman, Ann. Rev. Nucl. Part. Sci. **29** (1979) 313.

[22] G. Steigman, "Particle Physics in the Early Universe," **Physical Cosmology**, (Les Houches, Session XXXII, North Holland Publisher, 1979).

[23] S. Frautschi, G. Steigman, and J. Bahcall, Astrophys. J. **175** (1972) 307.

[24] G. F. Chapline, Nature **261** (1976) 550.

[25] H. Goldberg, Phys. Rev. Lett. **48** (1982) 1518.

[26] A. D. Dolgov and Ya. B. Zel'dovich, Rev. Mod. Phys. **53** (1981) 1.

[27] A. F. Hebard and W. M. Fairbank, "Search for Fractional Charge (Quarks) Using a Low Temperature Technique," **Proceedings of the 12th International Conference on Low Temperature Physics, Kyoto**, (Tokyo, Keigaku Publishing Co., 1970).

[28] G. S. LaRue, W. M. Fairbank, and A. F. Hebard, Phys. Rev. Lett. **38** (1977) 1011.

[29] G. S. LaRue, W. M. Fairbank, and J. D. Phillips, Phys. Rev. Lett. **42** (1979) 142.

[30] G. S. LaRue, J. D. Phillips, and W. M. Fairbank, Phys. Rev. Lett. **46** (1981) 967.

[31] K. S. Lackner and G. Zweig, Searching for Fractionally Charged Particles, California Institute of Technology, Pasadena, California (June, 1981), Talk given at the First International Quark Searchers Conference, San Francisco.

[32] Y. S. Kim and N. Kwak, Fields and Quanta **3** (1972) 81.

[33] Y. S. Kim, Contemp. Phys. **14** (1973) 289.

[34] L. W. Jones, Rev. Mod. Phys. **49** (1977) 717.

[35] L. W. Jones, Phys. Rev. **D17** (1978) 1462.

[36] A. Piccard and E. Kessler, Arch. Sci. Phys. Naturelles **7** (1925) 340.

[37] A. M. Hillas and T. E. Cranshaw, Nature **184** (1959) 892.

[38] J. G. King, Phys. Rev. Lett. **5** (1960) 562.

[39] J. C. Zorn, G. E. Chamberlain, and V. W. Hughes, Phys. Rev. **129** (1963) 2566.

[40] L. Lyons, Progress in Particle and Nuclear Physics (1981) .

[41] D. D. Cook, G. DePasquali, H. Frauenfelder, R. N. Peacock, F. Steinrisser, and A. Wattenberg, Phys. Rev. **188** (1969) 2092.

[42] W. A. Chupka, J. P. Schiffer, and C. M. Stevens, Phys. Rev. Lett. **17** (1966) 60.

[43] C. K. Jørgensen, Structure and Bonding **34** (1978) 19.

[44] C. K. Jørgensen, Naturwissenschaften **67** (1980) 188.

[45] K. S. Lackner and G. Zweig, The Chemistry of Free Quarks, California Institute of Technology, Pasadena, California (1980), CALT-68-781.

[46] K. S. Lackner and G. Zweig, Introduction to the Chemistry of Fractionally Charged Atoms – Electronegativity, California Institute of Technology, Pasadena, California 91125 (1981), CALT 68-865.

[47] K. S. Lackner and G. Zweig, Lettere Al Nuovo Cimento **33** (1982) 65.

[48] K. S. Lackner and G. Zweig, Oxidation Numbers of Fractionally Charged Atoms, Los Alamos National Laboratory, Los Alamos, New Mexico (To be published).

[49] K. S. Lackner and G. Zweig, Crystal Radii of Fractionally Charged Atoms, Los Alamos National Labratory, Los Alamos, New Mexico (To be published).

[50] R. A. Millikan, **The Electron**, (Chicago, Illinois, University of Chicago, 1917).

[51] D. M. Rank, Phys. Rev. **176** (1963) 1635.

[52] J. W. Elbert, A. R. Erwin, R. G. Herb, K. E. Nielsen, M. Petrilak, Jr., and A. Weinberg, Nuc. Phys. **B20** (1970) 217.

[53] M. R. C. McDowell and J. B. Hasted, Nature **214** (1967) 235.

[54] G. Morpurgo, G. Gallinaro, and G. Palmieri, Nucl. Instr. Meth. **79** (1970) 95.

[55] V. B. Braginskii, Ya. B. Zel'dovich, V. K. Martynov, and V. V. Migulin, Soviet Physics JETP **27** (1968) 51.

[56] C. D. Hendricks and G. Zweig, Detection and Enrichment of Fractionally Charged Particles in Matter, , (1982), Proposal approved by the Division of Advanced Energy Projects, Office of Basic Energy Sciences, Department of Energy..

[57] C. A. Barnes, B. Cooper, and R. D. McKeown, ., California Institute of Technology, Pasadena, California (1982), Proposal to the National Science Foundation for Grant to Support Search for Fractional Charges.

[58] K. H. Chang, A. E. Litherland, L. R. Kilius, R. P. Beukens, W. E. Kieser, and E. L. Hallin, , Department of Physics, University of Toronto, Toronto, Canada (June, 1981), Paper presented at the First International Quark Searchers Conference, San Francisco.

[59] T. Gentile, H. Kagan, S. L. Olsen, and D. Elmore, , Department of Physics and Astronomy, University of Rochester, Rochester, New York (June, 1981), Paper presented at the First International Quark Searchers Conference, San Francisco.

RECENT RESULTS FROM THE SAN FRANCISCO STATE QUARK SEARCH

Peter C. Abrams, Kimberly C. Coburn, David C. Joyce, Maureen L. Savage, Frederick Wm. Walters, Betty A. Young, Roger W. Bland, and Robert T. Johnson
San Francisco State University, San Francisco, Ca. 94132

Christopher L. Hodges
San Francisco State University and University of California, Riverside, California 92521

ABSTRACT

In this paper we describe an automated Millikan-type apparatus used to test small liquid drops for fractional electric charge. Results of recent measurements are given, including the first direct observations with sea water. In 0.36 μg of sea water, we have observed no fractional charges. We have measured 500 times more water than Millikan, and it seems likely that an uncertain measurement of fractional charge reported by Millikan should be discounted.

OTHER SEARCHES FOR FRACTIONAL CHARGE

Robert A. Millikan was the first person to directly measure single electronic charges.[1] He found the net charge of small oil drops to be an integral multiple of a value which he showed to be the electronic charge. Before the oil-drop measurements, Millikan had made charge measurements on water drops. In reporting these results, Millikan mentioned[2] a single non-integral charge measurement. The measurements with water were made by applying a fixed electrical field to a cloud of droplets formed after decompression of saturated air. A drop which was nearly motionless in this field was selected, and the field adjusted until the drop was perfectly motionless. Then the field was removed, and the drop was timed in free fall. From the free-fall velocity the mass was determined, and from the value of the electric field the charge was calculated. In reporting these results, Millikan described the single non-integral charge measurement as "one uncertain and unduplicated observation apparently upon a singly charged drop, which gave a value of the charge on the drop some 30 per cent lower than the final value of e.[2]" The mass of all the water drops measured was

0094-243X/82/930015-09$3.00

about 7 x 10^{-10} grams.

Millikan subsequently abandoned measuring water drops in favor of oil drops, because of difficulties associated with evaporation of the water. Even in saturated air, small water drops experience a net loss of mass due to more rapid evaporation from their curved surface. This means that measurements on small water drops must be made rapidly, and some error is unavoidable due to mass loss during the free-fall measurement. A long series of repeated measurements on single water drops was not possible by this method. Oil drops, however, could be measured repeatedly over a long period of time, permitting more accurate and reliable measurements.

Many experimental searches for fractional charge were taken up after the proposal of quarks in 1963 by Gell-Mann and Zweig. The Millikan technique was used on large oil drops[3] and on sea-salt crystals.[4] Various ingenious experiments were performed involving concentrating samples in quarks before measurement; these experiments are all limited to some degree by the properties of quarks assumed in the concentration and detection schemes. These experiments were reviewed by Lawrence W. Jones;[5] none of them gives evidence for fractional charge. The most sensitive direct measurements are the levitometer experiments of Gallinaro, Marinelli, and Morpurgo[6] and of LaRue, Hebard, Phillips, and Fairbank[7]. Gallinaro *et al* find no fractional charges in 3.7 mg of iron, whereas LaRue *et al* observe a third-integral charge in about a third of their measurements on thirteen 90-microgram niobium spheres. The results of LaRue *et al* have not been confirmed by an independent experiment.

THE SAN FRANCISCO STATE QUARK SEARCH EXPERIMENT

The SFSU quark search uses a modified Millikan apparatus. Small drops (5 to 15 microns in diameter) are introduced one at a time into a measuring chamber, where their drift velocity in air is measured in an electric field. Switching the field polarity in mid-measurement permits determining the mass and charge of the drops. The measurement is highly redundant, and various selection criteria (unbiassed as to fractional charge) reject about half of the measurements. Drops which change charge in flight are identified and rejected. Drops can be measured at a rate of about one a second, and we measure about five micrograms in an hour of ideal running.

Some details of the apparatus are shown in figure 1. A single drop is ejected on command from a

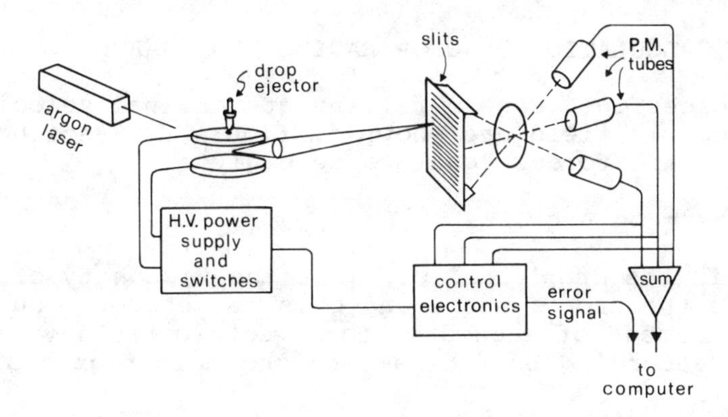

Figure 1. Diagram of the apparatus.

piezoelectric drop ejector, and falls into the space between the plates. There it is illuminated by an argon laser, with the scattered light imaged on a plane of slits. Light passing through the slits is collected by three photomultiplier tubes. One tube monitors the first two slits, a second monitors eighty-eight central slits, and a third monitors the last two slits. The control electronics switches the electric field as the drop image passes predetermined slit numbers. The electronics also checks the number of slits counted by each tube, with an error signal sent to the on-line computer. This computer digitizes the summed signal from the three photomultipliers, searches the digitizations for peaks, and calculates the time of each slit crossing. From these times the velocities under the different field polarities are determined. The computer also monitors the field-switching signals, and checks extensively for errors in numbers of peaks, field-switching times, etc.

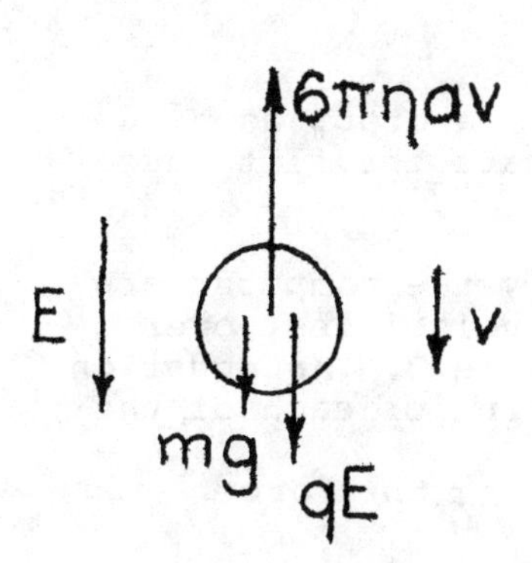

Figure 2. Forces acting on a sphere falling through air in the presence of an electric field.

Charge, mass, and other quantities of interest are calculated on line for each drop, and histograms of these quantities are stored in memory. At regular intervals (about once an hour) run summaries are sent to a second computer for display, storage, and further analysis of the data.

CALCULATION OF DROP RADIUS AND CHARGE

The forces on a sphere falling at terminal velocity in an electric field are shown in figure 2. Setting the sum of the forces equal to zero gives

$$6\pi\eta a v = mg + qE, \qquad (1)$$

where *a* is the drop's radius, η is the viscosity of air, *m* is the drop's mass, and *q* is its charge. During the measurement of each drop the electric field is switched according to the pattern shown in figure 3.

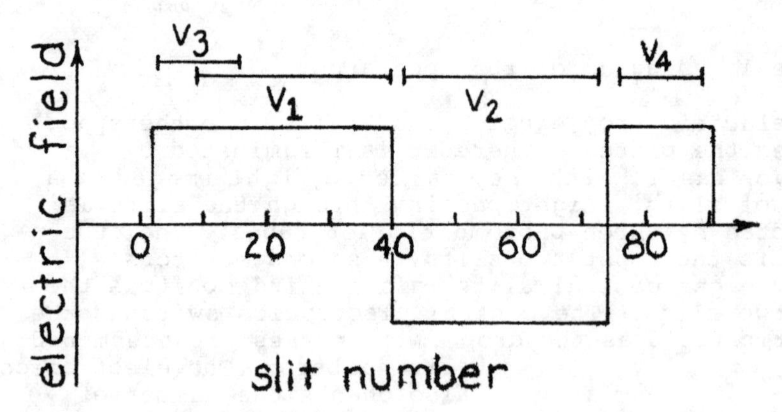

Figure 3. The electric field seen by a drop, as a function of slit number. Bars indicate the slits used to calculate velocities v_1 to v_4.

The slit-crossing times determined by the computer are then used to calculate four average velocities, over the groups of slits indicated in figure 3. An equation similar to equation (1) can be written for each of the velocities v_1 to v_4.

Taking the sum and the difference of the force equations for v_1 and v_2 leads to

$$6\pi\eta a\left(\frac{v_1+v_2}{2}\right) = mg, \qquad (2)$$

$$6\pi\eta a\left(\frac{v_1-v_2}{2}\right) = qE. \qquad (3)$$

Using $m = 4/3\ \pi a^3$, the first equation gives the radius *a*. The second then gives the charge *q*. Velocities v_3

and v_4 are used to identify occasional drops whose charge changes during the measurement. In those cases, v_3 and v_4 will not be equal; the change in charge is given by

$$dq = 6\pi\eta a\left(\frac{v_4 - v_3}{2}\right)/E. \qquad (4)$$

CHARGE CHANGES

The distribution of charge-change values for well-measured events from several one-hour data runs is shown in figure 4. These runs were atypical, chosen

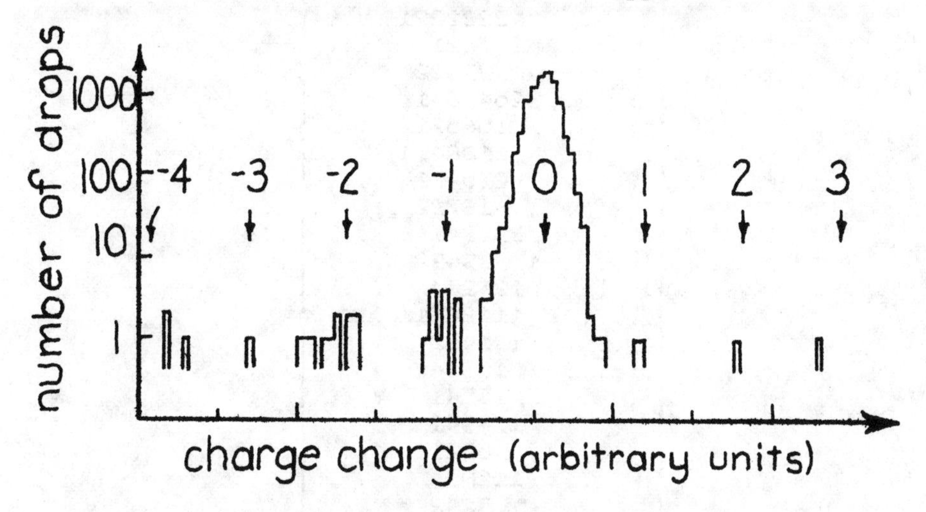

Figure 4. Distribution of dq for several runs of mercury-drop data.

for their high frequency of charge changes. In figure 4 one sees, in addition to a dominant peak at zero charge change, peaks at several other integral values. The resolution is evidently sufficient that most charge changes can be rejected, while leaving most drops which did not change charge. In the water measurements reported here, and in most of our previously reported results, drops with $|dq| \geq e/2$ are excluded from the data sample.

During best operation of our apparatus about one mercury drop in three or four thousand showed a charge change. When the chamber was dirty or when a deposit built up on the bottom of the chamber, however, the rate of charge changes increased dramatically. We attribute this to field emission of electrons which subsequently change the drop's charge. To avoid contaminating our sample with drops having double charge changes or other complications, we

```
..............111111.21.......|.....
....|.......11422441........|.....
....|.............412.............
....|.........1..4831.........|.....
....|........1.2471.3.........|.....
....|.........1.3221.1........|.....
....|.........211421.2........|.....
....|............2412.2.1.....|.....
....|...........24462.........|.....
....|...........223621........|.....
....|...........2222221.......|.....
....|.........1.7221..........|.....
....|............23422........|.....
....|.........254.641.........|.....
....|...........156521........|.....
....|...........118A411.......|.....
....|...........235432........|.....
....|............18722........|.....
....|.........332341..........|.....
....|........1.233432.........|.....
....|.........12.254..........|.....
....|.........11126.11........|.....
....|.........131432..........|.....
....|.........1183141.........|.....
....|...........15541.........|.....
....|.........1243411.........|.....
....|............5321.........|.....
....|.........112A1...........|.....
....|.........1.3242..........|.....
....|.........11151...........|.....
....|...........241.21........|.....
....|.
```

Figure 5. The charge distribution for one good run of sea water.

rejected entire data runs (of about an hour's duration) which had more than two charge changes per 1000 drops.

The charge histogram for our one good run of sea water is shown in figure 5. Here the contents of each of the 1024 bins of the histogram is represented by a single character; a dot represents zero, and 1-9 and A-T indicate 1 to 29 events in a bin, respectively. The clusters of characters thus represent clustering of charge measurements near certain values, taken to be integral multiples of e. (The analysis program has chosen the average length for a line of printout which makes the peaks line up vertically.) A third-integral charge value would appear near one of the dashed lines on the left- or right-hand side of the figure.

Figure 6a gives the distribution of residual charge

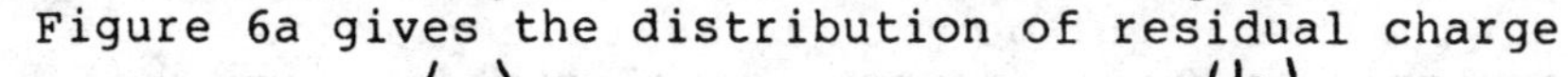

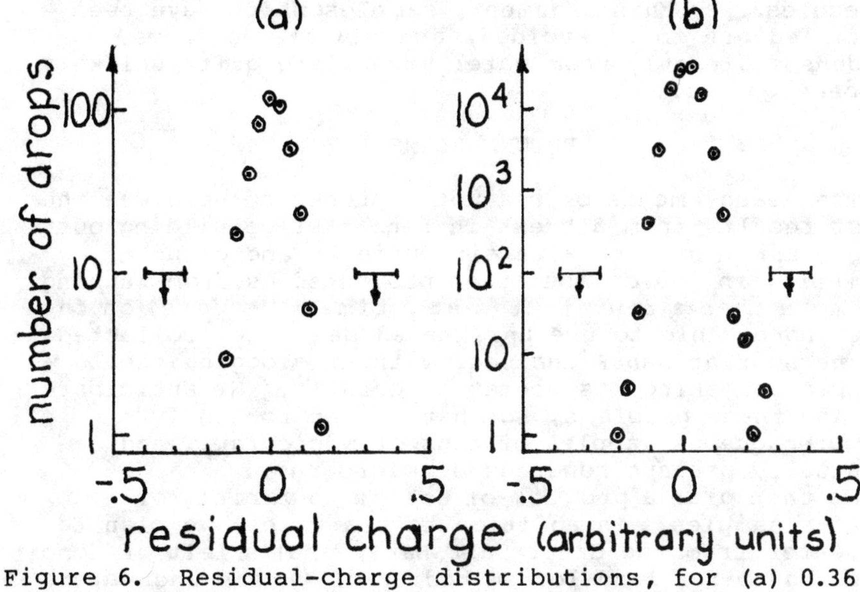

Figure 6. Residual-charge distributions, for (a) 0.36 micrograms of sea water, and (b) for 175 micrograms of native mercury.

for the same data. This is a vertical projection of figure 5. The integral peaks combine to give a single peak with a standard deviation $\sigma = 0.051e$, well approximated by a Gaussian. This graph represents 0.36 micrograms of sea water. In figure 6b we show a similar plot for all of the mercury we have measured, 60 micrograms of distilled mercury and 115 grams of native mercury. (The mercury measurements have been reported previously.[8]) The positions where third-integral values would fall are marked by bars four bins wide. These bars, about +/- 1.8 σ for the mercury data and about +/- 1.3 σ for the sea-water data, represent our fiducial regions for fractional charge. To date we have no measurements in these fiducial regions.

The mass of sea water which we measured is 0.36 x 10^{-6} grams. This is the first sample of sea water that we know to have been tested by direct measurment for fractional charge. It is also to our knowledge the first direct measurement of any water sample reported since Millikan's experiment. Our sample is 500 times larger than Millikan's water sample.

Sea water is an especially interesting material in which to look for fractional charges. Charged molecules should be inhibited from evaporation by an image-charge force which neutral molecules do not see,[9] and so the sea might be rich in soluble quarked molecules. By this argument, samples which have been distilled are to be avoided, and Millikan's drops, condensed directly from water vapor, are quite unlikely to bear quarks.

FUTURE PLANS

The measurements of sea water presented here are the first results from a break-in run after rebuilding our apparatus. Problems with air currents and cloud formation in the chamber then prevented us from running for more than a few minutes at a time. We were, on the other hand, able to use unfiltered sea water collected in the surf at San Francisco, without clogging the dropper or making its operation erratic. We anticipate solving these problems, so that we can routinely measure aqueous samples of tens of micrograms, and eventually perhaps hundreds of micrograms.

We then plan a program of testing a variety of aqueous samples. In addition to sea water, we plan to use water from the Great Salt Lake, what's left of Mono Lake, and other land-locked bodies of water; and water having passed through various geological environments, such as bottled mineral water, water from limestone caverns, and water from geysers. We hope to be able to process homogenized cells from plants and animals. We are also attempting to measure dilute solutions of niobium pentaethoxide, liquid at room temperature, and niobium pentafluoride,[10] which becomes liquid at 72 degrees Celsius.

CONCLUSION

We have measured 0.36 micrograms of sea water by an automated Millikan technique, finding no fractional charges. This is the first direct measurement to be made on sea water, and conflicts with the interpretation of a measurement made by Millikan on a water drop as evidence for fractional charge. In the future we expect to increase the size of the sample, and to measure a variety of aqueous samples.

This work was supported by the U.S. Department of Energy under Contract No. DE-AC03-81ER40009, and early phases of the work were supported by a grant from the Research Corporation.

1. R.A. Millikan, <u>The Electron</u> (University of

Chicago Press, 1963), and references cited therein.
2. R.A. Millikan, "A New Modification of the Cloud Method of Determining the Elementary Electrical Charge and the most Probable Value of that Charge," Phil. Mag. S. 6. Vol. 19 (1910), pp. 220 and 223.
3. D.M. Rank, Phys. Rev. 176, 1635 (1968)
4. Yasuhiro Mitsuhashi, Eiichi Goto, and Ryo Kuroda, Phys. Soc. Japan 40, 613 (1976)
5. Lawrence W. Jones, Rev. Mod. Phys. 49, 717 (1977)
6. G. Gallinaro, M. Marinelli, and G. Morpurgo, Phys. Rev. Letters 38, 1255 (1977); M. Marinelli and G. Morpurgo, Physics Reports 85, 162 (1982), and references cited therein.
7. G. LaRue, W.M. Fairbank, and A.F. Hebard, Phys. Rev. Letters 38, 1011 (1977); G. LaRue, W.M. Fairbank, and J.D. Phillips, Phys. Rev. Letters 42, 142 (1979); G. LaRue, J.D. Phillips, and W.M. Fairbank, Phys. Rev. Letters 46, 967 (1981)
8. C.L. Hodges, P. Abrams, A.R. Baden, R.W. Bland, D.C. Joyce, J.P. Royer, R.Wm. Walters, E.G. Wilson, P.G.Y. Wong, and K.C. Young, Phys. Rev. Letters 47, 1651 (1981)
9. G. Zweig, Science 201, 973 (1978)
10. This compound was suggested to us by Herman H. Hobbs, Department of Physics, George Washington University, private communication, submitted to Phys. Rev. Letters as "Absence of Fractional Charges on Mercury Drops: A Decisive Result?"

SEARCH FOR FREE QUARKS AT PEP*

S. J. Freedman[†] [††]
Stanford University, Stanford, CA 94305
and
Argonne National Laboratory, Argonne, IL 60439**

ABSTRACT

The results of a search for fractionally charged particles produced in e^+e^- annihilation at 29 GeV/c^2 are discussed. Results from cosmic ray searches for fractional charged particles, tachyons, and massive particles using the same detector are also presented.

INTRODUCTION

With the commissioning of each new high energy accelerator a new chapter in the search for free quarks is begun. Inevitably the new possibilities opened up by increased energies or luminosities entices a few brave souls to search for the tantalizing particles with third integer charges proposed eighteen years ago as the ultimate constituents of matter. I will present some results of a search at PEP[1] during its first year of operation which followed this tradition. The experiment was sensitive to quarks produced in e^+e^- annihilation at 29 GeV center of mass energy, either exclusively in pairs ($e^+e^- \rightarrow q\bar{q}$) or inclusively along with hadrons ($e^+e^- \rightarrow q\bar{q}\ X$). The detector was also exploited to search for exotic particles, including quarks, in cosmic rays and I will present these results as well. The experimental work was completed in the summer of 1981 but the analysis is just now being completed and some of the results are still preliminary.

Modern searchers for free quarks must somehow remain enthuiastic in the face of the widely held theoretical conjecture that free quarks will never be discovered becasue they are permanently confined within hadrons.[2] From an experimental point of view the idea of quark confinement must be rigorously tested despite theoretical bias. This is especially true in light of the positive evidence for fractionally charged matter reported by Fairbank et al.[3] Free quarks with $Q = \pm 1/3$ or $Q = \pm 2/3$ are obvious candidates to explain Fairbank's result but free fractional charge does not necessarily mean free quarks. Indeed the existence of fractionally charged leptons or fractionally charged hadrons made up of integrally- and fractionally-

---

*Research supported in part by the U.S. Department of Energy and I.N.F.N. (Italy).

†A. P. Sloan Foundation Fellow.

††Representing the PEP 14 Collaboration, see Ref. 1.

**Present Address.

0094-243X/82/930024-15$3.00

charged quarks are not inconsistent with color confinement. Most free quark searches (including the ones I will describe) are also sensitive to other possible fractionally-charged particles.

Apart from the stable matter searches, most searchers for free quarks have concentrated on particles produced in cosmic rays or in hadronic collisions at accelerators. When the present experiment was proposed the possibility that quarks might be produced electromagnetically had been largely neglected. The effects of the existence of quarks in particularly dramatic in e^+e^- annihilation and this process is the best available method for concentrating energy in space. When compared to the "softer" and less well understood collisions of hadrons it may be a more likely place to look for liberated quarks.

EXPERIMENTAL METHOD

The FQS detector was designed to measure charge. The method was simple and there was no magnetic field. The charges of particles were determined from measurements of ionization energy loss (dE/dx) in plastic scintillators and velocity ($\beta = v/c$) by time-of-flight (TOF). To reduce the effect of the Landau tail and to improve resolution the most probable dE/dx was estimated with a truncated average of many individual dE/dx measurements. The charge (Q) was determined from the Bethe-Bloch formula.

$$dE/dx = (Q^2/\beta^2)\ f(\beta).$$

where $f(\beta)$ is a slowly varying function of β. To a good approximation,

$$dE/dx \approx Q^2/\beta^2\ (dE/dx)_{MIP},$$

where $(dE/dx)_{MIP}$ is the ionization loss for a relativistic (minimum ionizing) Q=1 particle.

A view of FQS detector along the beam direction is shown in Fig. 1. The detector solid angle was $1/3 \times 4\pi$ sr.

The detector consisted of two identical arms. Each arm was made up of 12 scintillation counter hodoscopes, 9 multiwire proportional chambers (MWPC), and a lucite Cherenkov counter hodoscope. Each counter had a photomultiplier (PM) at each end. A subset of 5 planes of scintillation counters (see Fig. 1) measured TOF over a 1.5 to 26 m flight path depending on the track angle.

The lucite Cherenkov counters gave a check on the velocity determination; they triggered on particles with $\beta \gtrsim 0.7$.

The chamber gas was 80% Ar - 20% CO_2. The inner 5 MWPC's were 1.6 cm thick and the outer MWPC's were 2.4 cm thick.

The inner 5 MWPC's measure dE/dx as well as position and in

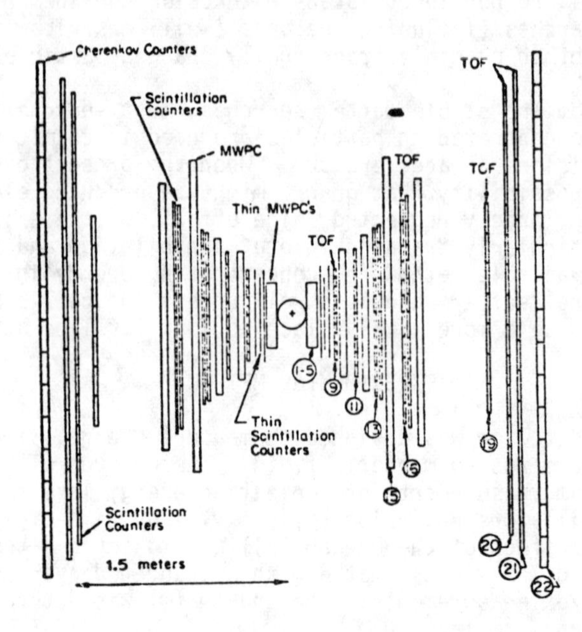

Fig. 1. Elevation view of the detector as viewed along the beam pipe. The elements are numbered sequentially from 1 to 22 moving outward from the IR (some of the layers are numbered in the figure). The "thin" MWPC's (layers 1 to 5) are not shown individually. Scintillation layers 9, 16, 19, 20, and 21 are equipped with TOF electronics.

conjunction with the first 3 layers of dE/dx counters they were used to make a highly-interacting quark search to be discussed below. Table I lists the detector components, their orientation and thickness in interaction lengths.

PERFORMANCE OF THE DETECTOR AND EXPERIMENTAL CHECKS

Cosmic rays were used to calibrate and align detector components. A serious consideration was to ensure sensitivity to lightly ionizing particles.

Scintillators: A cosmic ray pulse height spectrum from a typical scintillator is shown in Fig. 2. Scintillator attenuation length effects were accounted for by using the geometric mean of the two PM pulse heights. The low pulse heights in the figure were due to particles which "clip" scintillator edges. In our analysis this effect was accounted for by adding in the corrected pulse height from the adjacent counter for tracks near edges. The typical counter

Layer	Detector Type	Orientation Angle*	# of Elements	Pulse Height	TOF	Cumulative Collision Lengths (Normal Incidence)
1	thin MWPC	+45°		Yes		
2	thin MWPC	-45°		Yes		
3	thin MWPC	90°		Yes		
4	thin MWPC	0°		Yes		
5	thin MWPC	0°		Yes		.005
6	thin scint.	90°	8	Yes		.008
7	thin scint.	0°	8	Yes		.011
8	thin scint.	90°	10	Yes		.014
9	scint.	0°	10	Yes	Yes	.031
10	MWPC	80°				.045
11	scint.	0°	10	Yes		.072
12	MWPC	0°				.086
13	scint.	0°	10	Yes		.116
14	scint.	0°	10	Yes		.145
15	MWPC	100°				.160
16	scint.	0°	10	Yes	Yes	.202
17	scint.	0°	10	Yes		.243
18	MWPC	0°				.258
19	scint.	0°	7	Yes	Yes	.304
20	scint.	0°	15	Yes	Yes	.351
21	scint.	0°	16	Yes	Yes	.398
22	Cerenkov	0°	16	Yes		.492

*For scintillation counters, 0° means the long axis is parallel to the beam line. For MWPC's, 0° means the anode wires are parallel to the beam line.

Table I. The Components of the Quark Detector.

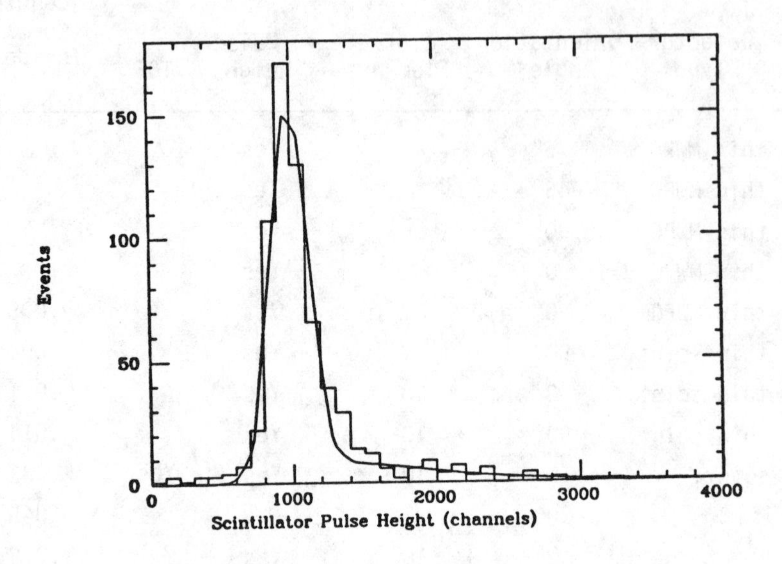

Fig. 2. A dE/dx Spectrum from cosmic rays from a typical scintillator.

resolution was $\approx$28% FWHM for $(dE/dx)_{MIP}$ corresponding to $\approx$150 photoelectrons. Discriminator thresholds for TOF were set at $\approx$130 of $(dE/dx)_{MIP}$. Counters used in the trigger were determined to be fully efficient for 1/10 x $(dE/dx)_{MIP}$ in tests which utilized the low pulse heights from edge hits.

After calibration and pulse height corrections the single counter time resolution for cosmic rays was $\sigma \approx 150$ psec. The resolution was slightly poorer for beam events, $\sigma \approx 180$ psec. The time resolution depends on pulse height and increases to $\sigma \approx 650$ psec for 1/9 x $(dE/dx)_{MIP}$.

MWPC: Thresholds were set at 1/50 of $(dE/dx)_{MIP}$. Lightly ionizing particles were simulated by observing cosmic rays with reduced chamber voltage and thus reduced gain. A check was made with a thinner chamber and different chamber gas (80% He - 20% CH_4) to verify that the procedure was valid including the effects of primary ionization statistics.[4]

COSMIC-RAY SEARCH EXPERIMENTS

The FQS detector provided several valuable features for cosmic-ray searches including high segmentation, high redundancy, good tracking, and excellent TOF resolution. In general it was much more elaborate than detectors previously employed. Exploiting these characteristics we conducted single-particle searches for fractional charge, tachyons, and massive particles in cosmic rays. A more detailed discussion of these cosmic-ray searches are found in Refs. 5 and 6.

Search for fractional charge in cosmic rays: For cosmic rays the acceptance of the FQS detector was at large zenith angles (45° to 90°). The geometric admittance was 4.0×10^3 cm^2sr. It has been noted[7] that high zenith angle searches may be more sensitive than vertical searches to penetrating fractionally charged produced in high energy air showers because less penetrating particles (presumably of lower mass) would be effectively dispersed or absorbed by the intervening atmosphere. However, very few high zenith angle searches have been made.[7,8,9]

Three fractionally-charged candidates with low velocity were reported in a previous vertical search.[10] Although unconfirmed by a second experiment[11] such low velocity particles would have been missed in previous experiments without TOF.

Cosmic-ray data for the fractional charge search were collected for $\cong 2.3 \times 10^6$ sec. About 2/3 of the data were obtained while the detector was enclosed in the $\cong 250$ g/cm^2-thick PEP shielding tunnel and the rest after it was removed. The trigger requirement was at least one hit in layer 9 on either arm and at least one hit in 16 on each arm. Using MWPC and scintillator information a loose tracking requirement identified $\approx 85\%$ of the $\approx 10^7$ triggers as single particle events. After a series of cuts to eliminate several systematic effects (primarily due to showering particles, interactions in the detector, and errors due to mistracking) no candidates for particles with $Q \leq 0.8$ were found. The analysis efficiency was 98% for Q = 1/3 and 95% for Q = 2/3 as determined by Monte Carlo. The final flux limits (90% C.L.) are shown in Fig. 3. Table II compares our limits with those of previous high zenith angle searches and previous searches with TOF.

Tachyon Search: Most cosmic ray searches for faster-than-light particle looked for particles preceding the relativistic front of air showers. From time to time positive evidence has been claimed but none have survived subsequent scrutiny.[12]

Only two previous experiments[13,14] measured particle velocity directly. Both used the TOF method and large plastic scintillators. Ashton[14] obtained a limit of $\Phi_t < 2.2 \times 10^{-5}$ cm^{-2} $sr^{-1}sec^{-1}$ for particles with $\beta > 1.6$ producing a signal corresponding to a dE/dx of > 4 MeV/cm in plastic scintillator.

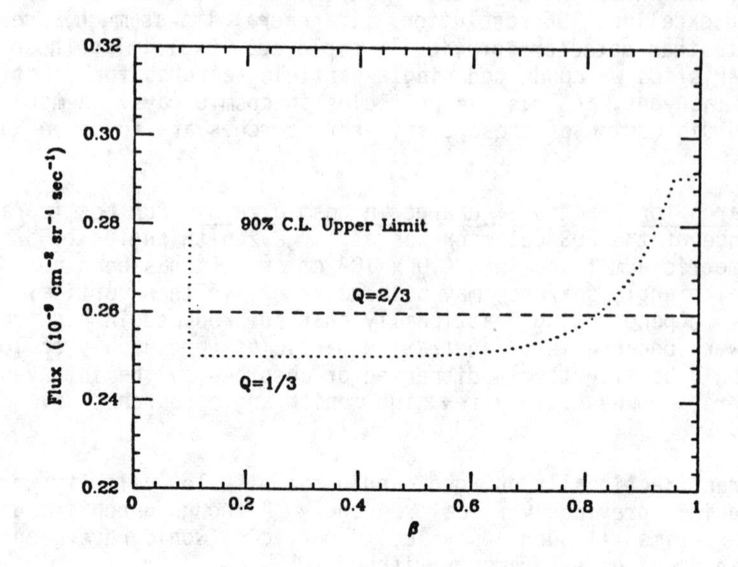

Fig. 3. Flux Limits from the present cosmic ray single particle fractional charge search as a function of β.

Expt.	Range of Zenith Angles	Flux Q = 1/3	β	Flux Q = 2/3	β
Ref. 8	$45° \leq \theta \leq 90°$	$\leq 2.3 \times 10^{-10}$	≈1	---	---
Ref. 7	$75° \leq \theta \leq 90°$	$\leq 1.7 \times 10^{-8}$	≈1	$\leq 1.7 \times 10^{-8}$	≈1
Ref. 9	$\theta \approx 84°$	---	---	$\leq 5.1 \times 10^{-8}$	0.5-0.9
This Expt.	$45° \leq \theta \leq 90°$	$\leq 2.9 \times 10^{-10}$	0.6-1.0	$\leq 2.6 \times 10^{-10}$	0.1-1.0
		$\leq 2.5 \times 10^{-10}$	0.1-0.6		

Table 2. Searches for Fractional Charge at Large Zenith Angles

In the present experiment the most serious systematic effect was from showering particles producing multiple hits in the long TOF counters, thus causing errors in the determined velocity. To reduce this background we required eight TOF measurements along the track and no more than two hit TOF counters not associated with the track. Figure 4 shows the velocity distribution for a one-hour run before and after this cut was applied.

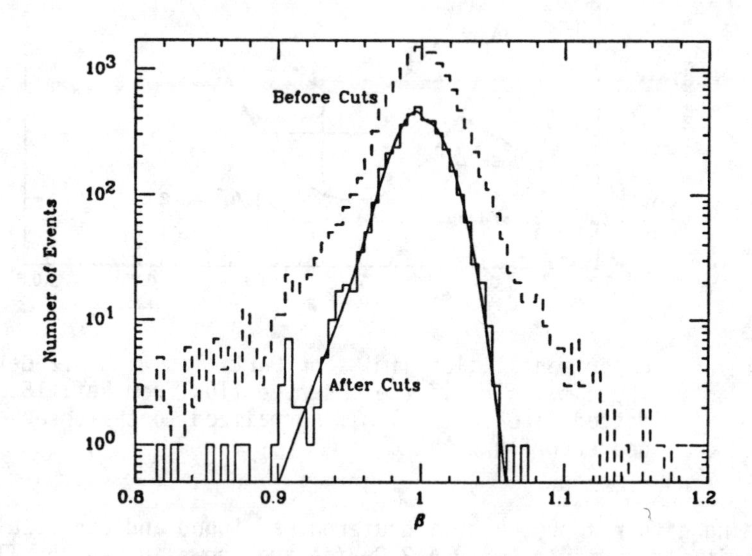

Fig. 4. Experimental velocity distribution before and after cuts to reduce the effects of showering particles. (See text).

For a run of 8×10^5 sec we observed six events above $\beta = 1.1$ and none above $\beta = 1.2$. The 6 events are above the 0.04 expected if the velocity resolution were Gaussian but they were not inconsistent with our estimate of undetected showers. Nevertheless, we state a limit only for $\beta \geq 1.2$, we obtain $\Phi_t < 2.4 \times 10^{-9}$ cm^{-2} sr^{-1} sec^{-1}$\beta >$ 1.2 (90% CL) for tachyons able to deposit at least 0.24 MeV/cm in plastic scintillator.

Massive Particle Search: Particles with masses of about 4.2 GeV/c^2 and unit charge have been suggested by several cosmic-ray experiments and a few accelerator searches.[15] We looked for heavy particles with the FQS detector after it was removed from PEP shielding tunnel. A steel absorber about 30 gm/cm^2 was placed between the two arms. We then searched for slowly-moving particles which slowed too little in the detector to be tritons. No particle inconsistent

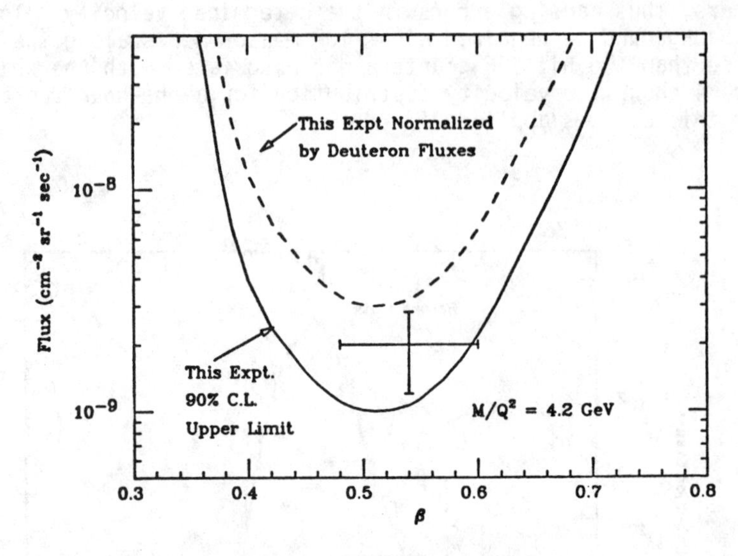

Fig. 5. Experimental flux limits on particles with 4.2 GeV/c^2. The data point is the observed flux from Ref. 15. The dotted line in the limit normalized to the observed deuteron flux.

with being either a proton or a deuteron was found and our resulting flux limit for particles with 4.2 GeV/c^2 are shown in Fig. 5. The positive result of a previous vertical search[15] is shown in the figure but since the range of zenith angles are different the two experiments are hard to compare directly.

Search for Fractional Charge at PEP

Colliding beam data were obtained at 29-GeV center-of-mass energy; the total luminosity was 16.0 pb^{-1} obtained from small-angle Bhabha scattering. The counters in the inner layers of the detector (scintillators between lay 9 and 17, see Fig. 1) were aligned in radial "roads" in each arm with 5 of the 6 counters hit. A separate fast trigger requirement was a hit in layer 9 and 16 in each arm. These conditions introduce a slight loss of efficiency for Q = 1/3 particles at relativistic velocities that is reflected in the final results.

QED Check: To gain confidence in the detector operation we measured the angular distribution of the reaction $e^+e^- \rightarrow e^+e^-, \mu^+\mu^-$.

Events were selected by requiring: (1) one and only one track in each arm, (2) tracks must come from the interaction region and correspond to particles produced in time with the beam crossing, and (3) collinearity of the tracks must be better than 4.0°.

After correcting for background the cross section agreed with the QED prediction and our measured luminosity to better than 5%. The angular distribution is shown in Fig. 6.

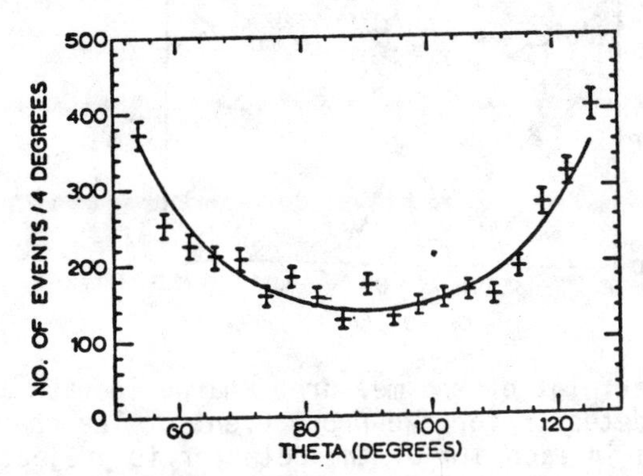

Fig. 6. The angular distribution for the events from the combined reactions $e^+e^- \rightarrow e^+e^-, \mu^+\mu^-$, plotted in terms of the production angle θ, relative to the beam direction. The curve shows the QED prediction, normalized to the total number of events.

<u>Exclusive Fractional Charge Production</u>: To search for $e^+e^- \rightarrow q\bar{q}$ we made the following selection: (1) and (2) above, (3) tracks back to back within 8° in the polar and azmuthal angles, and (4) at least three times must be measured along a track and the assumption of a constant velocity and production at the beam crossing time must yield a fit to the TOF data with $\chi^2 < 9$ per degree of freedom. These cuts reduced 1.5×10^6 triggers to 1.3×10^4 events. Figure 7 shows a scatter plot of the charge for each arm for the selected events. We required $Q \leq 0.8$ for both tracks and thus no candidates were found as seen in Fig. 7. We converted our limits on the differential cross section to total cross section limits by making radiative corrections and using the expected angular distribution for point-like fermions. Figure 8 shows the limits from this experiment as well as the recent limits for $Q = 2/3$ exclusive production from the MARK II[16] and JADE[17].

Present limits are now sufficient to rule out fractionally charged lepton production with either $Q = 1/3$ or $Q = 2/3$ up to masses of $\sim$ 14 GeV/c^2. A more detailed discussion of this exclusive quark search is found in Ref. 18.

Inclusive Fractional Charge Production

The process $e^+e^- \rightarrow q\bar{q} + X$ is probably more likely to produce free

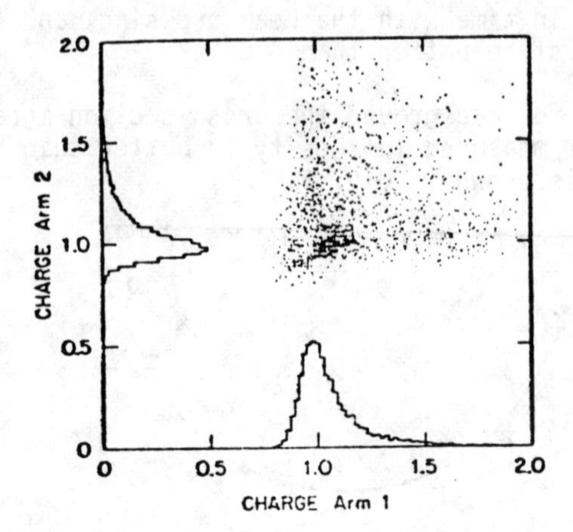

Fig. 7. A scatter plot of the measured charge in each arm of the detector for two-prong events. The charge measured in each arm of the detector is projected into a histogram illustrating the overall charge resolution. There are about 13,000 events in the figure.

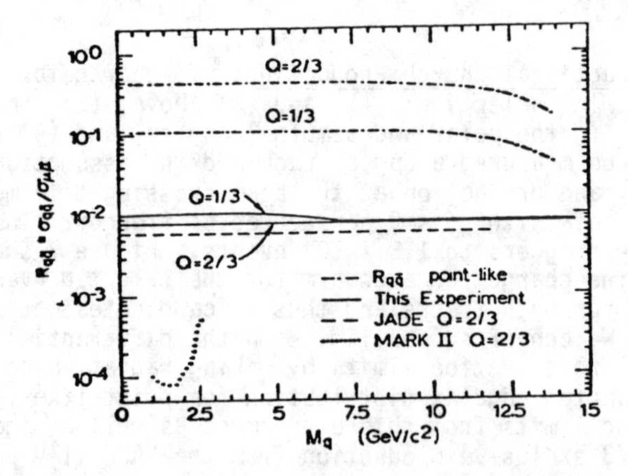

Fig. 8. Limits (90% confidence level) on exclusive quark production in e^+e^- annihilation. The limits for JADE are from Ref. 5 and the limits from MARK II are from Ref. 4.

quarks. In multiparticle final states we lost efficiency for measuring charge when more than a single particle hit a counter.

The following cuts reduce the 1.5×10^6 triggers to 6.1×10^3 events. (1) Events must be in time with the beam crossing and the particles must move out from the interaction point. (2) Three or more tracks are required. (3) To reduce backgrounds from accidental tracks at least two tracks are required to have dE/dx greater than 25% of normal. This last cut reduced the efficiency for those events in which there were 2 relativistic Q = 1/3 quarks and only one other particle within the acceptance.

The charge was calculated for tracks satisfying the following two conditions: (1) at least 5 of the innermost "road" dE/dx counters had signals above 3% of $(dE/dx)_{MIP}$ and (2) the track must have intersected at least two TOF counters with valid TOF hits. Figure 9 is a plot of $1/\beta$ vs dE/dx for the selected tracks. The region below the Q = 0.5

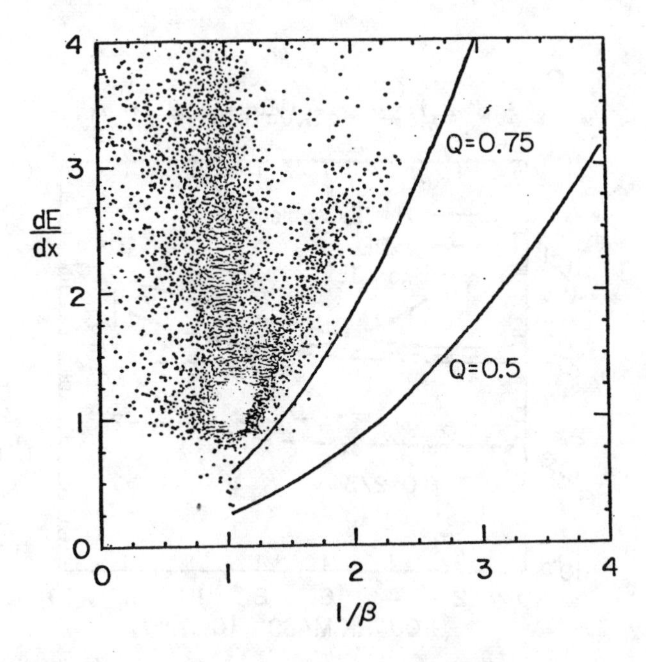

Fig. 9. dE/dx vs. $1/\beta$ for tracks (13840) in events which pass the selection criteria of the search for charge 1/3e particles. Q = 0.5 and Q = 0.75 contours are also shown. Tracks with low $1/\beta$ are accidental tracks which use very early time values in the outer counters. The tracks with high dE/dx and $\beta \approx 1$ are due to more than one particle in the road counters. Candidates are discussed in the text.

curve would contain > 95% of any charge Q = 1/3 particles. There is one point in this region with dE/dx = 0.9 and $\beta = 0.4$. This event was traced to a single incorrect TOF value along the track. Eliminating this value gave $\beta \sim 1$. This hypothesis was verified by a hit Cherenkov counter along the track and the candidate was discounted.

Similarly > 95% of the Q = 2/3 particles would fall below the line marked Q = 0.75 in Fig. 9. There are 16 candidates. These candidates were examined individually and could be attributed to two causes: accidental tracks due to low-energy photons (7 events) and tracks which passed along the edges of "road" counters depositing some of their energy in the counter wrappings. Additional cuts eliminated all of these tracks but only 35% of the "normal" tracks. The final limits are shown in Fig. 10 along with the limits from the MARK II[16] and JADE[17].

These results are preliminary but a paper is being prepared for Physics Letter.[19]

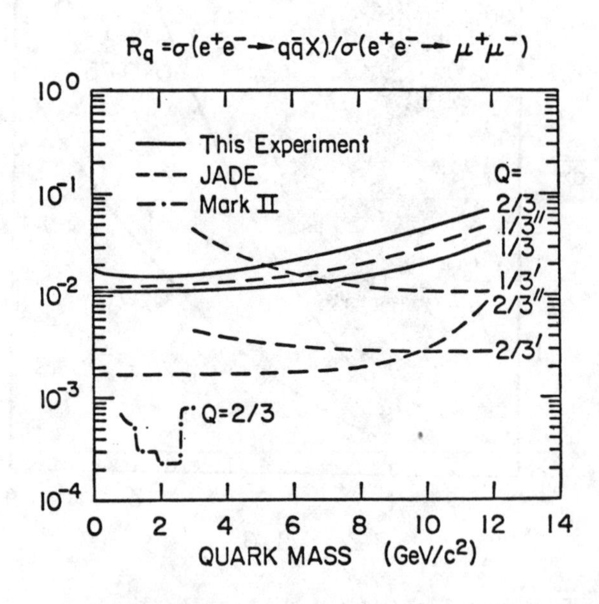

Fig. 10. Upper limits on quark production ($R_q = \sigma_{qq}\text{-}X/\sigma_{\mu\mu}$). Results from MARK II and JADE are included. In the JADE data analysis two hypothetical quark momentum spectra were used. The results from both of them are shown here. We used a Monte Carlo most similar to the JADE model identified by the double primes in the figure.

Highly Interacting Quark Search

One interesting possibility for why quarks have not yet been discovered at accelerators is that they have anomously large interaction cross sections and thus they interact before they reach the detector.[20,21] To explore this possibility the inner 5 MWPC's of the FQS detector were equipped with read-out for dE/dx. This system was capable of detecting quarks with 100 times normal hadronic cross sections. This data is still being analyzed but preliminary indications are that no quarks have been seen. The final results will be presented soon.

Conclusion

The most likely result of looking for something that is not expected to be there is a null result but if you don't look you will certainly never find anything.

References

1. The members of the PEP14 collaboration are: A. Marini, I. Peruzzi, M. Piccolo, F. Ronga--Laboratori Nazionali di Frascati dell' INFN. D. M. Chew, R. P. Ely, T. P. Pun, V. Vuillemia--Lawrence Berkeley Laboratory. R. Fries, B. Gobbi, W. Guryn, Donald H. Miller, M. C. Ross--Northwestern University. D. Besset, S. J. Freedman, A. M. Litke, J. Napolitano, T. C. Wang--Stanford University. Frederick A. Harris, I. Karliner, Sherwood Parker, D. E. Yount--University of Hawaii.

2. For a recent review of the status of the confinement hypothesis see M. Bander, Phys. Rep. 75, 205 (1981).

3. G. S. Larue, W. M. Fairbank, and A. F. Hebard, Phys. Rev. Lett. 38, 1011 (1977); G. S. Larue, W. M. Fairbank, and J. D. Phillips, Phys. Rev. Lett. 42, 142 (1979); and G. S. Larue, J. D. Phillips, and W. M. Fairbank, Phys. Rev. Lett. 46, 967 (1981).

4. S. I. Parker et al., Phys. Scr. 23, 4, 658 (1981).

5. J. Napolitano et al., Phys. Rev. D25, 2837 (1982).

6. A. Marini et al., Phys. Rev. D (in press).

7. R. B. Hicks et al., Il Nuov. Cim 14A, 65 (1973).

8. T. Kifune et al., UPSJ, 36 (1973).

9. P. Franzini and S. Shulman, Phys. Rev. Lett. 21, 1013 (1968).

10. P. C. M. Yock, Phys. Rev. D18, 641 (1978).

11. P. C. M. Yock, Phys. Rev. D22, 1 (1980).

12. For a review of tachyon and quark searches see L. W. Jones, Rev. Mod. Phys. 49, 717 (1977).

13. H. Hanni and E. Hugentobler, in Tachyons, Monopoles and Related Topics, E. Recami, ed; North-Holland Publishing Company (1978).

14. F. Ashton et al., Nucl. Instrum. Methods 93, 349 (1971).

15. Ref. 11, P. C. M. Yock, Phys. Rev. D23, 1207 (1981). See the references cited in these papers for previously reported evidence for massive particles.

16. J. M. Weiss et al., Phys. Lett. 101B, 439 (1981).

17. J. Burger, Proceedings, 10th International Symposium on Lepton and Photon Interactions at High Energies, Bonn 1981.

18. A. Marini et al., Phys. Rev. Lett. 48, 1649 (1982).

19. M. Ross et al., Phys. Lett. (to be published).

20. J. Orear, Phys. Rev. D18, 3504 (1978).

21. A model of modified QCD that indicates highly interacting free quarks is described in A. DeRujula, R. C. Giles, and R. L. Jaffe, Phys. Rev. D22, 227 (1980).

IS FRACTIONAL ELECTRIC CHARGE PROBLEMATIC FOR QCD?*

R. Slansky
Theoretical Division, Los Alamos National Laboratory**
University of California, Los Alamos, New Mexico 87545

ABSTRACT

A model of broken QCD is described here; SU_3^c is broken to SO_3^g ("g" for "glow") such that color triplets become glow triplets. With this breaking pattern, there should exist low-mass, fractionally-charged diquark states that are not strongly bound to nuclei, but are rarely produced at present accelerator facilities. The breaking of QCD can be done with a $\underline{27}^c$, in which case, this strong interaction theory is easily embedded in unified models such as those based on SU_5, SO_{10}, or E_6. This work was done in collaboration with Terry Goldman of Los Alamos and Gordon Shaw of U.C., Irvine.

The work described in this talk was done with Terry Goldman and Gordon Shaw.[1] In it we took seriously the growing experimental evidence that a tiny portion of the charged particles in Nature have fractional electric charge, $\pm \frac{1}{3}$ + integer.[2] There has been more than the usual resistance to accepting this evidence, and not all of the grumbling is based on experimental objections, but rather on the understanding that the observation of fractional charge would likely require a major revision of QCD. Although it is not yet proven that unbroken QCD must confine, there is some good evidence for it. Thus if fractional charges do exist and are related to quarks, the subnuclear fermions hypothesized by Gell-Mann and Zweig to be the basic constituents of hadrons,[3] then SU_3^c must be broken. There is an example of broken QCD,[4] but the breaking mechanism described there requires other major modifications of the standard model: the breaking is done with color-triplet scalars. Usually $\underline{3}^c$ particles of any kind must have fractional charge, although these $\underline{3}^c$ are electrically neutral. Moreover, the origin of additional symmetries imposed on their model is not clear. So, some resistance to accepting Ref. 2 is based on the belief that a major revision of the theory is necessary to accommodate fractional charge. However, in the final analysis, we must await further experimental confirmation of those observations.

---

*Presented at Novel Results in Particle Physics, 5th International Conference, Vanderbilt University, May 24-26, 1982.

**Work supported by the U.S. Department of Energy.

Our object is to show that the observation of fractional charge does not require such drastic modifications of the standard model, but can be achieved by extending slightly the symmetry-breaking scheme of the theory. Before proceeding, several comments are in order. What we have tried to do is construct a gauge-theoretic phenomenology that allows for fractional charge, and at the same time disturbs the standard model very little. In a gauge-theory framework there are many restrictions, and so the model has many predictions for the properties and interactions of fractionally-charged particles. However, there are other ways that fractional charge could appear in a theory. For example, in many models we may simply modify the electric-charge operator. There could be color nonsinglets with charge assignments outside the usual triality-electric charge constraints such as a $\underline{3}^c$ with charge +1/3, so that composite color singlets containing such a constituent would be fractionally charged. In the same vein, there could be color singlet "leptons" with fractional charge. It is possible in some unified models to construct such unusual electric charge operators, although it is not possible for theories based on the exceptional groups. It appears that these schemes would[5] imply an unacceptably large abundance of relic fractional charge left from the "big bang."

The widely-held belief that unbroken nonabelian gauge theories confine the charges of the local-symmetry group is accepted here. We also assume that the fractionally-charged particles have constituents that are not color singlets, so that they are strongly interacting. We maintain the usual connection between charge and $SU_3{}^c$ triality. Consequently, in the context of QCD we must find a symmetry-breaking pattern that allows for unconfined fractional charge. Thus, our model is constructed in the spirit of that of De Rújula, Giles, and Jaffe (DGJ),[4] but we propose a different symmetry-breaking scheme for $SU_3{}^c$, and as a result, the implications of the model are quite different. The most important differences are, phenomenologically, that some fractionally-charged states (called diquarks) need not be very massive and need not be strongly bound to nuclei, although their production at present machine energies is suppressed; theoretically, the breaking scheme is easily embedded into unified models and requires neither extra symmetries nor abnormal electric-charge assignments. After presenting the salient features of our version of broken QCD including the spectrum of fractionally charged particles, we discuss in the following order some questions regarding the symmetry breaking, masses, production cross sections, the formation of bound states with nuclei, and the highly speculative yet enormously important application of using the diquarks as a catalyst in fusion reactors.[6]

SU_3 has two maximal subgroups, SO_3 and $SU_2 \times U_1$, and several subgroups that are not maximal, including SU_2, $U_1 \times U_1$, U_1, and nothing. We assume that SU_3^c is broken to its maximal subgroup SO_3^g (g for "glow"); this embedding is defined by the branching rule $\underline{3}^c = \underline{3}^g$. (Mathematically this is the same embedding that is used in nuclear physics.)[7] In breaking SU_3^c to SO_3^g, three gluons are left massless and the other five gluons get equal masses μ. In the short distance limit $\ll 1/\mu$, this theory retains the features of unbroken QCD. At long distances all eight gluons are confined, since all have nonzero glow, the SO_3^g charges.

We now describe the spectrum of multiquark states, assuming that glow is confined.[8] One of the original motivations for selecting SU_3 as the color group was the statistics problem encountered in forming the observed $\underline{56}$, $\ell = 0$ of low-mass baryons from three quarks.[9] The $\underline{56}$ is formed from the symmetrized product, $(\underline{6} \times \underline{6} \times \underline{6})_s$ of SU_6, where SU_6 contains the $SU_3^e \times SU_2^s$ of the Eightfold Way (e) and rotational symmetry as $\underline{6} = (\underline{3}^e, \underline{2}^s)$, so $(\underline{6}^3)_s = (\underline{8}^e, \underline{2}^s) + (\underline{10}^e, \underline{4}^s)$. (All irreps, including those of SU_2, are labeled by their dimensions.) With the addition of the color quantum numbers the group becomes $SU_6 \times SU_3^c$, and the triquark states satisfying the exclusion principle are

$$[(\underline{6},\underline{3}^c)^3]_a = (\underline{56},\underline{1}^c) + (\underline{70},\underline{8}^c) + (\underline{20},\underline{10}^c) \quad . \tag{1}$$

The SO_3^g singlets exactly coincide with the SU_3^c ones, since the branching rules are $\underline{8}^c = \underline{3}^g + \underline{5}^g$ and $\underline{10}^c = \underline{3}^g + \underline{7}^g$. For any other subgroup of SU_3^c there are $\underline{70}$'s, $\underline{20}$'s, or both that are unconfined, since the $\underline{8}^c$, $\underline{10}^c$, or both will have subgroup singlets. The SO_3^g breaking scheme is the only one that does not disturb the spectrum of triquark singlets.

One objection in the past to an SO_3^g "color" group is the existence of diquark singlets: the difermions are

$$[(\underline{6},\underline{3}^c)^2]_a = (\underline{21},\bar{\underline{3}}^c) + (\underline{15},\underline{6}^c) \quad , \tag{2}$$

and the $(\underline{15},\underline{6}^c)$ has a glow singlet, since $\underline{6}^c = \underline{1}^g + \underline{5}^g$. The $SU_3^e \times SU_2^s$ content of the $\underline{15} = (\underline{6}^e,\underline{1}^s) + (\bar{\underline{3}}^e,\underline{3}^s)$. The Q diquarks may have masses quite similar to the $q\,\bar{q}$ color-singlet masses.

The spectrum of "low-mass" observable scalar diquarks in Eq. (2) consists of a strong isotriplet with charges of uu, ud, and dd, and so forth. The three axial-vector diquarks have the SU_3^e quantum numbers of $\bar{u}$, $\bar{d}$, and $\bar{s}$.[10] The relative mass of the scalar and axial vector states is model dependent. If the average mass difference is dominated by the color magnetic splitting, then its value is -1/4 that of the $\underline{1}^c$ $q\bar{q}$ system. However, if only the glow-magnetic spin splitting dominates, then the scalars are lower in mass. Thus, it is possible that the charge 4/3 scalar is the least massive and is the only stable diquark state. (For example, M_{Σ^+} = 1189.3 MeV, M_{Σ^o} = 1192.5 MeV, and M_{Σ^-} = 1197.3 MeV suggests this pattern may be correct, at least if the gluon field, which is flavor independent, is ignored.) Of course, the theory itself does not depend on the speculation that this is the correct ordering of masses.

The $q\bar{q}$ color-singlet states coincide with the glow singlets. The diquark-antiquark states do include glow singlets:

$$(\bar{\underline{6}},\bar{\underline{3}}^c) \times [(\underline{6},\underline{3}^c)^2]_a = (\underline{6},\bar{\underline{6}}^c) + (\underline{120},\bar{\underline{6}}^c) + (\underline{6},\underline{3}^c) + (\underline{6},\underline{3}^c) + (\underline{84},\underline{3}^c) + (\underline{120},\underline{3}^c) + (\underline{6},\underline{15}^c) + (\underline{84},\underline{15}^c). \quad (3)$$

The $\underline{15}^c$ and $\underline{3}^c$ contain no glow singlets, but $\underline{6}^c$ does; thus the observable states, classified by $SU_3^e \times SU_2^s$ are $\underline{6} = (\underline{3}^e,\underline{2}^s)$ and $\underline{120} = (\underline{3}^e,\underline{2}^s) + (\bar{\underline{6}}^e,\underline{2}^s) + (\underline{15}^e,\underline{2}^s) + (\underline{3}^e,\underline{4}^s) + (\underline{15}^e,\underline{4}^s)$. These are expected to be more massive than diquarks.

In a sense, the unbroken SO_3^g theory cannot confine quarks because quark-gluon bound states always have a glow singlet. Thus a quark in $\underline{3}^c$ can bind with a gluon in $\underline{8}^c$ to form states in $\bar{\underline{6}}^c + \underline{3}^c + \underline{15}^c$, where the $\bar{\underline{6}}^c$ has a glow-singlet piece that may be indistinguishable from a quark. However, in analogy with QCD, where glueball masses are expected to be around 1.6 GeV (perhaps), the masses of the quark-glue states should be at least as large as typical baryon masses and significantly larger than the diquark masses.[11]

The breaking of SU_3^c can be described in terms of an irrep or a conjugate pair of complex irreps of (effective) Higgs spinless bosons. If the spinless bosons satisfy the usual connection between SU_3^c triality and electric charge, the breaking representation must have zero triality: $\underline{8}^c$, $\underline{10}^c+\overline{\underline{10}}^c$, $\underline{27}^c$, $\underline{28}^c+\overline{\underline{28}}^c$, $\underline{35}^c+\overline{\underline{35}}^c$, $\underline{55}^c+\overline{\underline{55}}^c$, $\underline{64}^c$, etc. However, before an irrep can break a group to a subgroup, it must have a subgroup singlet, and of these, only the $\underline{27}^c$ and $\underline{28}^c$ $+\overline{\underline{28}}^c$ have SO_3 singlets. (The $\underline{28}$ is formed irreducibly as $[\underline{3}^6]_s$. The $\underline{27}^c$ has the $SO_3{}^g$ branching rule $\underline{27}^c = \underline{1}^g + \underline{5}^g + \underline{5}^g + \underline{7}^g + \underline{9}^g$.) The simplest candidate for breaking color to glow is a $\underline{27}^c$ of completely flavorless spinless bosons; the Lagrangian mass is expected to be $O(\mu)$. Production of the unconfined $\underline{1}^g$ "Higgs" is suppressed as are all $\underline{1}^g$ that are not $\underline{1}^c$ states. The radiative corrections to the Higgs mass could be large enough that its experimental signature would be similar to that of a glueball.

The (effective) Higgs potential depends on the second-order invariant in $(\underline{27}^2)_s$ (defined as $|\underline{27}|^2$), two third-order invariants in $(\underline{27}^3)_s$, and four independent fourth-order invariants in $(\underline{27}^4)_s$ [including $(|\underline{27}|^2)^2$].[12] There are ranges of parameters for which the $\underline{27}$ can break $SU_3{}^c$ either to SO_3 or to $SU_2 \times U_1$, which are the two maximal little groups of the $\underline{27}$. If we assume that the fourth-order Higgs potential provides an exhaustive description of the possible breaking directions, then by Michel's conjecture these are the only two possible breaking directions for a single $\underline{27}^c$ with a nonzero vacuum expectation value.[13] We may select the parameters in a range where the breaking is to $SO_3{}^g$.

An important advantage of the $\underline{27}^c$ breaking scheme is the ease with which it can be embedded into unified models of electromagnetic, weak, and strong interactions. In most models the electric charge of a state is related to the triality of its color representation; only triality-zero irreps have neutral states. Also, the $\underline{27}^c$ is the irrep of highest weight in adjoint times adjoint; the (effective) scalars could conceivably be formed as a set of bound states. This possibility is easily generalized to unified models.

In particular, for SU_5 the irrep with highest weight in $\underline{24} \times \underline{24}$ is the $\underline{200}$, which contains the $SU_2{}^w \times SU_3{}^c$ ($\times U_1$) irreps $(\underline{1},\underline{1}^c)$ $+ (\underline{3},\underline{1}^c) + (\underline{5},\underline{1}^c) + (\underline{2},\underline{3}^c) + (\underline{2},\overline{\underline{3}}^c) + (\underline{4},\underline{3}^c) + (\underline{4},\overline{\underline{3}}^c) + (\underline{3},\underline{6}^c)$ $+ (\underline{3},\overline{\underline{6}}^c) + (\underline{1},\underline{8}^c) + (\underline{3},\underline{8}^c) + (\underline{2},\underline{15}^c) + (\underline{2},\overline{\underline{15}}^c) + (\underline{1},\underline{27}^c)$, where the U_1 charges of all irreps with zero $SU_3{}^c$ triality are zero[14]. Thus the $(\underline{1},\underline{27}^c)$ set of scalars carries no flavor at all. The one glow singlet in the $\underline{27}^c$, although strongly interacting, is rarely produced and would be difficult to detect; the rest of the $\underline{27}^c$ is confined. (We do not analyze the bound states with a scalar here: the quantum-number analysis is trivial and the mass estimates unreliable.) The $\underline{200}$ is by far the smallest irrep of SU_5 with a $\underline{27}^c$.

A similar discussion can be made for SO_{10} and E_6. In SO_{10} the $\underline{27}^c$ breaking is in the $\underline{770}$ only, which is the highest weight in $\underline{45} \times \underline{45}$; the only $\underline{27}^c$ is in the SU_5 $\underline{200}$ piece of the $\underline{770}$. In E_6 the irrep of highest weight in adjoint times adjoint is the $\underline{2430}$ and the only $\underline{27}^c$ is in the SO_{10} $\underline{770}$. After spontaneous symmetry breaking the exact local symmetry of these theories could easily be $U_2 \sim SO_3{}^g \times U_1{}^{em}$, not $U_3 \sim SU_3{}^c \times U_1{}^{em}$, as is supposed in the usual approach.

Recall that in the breaking pattern suggested by DGJ, all eight gluons have equal nonzero masses and the vacuum has a global SU_3 symmetry. However, $SU_3{}^c$ is completely broken so that all the multiquark states in $[(\underline{6},\underline{3}^c)^k]_a$ (k = 1, 2, ...) would be observable. DGJ expect from bag-model considerations that a gluon cloud of extension $1/\mu$ forms around any state that is not a color singlet. The self energy of the gluon field contributes to the mass, so the color nonsinglet states are much heavier than the color singlets. It is μ and not the QCD scale parameter that determines the mass of color nonsinglets; for small μ all fractionally charged states should be very heavy (of order $1/\mu$) in their model.[4,15]

If $SU_3{}^c$ is broken to $SO_3{}^g$, the situation is very different. Of the eight colored gluons, the $\underline{3}^g$ is massless and the $\underline{5}^g$ has mass μ. It is important to keep in mind that all gluons carry glow, so the volume of the gluon cloud may be determined by the confinement mechanism of glow. However, the diquarks carry $\underline{6}^c$ of color, and the shielding of $\underline{5}^g$ gluons may require a slightly larger hadronic volume, making mass estimates difficult. For the $\underline{1}^g$ Higgs in the $\underline{27}^c$, this could be a significant contribution to the physical mass. For the moment we assume the $\underline{5}^g$ gluon contribution to the mass to be negligible, consistent with earlier comments on the bound-state mass splitting. Then, the glow-singlet masses should be controlled by the scale parameter of $SO_3{}^g$, which must be equal to the QCD scale parameter for $Q^2 >> \mu^2$, since the gauge coupling of $SU_3{}^c$ is the coupling of any of its subgroups in this domain. Consequently we expect the glow-singlet diquarks in Eq. (2) to have masses of order of the $q\,\bar{q}$ mesons, and the $q\,q\,\bar{q}$ glow singlets in Eq. (3) to have typical baryon masses. Estimates of quark-glue masses lie between baryon masses and glueball masses of QCD. Note that the proton mass is below the quark-glue-diquark threshold. If the $\underline{5}^g$ field configuration is similar to that supposed by DGJ, then the diquark mass should go as $1/\mu$ as $\mu \to 0$. However, Georgi has argued that QCD should be broken in a first-order phase transition, so μ cannot be arbitrarily small.[16] Then the diquark mass spectrum is discontinuous, and there is no small μ limit. Nevertheless, we assume that μ is small enough that production processes are suppressed.

Although these glow-singlets appear in many ways as ordinary hadrons, their production cross sections are severely suppressed. As a typical example, a rough estimate of the production of light diquarks in $e^+ e^-$ annihilation can be obtained from the linear-potential model of confinement in QCD.[17] At distances less than $1/\mu$ ($\cong$200 fermi/μ, where μ is now in MeV), all eight gluons participate in the confinement of color nonsinglets and the potential rises linearly to $\sim$200 GeV/μ. At distances greater than 200 f/μ, the channel-dependent potential between states with glow continues to rise, but that between glow singlets drops rapidly to zero due to the finite range of the $\underline{5}^g$ gluons and the low mass of the free diquarks. In this picture a diquark must tunnel through the barrier or be energetic enough to go over the top of the barrier to reach the separation of 200 f/μ needed to cut off the confinement contribution of the $\underline{5}^g$ gluons. In addition, the production is suppressed by its form factor, since the diquark is a composite particle. A gluon mass of order 50 MeV may be small enough to explain the experimental dearth of diquarks at PETRA. Similar arguments may be made for other scattering processes. Since μ can be large (but less than Λ), the gluon-cloud contribution to the diquark mass $[O(\Lambda^2/\mu)]$ may be less than 1 GeV.

Copious production of diquarks might be achieved at lower energies in heavy-ion collisions. Let us suppose the interaction region can be viewed as a quark gas at an elevated temperature in approximate thermodynamic equilibrium. Then the effective confining potential of QCD may be constant over the interaction region, and the probability of finding glow singlets (that are not color singlets) separated by 200 f/μ or more should be greatly enhanced.

The diquark Q is a strongly interacting particle with interaction range similar to the color-singlet hadrons. The QCD description of the collision region should be unaltered, since $\mu < 200$ MeV. Neither the color-singlet states formed during the final stages of the collision nor the final-state (glow-singlet) diquark couple to the $\underline{5}^g$ gluons, and so the character of the final-state interactions should be similar to those in meson-nucleon scattering. If the mass estimates above are also qualitatively correct, we should not expect any deeply bound states.

The stability of the triality ±1 states, such as the quark-glue or antiquark-diquark states, depends on mass relations. In this regard $\bar{Q}$ N (or $\bar{Q}$ nucleus) scattering is of special interest. If the $\bar{Q}$ N threshold is above the lowest-mass triality-one state, then the $\bar{Q}$ N may annihilate to it plus photons or mesons. The lowest-mass triality-one state would then be nearly stable, since its decay to diquarks would have to proceed by interactions similar to those responsible for proton decay in unified models.

The more exciting possibility technologically is for the lowest-mass triality-one state to have a mass greater than the $\bar{Q}$ N threshold. Then the lowest-mass $\bar{Q}$ state is not only stable against weak decay, but it is not annihilated in interactions with nucleons. If this state is the $\bar{u}\,\bar{u}$ antidiquark as speculated above, then it should form long-lived $\bar{Q}$-nucleus "atoms" of size around 50 fermi. An anomaly without fractional charge could involve a glow singlet such as the Higgs particle and a nucleus. Such composites would have shortened mean free paths.

Finally, for such charge -4/3 $\bar{u}\,\bar{u}$ diquarks, we stress an enormously important application. In the event that they can be mined or produced in high-energy collisions and are not annihilated in collisions with nucleons, these $\bar{Q}$'s can be used as a catalyst in fusion reactors. As detailed by Zweig,[6] the perfect catalyst is a relatively light, stable, charged -4/3 "quark" that does not bind very strongly to nuclei. A charge less than -1 greatly enhances the binding in, for example, D-D or D-T molecules. For $\bar{Q}$ masses of order 600 MeV, the molecular size is small and the fusion rapid.[6] Typical temperatures projected for fusion reactors are ~10 KeV, which should be sufficient to free a large fraction of the $\bar{Q}$'s from atomic states in the resultant fusion products.

We have enjoyed helpful conversations with G. Zweig, M. Bander, G. Chapline, and C. Thorn.

REFERENCES AND FOOTNOTES

1. R. Slansky, T. Goldman, and G. Shaw, Phys. Rev. Lett. 47, 887 (1981).
2. G. S. LaRue, W. M. Fairbank, and A. F. Hebard, Phys. Rev. Lett. 38, 1011 (1977); G. S. LaRue, J. D. Phillips, and W. M. Fairbank, ibid, 42, 142 and 1019E (1979); and ibid, 46, 967 (1981).
3. M. Gell-Mann, Phys. Lett. 8, 214 (1964); G. Zweig, CERN Report Th. 412, (1964), unpublished.
4. A. De Rújula, R. C. Giles, and R. L. Jaffe, Phys. Rev. D17, 285 (1978); ibid, D22, 227 (1980).
5. E. W. Kolb and R. Wagoner (private communications) say that primordial production is suppressed in our model.
6. G. Zweig, Science 201, 973 (1978).
7. J. P. Elliott, Proc. Roy. Soc. A245, 128 (1958).
8. R. Barbieri, L. Maiani, and R. Petronzio, Phys. Lett. 96B, 63 (1980).
9. H. Fritzsch and M. Gell-Mann, in Proceedings of the XVI International Conference on High-Energy Physics, Vol. 2 (National Accelerator Laboratory, 1973), p. 135.
10. It is a trivial matter to add other quarks to this classification scheme.
11. The derivation of more quantitative estimates is hindered by the lack of experimental guidelines; this problem can be solved from first principles about as easily as the color-singlet mass spectrum problem.
12. A fairly simple way to derive these results is to use the maximal subgroup embedding, $E_6 \supset SU_3$, with $\underline{27} = \underline{27}$, compute the E_6 tensor products $(\underline{27}^3)_s = \underline{1} + \underline{650} + \underline{3003}$ and $(\underline{27}^4)_s = \underline{27} + \underline{351}' + \underline{7722} + \underline{19305}'$, and count the SU_3 singlets from the $E_6 \supset SU_3$ branching rules. Ref. 14 is an aid to this last step, but a few more tensor products are needed to count the singlets in the $\underline{19305}'$, which has just two singlets.
13. L. Michel, CERN-TH-2716 (1979), Contribution to the A. Visconti Seminar.
14. W. G. McKay and J. Patera, "Tables of Dimensions, Indices, and Branching Rules for Representations of Simple Lie Algebras" (Dekker, New York, 1981).
15. If the gluon mass μ is much less than the QCD scale, the gluon force becomes very strong before its range $1/\mu$. Thus the bag model may misestimate the hadronic size for small μ. Nevertheless, it is attractive to impose confinement by supposing the bag size is $O(1/\mu)$ as $\mu \to 0$.
16. H. Georgi, Phys. Rev. D22, 225 (1980).
17. For a review, see M. Bander, Phys. Rep., in press, especially, Fig. 10.1.

OBSERVATION OF FRACTIONAL CHARGE OF 1/3e ON MATTER

J. D. Phillips, W. M. Fairbank, and C. R. Fisel
Stanford University, Stanford, CA 94305

ABSTRACT

Measurements on niobium spheres which show unambiguously the existence of fractional charges of 1/3e are reported. Charge changes of 1/3e on particular spheres when they contact other surfaces continue to be observed. Of six new measurements, four have the same residual charge, one has a residual charge of +1/3e with respect to the four, and one of -1/3e. Extensive measurements and critical analyses have assured us that the background forces are either negligible or have been measured and taken fully into account.

INTRODUCTION

In previous publications[1-4] we have presented results of a superconducting magnetic levitation experiment giving evidence for the existence of fractional charge. Since then we have continued to modify and improve the experiment and we report here six new measurements. These and earlier measurements are shown in Fig. 1.

We can measure the force on the ball to within 0.01e E_A in several hours. By taking the background forces fully into account the residual charge can be determined to within the errors shown in Fig. 1. We are forced to conclude that fractional charges with magnitude 1/3e must exist.

Out of 29 repeat measurements, we have observed 12 residual charge changes, in every case of $\pm$1/3e. These changes occurred only between levitations, when a ball contacted other surfaces. All 13 measurements of ball 6 have been consistent with 0e or $\pm$1/3e. All eight measurements of ball 10 have been consistent with 0.

APPARATUS

The experimental configuration is shown schematically in Fig. 2. A superconducting niobium sphere, up to 280 μm in diameter, is supported against gravity in a vacuum at 4.2 K on the magnetic field produced by underlying coaxial superconducting coils. The suspension system is extremely stable because zero resistance supercurrents flow both on the surface of the sphere and in the support coils. An alternating electric field E_A is applied by capacitor plates above and below the ball. The potential applied to the plates is generated with the proper phase at the frequency of the vertical mechanical oscillations of the sphere, 0.9 Hz. This electric field provides a vertical force on the sphere which is directly proportional to its charge. The rate of change of the amplitude of the ball's oscillations is proportional to the force on the ball. A sensitive SQUID magnetometer located immediately above the top capacitor plate is used to detect the ball's motion.

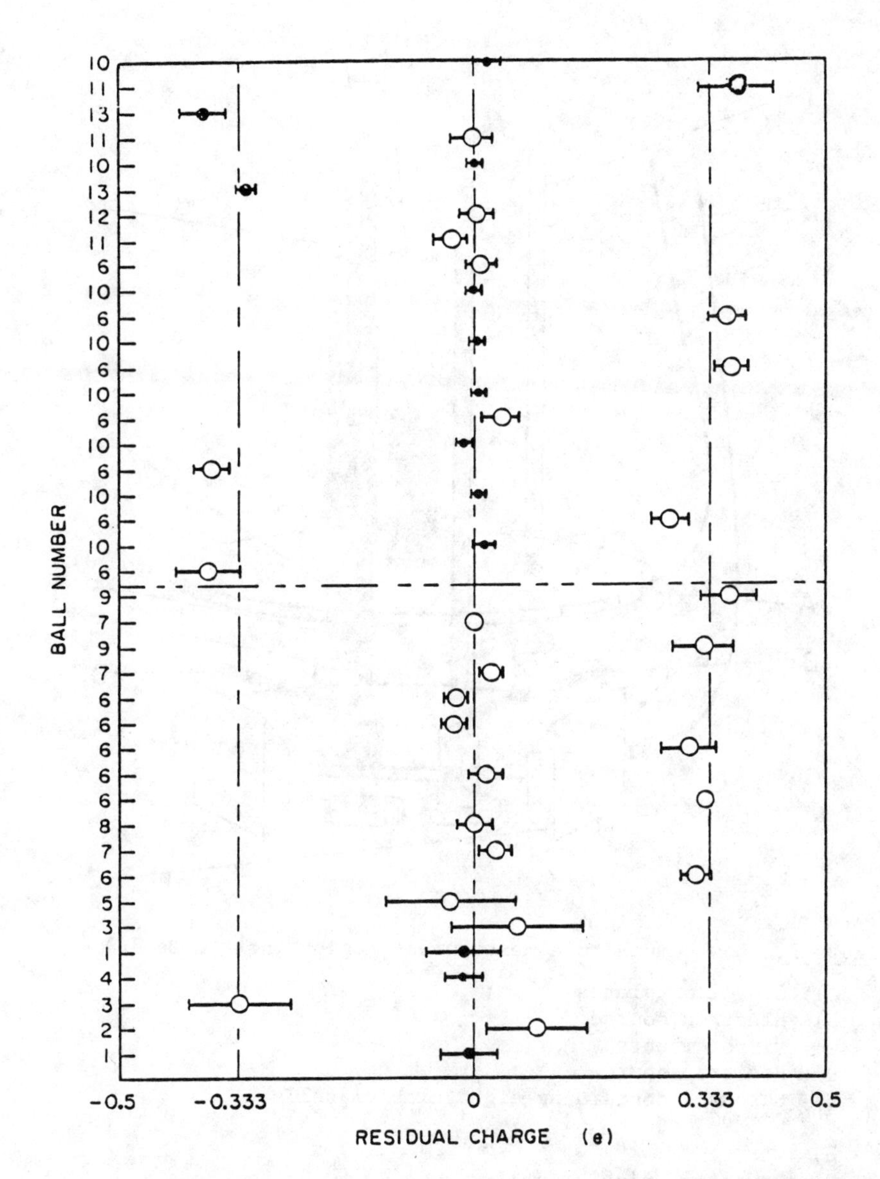

Figure 1. Histogram of all data taken up to March 1980. Residual charge differences (not shown here) from the September 1980 run are 0.000e, -0.003e, -0.002e, -0.256e, +0.259e, +0.004e ± 0.005e (statistical) ± 0.07e (systematic).

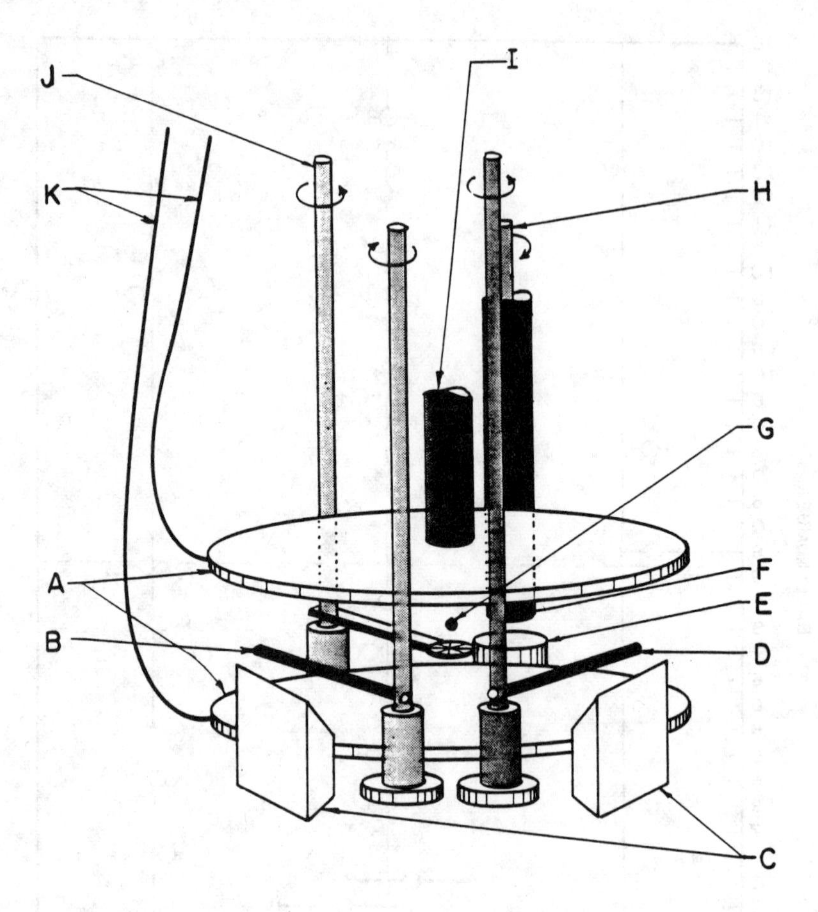

Figure 2. Low temperature measurement region (not to scale)

A) Capacitor plates
B) Electron source
C) Mirror mounts
D) Positron source
E) Carousel containing eight niobium balls
F) " " " " "
H) " " " " "
G) Levitated niobium ball
I) Magnetometer input coil
J) Ball arm
K) High voltage leads

The initial charge on a levitated sphere is typically on the order of $\pm 10^5 e$ and therefore must be neutralized to within a few unit charges of zero by exposure to electron or positron sources (see Fig. 2). This experiment does not measure the absolute charge on the sample as did Millikan's experiment, but rather the residual charge calibrated with respect to single electron charge changes. The charge is determined modulo one and it is therefore impossible to tell the difference between +1/3 and -2/3 of an electron charge.

FORCES ON THE BALL

In order to obtain the residual charge from the ball's motion, we must take account of background forces. The alternating force on the ball is

$$F_A = (q_r + ne)E_A + \vec{P}_A \cdot \vec{\nabla} E_F + \vec{P}_F \cdot \vec{\nabla} E_A + F_M + F_Q \tag{1}$$

where

- q_r = residual charge
- n = integer charge number
- E_A = z component of $\vec{E}_A$
- $\vec{P}_A$ = induced electric dipole of the ball
- E_F = z component of the fixed electric field arising from contact potential variations on the plates near the ball.
- $\vec{P}_F$ = permanent electric dipole of the ball
- F_M = magnetic force
- F_Q = quadrupole force

The measured residual force F_A^r is defined as the value of F_A for $n = 0$ in (1). It can be measured to within 0.01 eE_A with the technique described above.

We have shown[1-4] by measuring $\vec{P}_F$ and $\partial E_A/\partial z$ and by other measurements and calculations, that all known electromagnetic multipole forces are negligible at z_o (where $\partial E_A/\partial z = 0$) except for the force due to DC electric field gradients from the plates. Eliminating these negligible forces, (1) becomes

$$\frac{F_A^r}{E_A} = q_r + R^3 \frac{\partial E_F}{\partial z}(z) + \left(\frac{P_{Fz}}{E_A}\right) \frac{\partial E_A}{\partial z}(z) \tag{2}$$

where R is the ball's radius. We measure F_A^r as a function of each ball's position. q_r is independent of position. The second term in (2) represents the interaction of the ball's alternating electric dipole moment $R^3 E_A$ with the DC electric field gradient $\partial E_F/\partial z$ due to contact potential variations on the plates.

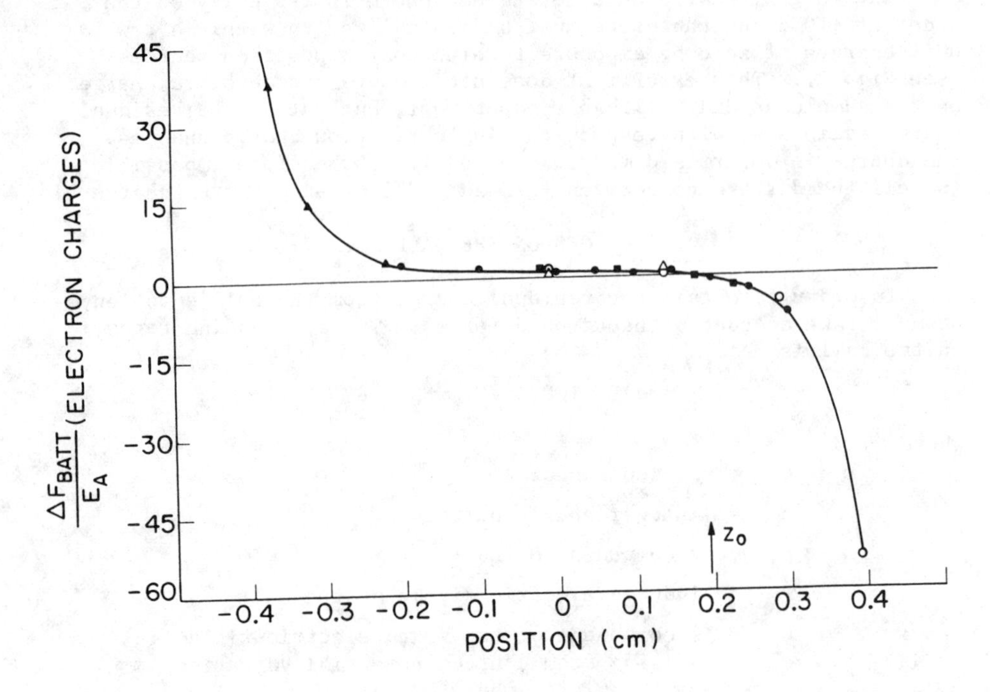

Figure 3. Measurements of the alternating electric field gradient $\partial E_A/\partial z$ as a function of ball position obtained by superimposing a DC ("battery") electric field on the AC field between the plates used to measure residual charge. The strong position dependence of these measurements, which are reproducible for all levitations, is used to determine accurately the relative positions of balls in different levitations. A DC electric dipole moment due to contact potential variations on the ball gives a force during residual charge measurements of 1.5% of that plotted here for the maximum vertical dipole usually observed, $P_{Fz} = 1.0 \times 10^{-7}$ esu cm. In more recent measurements with optically flat capacitor plates, the position at which $\partial E_A/\partial z = 0$, z_o, has been within 0.04 cm of the center of the plates, 0 cm.

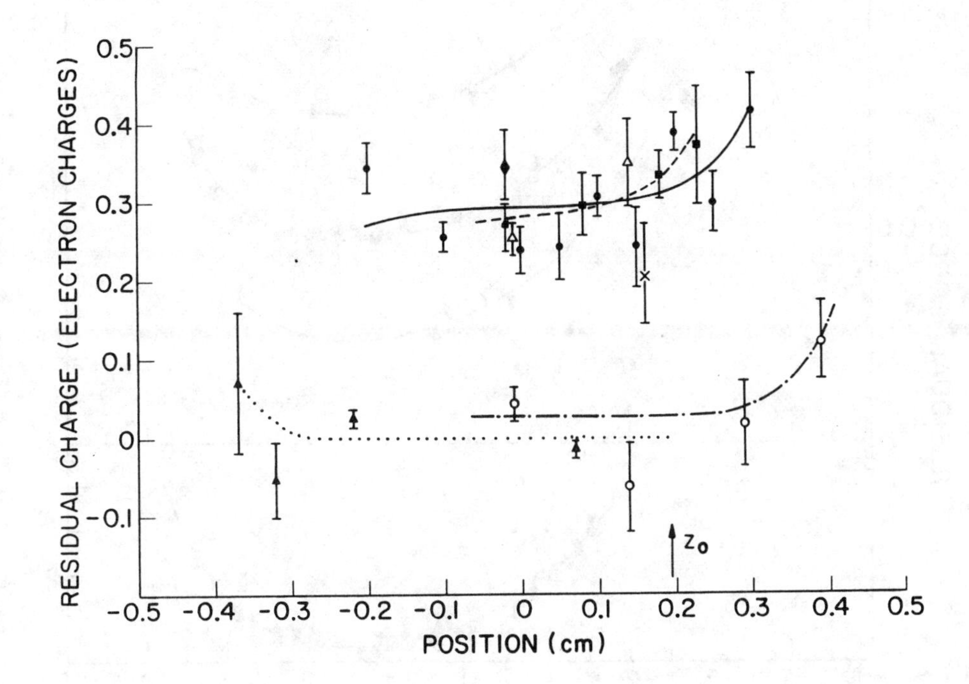

Figure 4. The position dependence of the residual force F^r_A/E_A for two runs early in 1977. The patch effect electric field gradient is 0 (case 1 in text). The curves are least squares fits to

$$\frac{F^r_A}{E_A} = q_r + P_{Fz}\left(\frac{\partial E_A}{\partial z}\right) .$$

The constant vertical difference between these curves, except for the upturns at the ends due to a DC electric dipole moment on the ball, indicates residual charges for the two upper curves of +0.313 $\pm$.019e and +0.344 $\pm$ .010e.

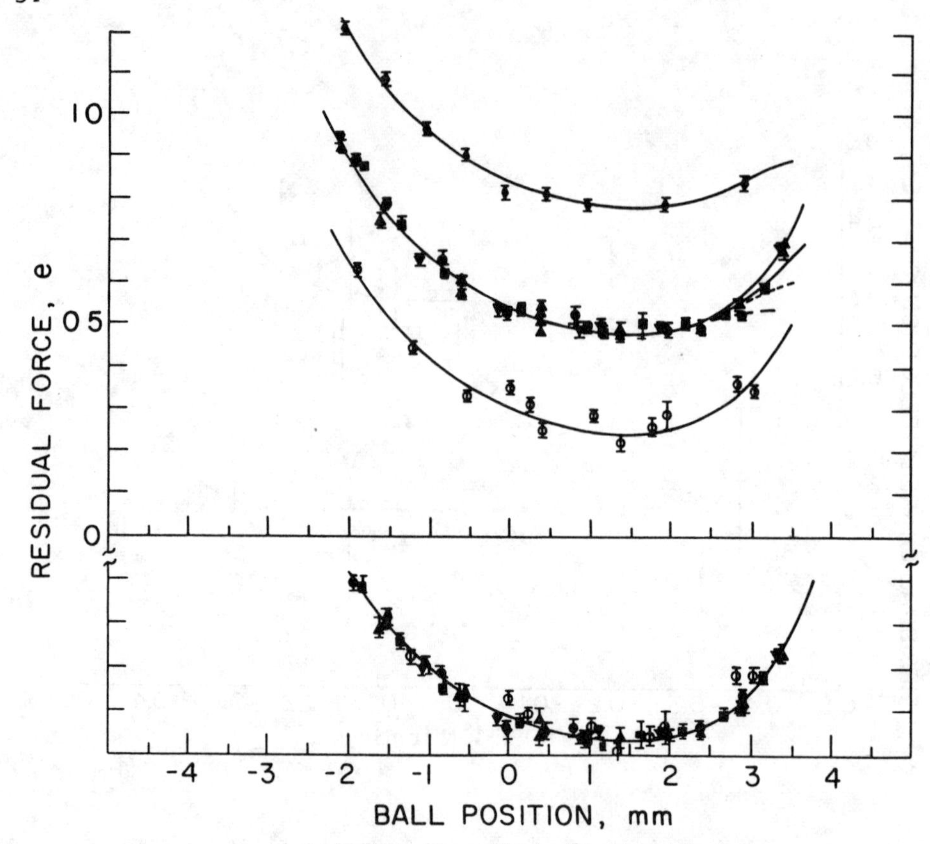

Figure 5. The position dependence of the residual force F_A^r/E_A for a run begun in September 1980. The patch effect electric field gradient for $\partial E_F/\partial z$ is not zero, and balls of the same radius were measured (case 2 in the text). The balls in the order in which they were measured are ball 14 (▲), 14 (▼), 15 (■), 16 (0), 15 (◆), and 14 (●). The curves in the upper half are least squares fit to

$$\frac{F_A^r}{E_A} = q_r + P_{Fz}\frac{\partial E_A}{\partial z} + R^3\frac{\partial E_F}{\partial z} ,$$

where a function representing $\partial E_F/\partial z$ is chosen to interpolate between the points. The constant differences between these curves indicate residual charge differences (not shown in Fig. 1 because no absolute charge determination could be made) for the upper and lower curves of +0.259e and -0.256e $\pm$0.005e (statistical) $\pm$0.07e (systematic).

In the lower curve, the data are offset by residual charge and corrected for DC electric dipole moment, showing that $\partial E_F/\partial z$ remained the same for these levitations.

The third term in (2) represents the interaction of the ball's dipole moment P_{Fz} with the alternating electric field gradient $\partial E_A/\partial z$ between the plates. The gradient $\partial E_A/\partial z$ is primarily due to the images of the ball's dipole moment in the plates, and is a calculable function of position. We measure $\partial E_A/\partial z(z)$ independent of the ball's charge by inducing a large DC dipole on the ball with a DC ("battery") potential difference between the plates. The measurements are plotted in Fig. 3. Both they and the third term in (2) are proportional to $\partial E_A/\partial z(z)$ and thus have the same position dependence. We extract the proportionality constant $P_{Fz}/(2 R^3 E_{Batt})$ by a least squares fit with a separate value of P_{Fz} for each levitation. If P_{Fz} is 1×10^{-7} esu cm (our largest dipole has been 2×10^{-7} esu cm), $(P_{Fz}/E_A)(\partial E_A/\partial z)$ is 1.5% of the battery curve at all positions z, and both vanish at $z = z_o$.

ANALYSIS

Now we have

$$\frac{F^r_A}{E_A}(z_o) = q_r + R^3 \frac{\partial E_F}{\partial z}(z_o) \tag{3}$$

From this point on the analysis falls into three cases listed below, depending on whether $\partial E_F/\partial z$ is zero, and on whether balls of different radius are measured. Of course, $\partial E_F/\partial z$ must remain constant for this analysis to be valid.

Case (1) - The patch effect field gradient $\partial E_F/\partial z$ is zero. Figure 4 shows measurements[2] of $F^r_A/E_A = q_r$ at z_o. These measurements show residual charges near zero and 1/3 e.

Case (2) - The patch effect field gradient $\partial E_F/\partial z$ is not zero and all balls measured have the same radius R. With a constant $R^3\, \partial E_F/\partial z$ at z_o, a difference in F^r_A/E_A for these balls is a difference in q_r. Data taken since our most recent publication[4] are shown in Figure 5. The differences of F^r_A/E_A of each curve with respect to the average of that of the four middle curves are ball 14: 0.000e, 14: -0.003e, 15: -0.002e, 16: -0.256e, 15: +0.259e, 14: +0.004e. The statistical errors for all are ±0.005e. There is a magnetic background force which is proportional to the angle between the magnetic levitation field and $\vec{g}$, which in the run shown here, begun in September 1980, could have contributed a systematic error up to ±0.07e. For the data above the dashed line in Fig. 1, taken before September 1980, and after the September 1980 run, this angle was checked and adjusted so that the error proportional to it was less than 0.01e. Other systematic errors, summarized in Ref. 3, are $\lesssim$0.03e. Ball 16 is 8% smaller than balls 14 and 15. This gives rise to an additional ±.03e uncertainty in its charge, depending on the value of $\partial E_F/\partial z$.

The lower curve in Fig. 5 shows

$$R^3 \frac{\partial E_F}{\partial z} = \frac{F^r_A}{E_A} - q_r - \left(\frac{P_{Fz}}{E_A}\right)\frac{\partial E_A}{\partial z}$$

for each measurement. Since these points all fall on the same curve, $\partial E_F/\partial z$ remained constant. The constancy of $\partial E_F/\partial z$ is crucial to subtracting out this background force.

The September 1980 run continued into 1981. The measurements were consistent with fractional charge, but the measured residual force sometimes drifted. This occurred over a period of six hours, the length of time for which a ball is usually left at one position between the plates while the residual force is measured. Remeasurements of the residual force at any particular plate position failed to repeat, once by as much as 0.2 eE_A. The drift was definitely not present in any run before 1981, so it was not present when any of the data shown in Fig. 1 was taken. We are continuing to investigate it.

Case (3) - The patch effect field gradient $\partial E_F/\partial z$ at z_o is not zero and a number of balls with different radius are measured. Figure 6a shows previously published data on five balls of different radius. If all balls have $q_r = 0$ then

$$\frac{F_A^r}{(E_A R_i^3)}(z_o) = \frac{\partial E_F}{\partial z}(z_o),$$

where R_i is the radius of ball i. This quantity should be the same for every levitation. The two curves shown differ by a constant, so they have different residual charges. Figure 6b shows these data adjusted for differences in radius, residual force and dipole moment. They all fit the same curve, implying that $\partial E_F/\partial z$ remained constant during the measurement.

Figure 6c shows $F_A^r/E_A(z_o)$ plotted against R^3. According to (3), the lines have slope $\partial E_F/\partial z(z_o)$ and intercepts which are the residual charges. Note that each of the points falls on one of these lines. Thus these data are consistent with four measurements of zero residual charge and one of -e/3 residual charge.

RESULTS

We have made 45 independent measurements on 16 balls. The results are shown in Figure 1 in chronological order from the bottom to the top. The last six measurements have been made since our most recent publication[4], and are not included in the figure because only residual charge differences were measured. The statistical errors represent 1 standard deviation. Out of 29 repeat measurements, we have observed 12 residual charge changes, in each case of $\pm 1/3e$. These changes have occurred only when a ball was brought into contact with other surfaces. We have made 14 independent measurements on ball 6 over a 3½ year period, every measurement consistent with 0 e or $\pm 1/3e$, and have observed 9 changes of 1/3e. For the 27 measurements since Ref. 4, the ball stayed at low temperature during any one cooldown. All of the balls measured since Ref. 2 were heat-treated on a tungsten plate at 1800° C for 17 hr in a vacuum of 10^{-9} torr.

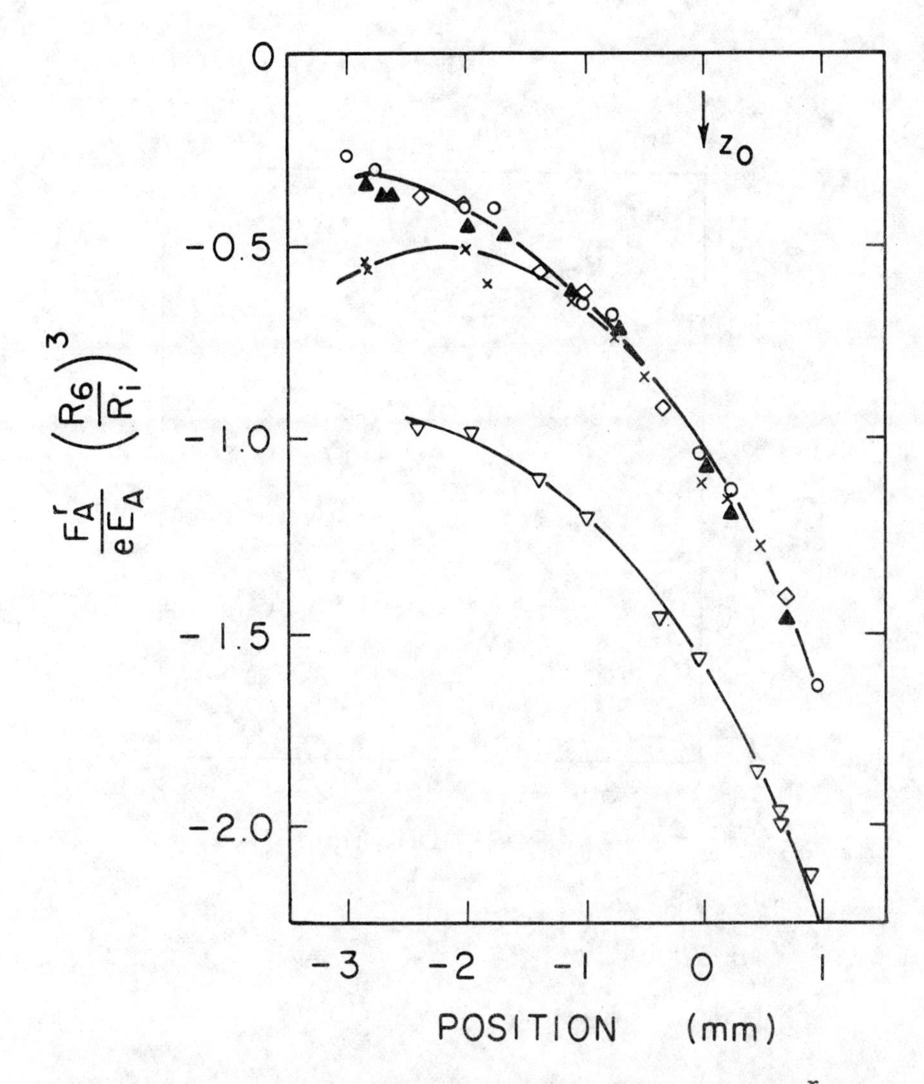

Figure 6a. The position dependence of the residual force F_A^r/E_A for a run begun in January 1980. The patch effect electric field gradient $\partial E_F/\partial z$ is not zero, and balls of differing radius were measured (case 3 in text). The curves are least squares fit to

$$\frac{F_A^r}{E_A}\left(\frac{R_6}{R_i}\right)^3 = q_r\left(\frac{R_6}{R_i}\right)^3 + P_{Fz}\frac{\partial E_A}{\partial z}\left(\frac{R_6}{R_i}\right)^3 + R_6{}^3\frac{\partial E_F}{\partial z}$$

(plotted in units of e), where a function to interpolate $\partial E_F/\partial z$ is chosen as in Fig. 4.

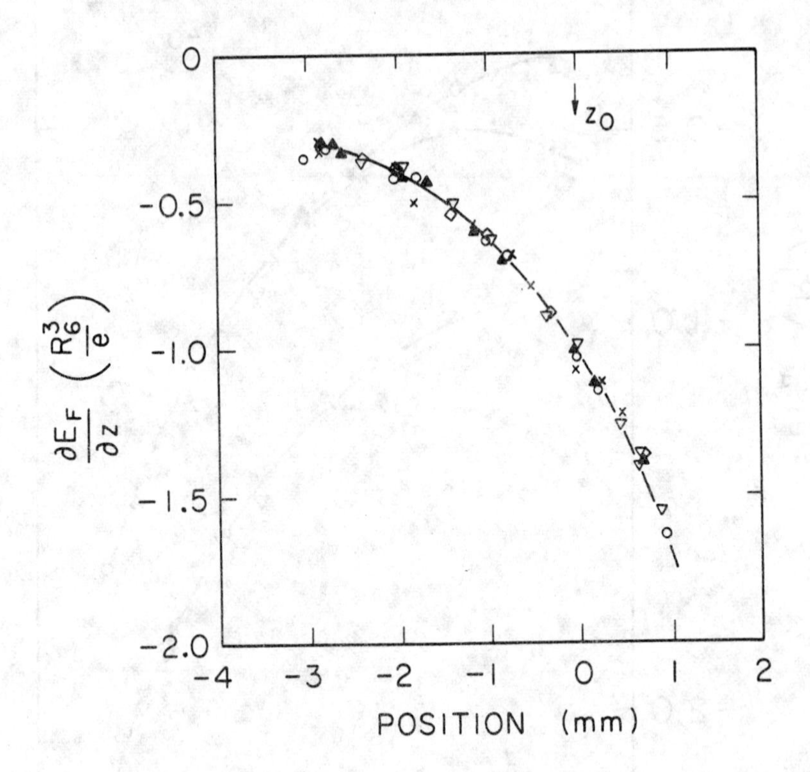

Figure 6b. The data of Fig. 6a offset by residual charge q_r, and corrected for permanent electric dipole P_{Fz} and radius, showing that the patch effect electric field gradient $\partial E_F/\partial z$ remained the same during all five levitations.

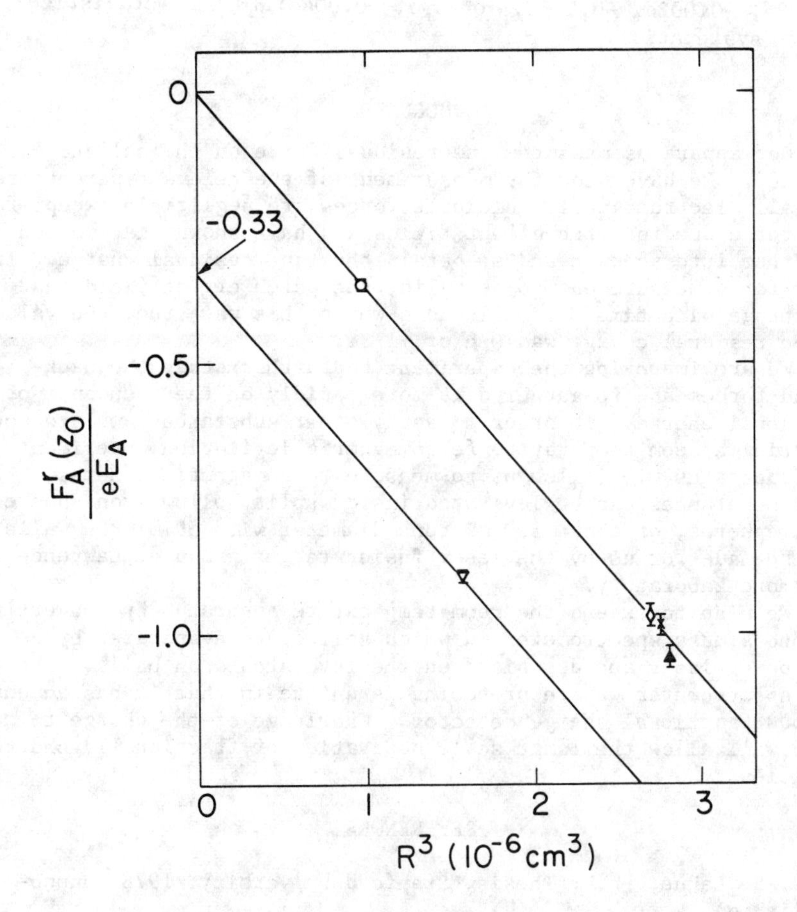

Figure 6c. The residual force of Fig 6a evaluated at z_o, where no correction is necessary for a permanent electric dipole moment on the ball, plotted against R^3. The lines have slope $\partial E_F/\partial z$ and intercept q_r, the residual charge. Ball 13, with $R^3 = 1.6 \times 10^{-6}$ cm^3, has a residual charge of $-0.324 \pm .014e$.

The residual charge differences reported here are: 0.000e, -0.003e, -0.002e, -0.256e, +0.259e, +0.004e; ±0.005 (statistical) ±0.07 (systematic).

CONCLUSION

Our apparatus measures the residual force on the ball to 0.01 eE_A. We have shown by measurement of the relevant parameters that all electromagnetic multipole forces are negligible except for that force due to patch effect fields and have shown that we can take them into account and so obtain the true residual charge. In order for a measurement to be valid, the patch effect field must not change with time. In all cases where this was true, the value of the residual charge was 0 e or ±1/3e.

We are improving the experiment to further study the background forces and to gather data more rapidly on the abundance of fractional charge. In order to assay other substances, we are constructing a room temperature ferromagnetic levitation experiment. This apparatus will allow us to measure ferromagnetic spheres. Other substances can be levitated inside split hollow iron spheres. These spheres, of about 300 microns diameter with 50 micron walls, will be made for us by the laser fusion target group at Lawrence Livermore Laboratory.

We plan to extend the room temperature apparatus by converting it into a mass spectrometer in which solids are evaporated by a high power laser and deposited on the levitated iron balls. The mass spectrometer we are proposing is unique in that it has an unambiguous fractional charge detector. Knowledge of the charge to mass ratio will allow the large scale separation of fractionally charged particles.

REFERENCES

1. G. S. LaRue, Ph.D. Thesis, Stanford University, 1978 (unpublished).
2. G. S. LaRue, W. M. Fairbank and A. F. Hebard, Phys. Rev. Lett. 38, 1011 (1977).
3. G. S. LaRue, W. M. Fairbank, and J. D. Phillips, Phys. Rev. Lett. 42, 142, 1019(E) (1979).
4. G. S. LaRue, W. M. Fairbank, and J. D. Phillips, Phys, Rev. Lett. 46, 967 (1981).

FREE QUARKS IN THE EARLY UNIVERSE

Jay Orear
Cornell University, Ithaca, NY 14853

ABSTRACT

Starting with the assumption that during the big-bang a small number of free quarks or other fractionally charged particles survive hadron formation, it is likely that within the first 60 minutes all the negative quarks will be captured by helium nuclei to form compact "particles" of charge +4/3. At the time of solar system formation only ~1% of these negative quarks will have been reprocessed in supernovas. Thus the free quark content of our planet should be ~50% positive quarks of charge +2/3, ~50% (^{4}He-$\bar{u}$) "particles" of charge +4/3, and ~1/2% negative $\bar{u}$ quarks bound to nuclei with $Z > 2$. It is shown that such a large surplus of positive over negative fractional charges is consistent with the niobium ball results of Fairbank, et al.

Recent experiments[1,2,3,4] suggest that free quarks may exist on our planet at low abundances. Wagoner and Steigman[5] have estimated that ~one free quark per 10^{20} nucleons could have survived the big-bang if the free quark rest mass is ~10 GeV or greater. While it is presently fashionable to postulate total confinement as an exact rule; as yet there is no proof that QCD requires total confinement. So in this paper, we shall explore the consequences of a weak violation of total confinement. In addition to postulating the existence of free quarks, we shall, in order to explain mass differences, assume that the u-quark has smaller rest mass than the other quarks. Then in a matter of seconds after the big-bang, all free quarks will have decayed via the weak interaction into equal amounts of $u(q = +2/3)$ and $\bar{u}(q = -2/3e)$.

0094-243X/82/930061-05$3.00

In this paper the term "quark" is used freely. It could just as well be a fractionally charged lepton or a fractionally charged color singlet hadron containing a colored quark of charge 0 or +1/3. Just to be specific, we shall use the u-quark as an example. In the following discussion it is further assumed that the (qq) and (qqqq) systems cannot be bound; i.e., the "strong" force between a free q or $\bar{q}$ and a nucleon is short range and not strong enough to bond a free quark to a proton or a neutron. This is consistent with the QCD calculations of Milton, Palmer, and Pinsky[6] who find a repulsive force between q and q but the usual binding between q and $\bar{q}$. This is at variance with the bag model of DeRujula, Giles, and Jaffe[7] where a massive free quark might capture several nucleons. Since so little is known about the quark-nucleon interaction, we shall calculate the effects of the coulomb interaction only. If there is an additional quark-nucleon interaction the main conclusions will be unaltered as long as positive free quarks cannot bind to a nucleon.

Four minutes after the big-bang about 25% of the baryons find themselves bound in ^{4}He, about 10^{-9} in ^{7}Li and about 10^{-12} in ^{6}Li.[8] The abundance of any other heavier nucleus at this time is much less than 10^{-12}. At t = 4 minutes just after the nuclei are formed, the free quarks have about 120 keV of thermal energy compared with a Coulomb binding energy of ~350 keV for the ^{7}Li-$\bar{u}$ system, 120 keV for the ^{4}He-$\bar{u}$ system, and 10.2 keV for the $\bar{u}$-proton system. (These Coulomb binding energies are for a $\bar{u}$ rest mass of 10 GeV and correctly take into account the reduced Coulomb field inside the nucleus.) As the universe continues to cool down the $\bar{u}$-quarks would freeze out on the ^{7}Li nuclei provided there is time enough to have a successful recombination collision with one of the rare ^{7}Li nuclei. The expansion time in going from 120 keV to 30 keV is about 60 minutes. The mean collision time for $\bar{u} + {}^7\text{Li} \rightarrow ({}^7\text{Li-}\bar{u})_{\text{bound}} + \gamma$ is $\tau = (N\sigma v)^{-1} = 1.3 \times 10^8$ sec where $N \approx 4 \times 10^{14}$ Li nuclei/cm^3, $\sigma \approx 1.5\text{x}10^{-29}$cm^2, and $v \approx 1.3\text{x}10^6$cm/sec. Thus only about $2\text{x}10^{-5}$ of the $\bar{u}$'s will freeze out on nuclei heavier than heli

The collision time for $\bar{u} + {}^4He \rightarrow ({}^4He\text{-}\bar{u})_{bound} + \gamma$ is 0.7 sec compared to the expansion time of ~4000 sec to go from 120 keV to 30 keV. Thus nearly all of the $\bar{u}$'s will freeze out on ${}^4He$ before they have reached a low enough temperature to freeze out on the protons. Even if the negative quarks are captured by hydrogen, the following recommendations on where to look for quarks still hold.

So the matter out of which stars were formed contained 50% positive u-quarks and 50% (${}^4He\text{-}\bar{u}$) bound objects of ~130 keV binding energy with 6.6×10^{-13} cm Bohr radius and charge +4/3. This binding energy is so much higher than chemical energies and the size so much smaller than atoms that such objects would behave as stable particles of charge +4/3. Most would capture an orbital electron and behave as a "hydrogen atom" of charge +1/3. The nuclear reactions in a stellar core would dislodge the $\bar{u}$'s and allow them to bind on to heavier nuclei. The gas and dust out of which the solar system condensed is estimated to contain about 1% of matter from supernova explosions with the remaining 99% being the primordial hydrogen and helium. Then the ratio of positive quarks plus the positive (${}^4He\text{-}\bar{u}$) "particles" to negative quarks bound in heavier nuclei would be ~200 to 1. The positive particles could bind to atomic hydrogen, but unlike ordinary hydrogen molecules, their unbalanced charge would make them quite sticky in collisions with dust particles. They would eventually find themselves bound to those dust particles with the highest hydrogen affinities and should survive the condensation into a planet. So even though very little of the primordial hydrogen and helium would find itself as constituents of the planet earth, we might expect most of the positive quarks and the (${}^4He\text{-}\bar{u}$) objects to occur in their original abundance. Then there would be a factor of ~200 advantage in searching for positive fractionally charged particles as opposed to negative. It might be possible to concentrate these positive fractionally charged particles on surfaces of metals such as niobium, but it would be very unlikely to find a negative quark bound to a heavy nucleus in a refined sample of metal. A niobium atom with a $\bar{u}$ in the nucleus no longer has the chemistry of niobium and neither would any other material. The refining process would remove such contaminants. However, there are at least 2 mechanisms which could accumulate positive quarks on the surface of the niobium balls used in the experiments of Fairbank, et al. In the manufacture of those balls, they are in contact with a reasonably large amount of various grinding compounds for many hours. Niobium with its large hydrogen affinity might be expected to collect hydrogen-like quark impurities. The other mechanism would involve transfer of positive quarks from a tungsten plate to the niobium balls during heat treatment. It is known that a large amount of hydrogen can be dissolved in a metal such as tungsten. If some of the hydrogen (along with positive quarks) is driven to the surface during heat treatment, then surface diffusion should allow them to preferentially collect on the region of highest hydrogen affinity; i.e., the niobium ball. This explanation is consistent with the combined measurments of Fairbank, et al.[1,2,3,4]. The 11 niobium balls which were heat treated on tungsten initially showed 3 balls with charge -1/3, 5 with charge 0, and 3 with charge 1/3. Repeated measurements on these 11 balls gave 12 with charge -1/3, 14 with charge 0, and 10 with charge +1/3 including some changes of charge between runs when the surfaces were disturbed. All directions of charge change were observed, i.e., -1/3 to 0, -1/3 to 1/3, 0 to -1/3, 0 to 1/3, 1/3 to 0, and 1/3 to -1/3. However the 5 balls which were heat treated on niobium initially gave 0 balls with -1/3, 5 with zero charge, and none with +1/3 charge. These 5 balls had no change of charge during repeated measurements. All the above phenomena are consistent with the hypothesis that each ball heated on tungsten picked up several quarks. Then even if all the quarks were of the same fractional charge, there would be equal distribution of -1/3, 0, and +1/3 residual charge. All 6 types of charge change would occur when one or more quarks would be lost from the surface.[10]

The author wishes to thank E. Salpeter, R. Wagoner, I. Wasserman, and Wm. Fairbank for helpful conversations. This research was supported in part by a grant from the U. S. National Science Foundation.

REFERENCES

1. LaRue, Fairbank, and Hebard, Phys. Rev. Lett. 38, 1011 (1977).
2. LaRue, Fairbank, and Phillips, Phys. Rev. Lett. 42, 142 (1979).
3. LaRue, Fairbank, and Phillips, Phys. Rev. Lett. 46, 967 (1981).
4. Wm. Fairbank, Proceedings of 5th International Vanderbilt conference, May 1982.
5. R. V. Wagoner and G. Steigman, Phys. Rev. D20, 825 (1979).
6. K. A. Milton, W. F. Palmer, and S. Pinsky, Phys. Rev. D25, 1718 (1982).
7. A. DeRujula, R. C. Giles, and R. L. Jaffe, Phys. Rev. D17, 285 (1978).
8. R. V. Wagoner Ap. J. 179, 343 (1973).
9. H. Bethe and E. Salpeter, "Quantum Mechanics of One- and Two-Electron Atoms" (Springer Verlag, Berlin, 1957) p. 322.
10. J. Orear, Phys. Rev. D20, 1736 (1979); Erratum D22, 233 (1980).

AXIONS: VISIBLE AND INVISIBLE*

Michael Dine
The Institute for Advanced Study, Princeton, N. J. 08540

ABSTRACT

The strong CP problem and the Peccei-Quinn solution are briefly reviewed. The properties of the Peccei-Quinn axion are derived, and shown to be phenomenologically unacceptable. A recently proposed generalization of the Peccei-Quinn solution is described, and shown to lead to a light, very weakly coupled axion.

I have been asked today to talk about axions. This is a rather amusing subject for an experimental meeting, since the major developments in this area over the last year or so have been in understanding why axions are not seen in experiments; moreover, this understanding suggests that axions will never be seen in experiments. I would like, today, to review these developments and convince you that nothing in our current understanding of particle physics forces observable axions upon us. But I should like to remind you, as well, that the physics of symmetry breakdown in weak interactions is at best dimly understood. Axion-like objects may well exist, and if such objects are found they will raise questions — and provide clues, about this physics. In particular, the kind of analysis I will discuss to determine the properties of these particles is not special to the strong CP problem. It is, in fact, just the current algebra analysis developed to understand the low-energy properties of π's and nucleons, and it is the appropriate tool for any discussion of spontaneous symmetry breaking.

Let me first spend a few minutes reviewing the strong CP[1] problem and the Peccei-Quinn solution.[2] From here, it will be a small step to understand the solution of the strong CP problem which has been suggested recently.[3] The basic problem is that QCD contains a hidden parameter. It is possible to add to the QCD Lagrangian a term[1,4]

$$L_\Theta = \frac{\Theta g^2}{32\pi^2} F^a_{\mu\nu}\tilde{F}^{a\mu\nu} \tag{1}$$

This term is a pure divergence:

$$F_{\mu\nu}\tilde{F}^{\mu\nu} = \partial_\mu \; 2\epsilon^{\mu\nu\lambda\sigma}\left(A^a_\nu \partial_\lambda A^a_\sigma - \frac{1}{3} f^{abc} A^a_\nu A^b_\lambda A^c_\sigma\right) \tag{2}$$

and one might be tempted to ignore it, since it does not affect the eqations of motion. However, in QCD, we know that it is important.

*Invited talk presented at the High Energy Physics Conference, Vanderbilt University, May 24-26, 1982.

This can be illustrated explicitly by semiclassical calculations in these theories.[1,4] One knows that when one tries to calculate transition amplitudes, etc., in this theory they are in fact sensitive to the choice of Θ. This is because configurations where $\int d^4x\, F\tilde{F}$ is non-zero are present in the theory (instantons). Unfortunately, these calculations are plagued with infrared divergences, and it is rarely possible to do an honest calculation which gives the Θ-dependence of a physical quantity.

There is, however, an alternative way to see that the $F\tilde{F}$ term plays a role in the strong interactions. This is by examining the U(1) problem,[1] the question of the ninth pseudoscalar meson. From now on, let me, for simplicity, focus on a world with just two quark flavors, u and d, of mass m_u and m_d, where these masses are much smaller than characteristic QCD scales such as m_N, f_π, etc. Then the current

$$j_5^\mu = \bar{u}\gamma_\mu\gamma_5 u + \bar{d}\gamma_\mu\gamma_5 d \tag{3}$$

satisfies

$$\partial_\mu j_5^\mu = m_u\bar{u}\gamma_5 u + m_d\bar{d}\gamma_5 d + N\frac{g^2}{32\pi^2}F\tilde{F} \tag{4}$$

where N is the number of quark flavors. The last term represents the Adler-Bell-Jackiw anomaly.[5] If we ignore the anomaly, then this is a partially conserved current just like the usual axial isospin current, and one can do the usual current algebra analysis with it.[6] In particular, one can show that, corresponding to this current there should be another massless particle with mass of order m_π. The resolution of this problem must lie in the $F\tilde{F}$ term.[1] Because of this term, this axial current is not, in fact, conserved. This can be seen, again, if we start computing appropriate matrix elements using instantons. Ignoring the quark mass terms, we can integrate (Eq. 4) from $t = -\infty$ to $t = +\infty$ and over all space to find

$$Q_5(+\infty) - Q_5(-\infty) = N\frac{g^2}{32\pi^2}\int d^4x\, F\tilde{F}\ . \tag{5}$$

In the presence of instantons the right-hand side takes on a non-zero, integer value. Now, as I said earlier, instanton calculations tend to be very divergent, so this argument is mainly suggestive: it tells us simply that the anomaly is important. The lesson we draw from this — and from the simple fact that there is no ninth light pseudoscalar — is that a current afflicted with an anomaly is not a conserved current.

There is, however, another side to this argument. If the $F\tilde{F}$ term plays a significant role in QCD, then adding the term L_Θ (Eq. 1) to the action has physical consequences. In particular, since this term is CP odd (it is just $\vec{E}\cdot\vec{B}$), it will induce real CP violation. In fact, our previous discussion may be turned around to predict the neutron electric dipole moment as a function of Θ. Baluni and Crewther et al. obtain the estimate[7]

$$d_n \cong 3\times10^{-16}\ \Theta \text{ e cm.} \tag{6}$$

Since it is known that[8]

$$|d_n| \lesssim 6\times10^{-25} \text{ e cm.} \tag{7}$$

we find

$$\Theta \leq 2\times10^{-9}. \tag{8}$$

The strong CP problem is the question: why is this quantity so small?

To see that we can't simply set $\Theta = 0$, we again examine the anomaly equation. Suppose both m_u and m_d had a common phase, α. This could be eliminated by redefining the fermion fields

$$u' = e^{i\alpha\gamma_5} u \qquad d' = e^{i\alpha\gamma_5} d\ . \tag{9}$$

But the anomaly tells us that under this transformation

$$\delta L = \frac{\alpha g^2}{32\pi^2} F\tilde{F} \tag{10}$$

Now, even if we start with $\Theta = 0$ and m_u, m_d real, weak interactions will induce such a phase, α, in the mass matrix. This phase will be logarithmically divergent, and require renormalization. So we are <u>forced</u>, in general, to include Θ (or α) in the tree level theory. There is no obvious reason that it should be small.

One possible solution to this problem is to set one quark mass to zero (say $m_u = 0$). Then we can always do a chiral rotation which eliminates Θ without changing anything else. This, however, the experts in current algebra (e.g. Weinberg[9]) assure us cannot be the case. A second solution was proposed by Peccei and Quinn.[2] Suppose the full theory of weak and electromagnetic interactions has an exact U(1) symmetry at the classical level, but that the corresponding current, j^μ_{PQ}, is afflicted with an anomaly, i.e.

$$\partial_\mu j^\mu_{PQ} = \frac{cg^2}{32\pi^2} F\tilde{F}\ . \tag{11}$$

Then we can again eliminate the Θ parameter without changing anything else. The easiest way to obtain such a symmetry, they noted, is to enlarge the Higgs sector of the standard model by adding a second Higgs doublet. Again, we confine ourselves to a theory with just u and d quarks. We let

$$Q_L = \begin{pmatrix} u \\ d \end{pmatrix}_L \tag{12}$$

denote the quark doublets, while we denote the singlets by u_R and d_R. The Higgs doublets have hypercharge +1 and -1, and we denote them by H_U and H_D, respectively. The Lagrangian for the Higgs sector then looks like

$$G_u \bar{u}_R H_U Q_L + G_d \bar{d}_R H_D Q_L + \text{H.C.} + V(H_U, H_D) \; . \tag{13}$$

With appropriate restriction on V, this is invariant under the transformation

$$\begin{aligned} H_U &\to e^{i\alpha} H_U \qquad & \bar{u}_R Q_L &\to e^{-i\alpha} \bar{u}_R Q_L \\ H_D &\to e^{i\alpha} H_D \qquad & \bar{d}_R Q_L &\to e^{-i\alpha} \bar{d}_R Q_L \end{aligned} \tag{14}$$

which we will call $U(1)_{PQ}$. Moreover,

$$\partial_\mu j^\mu_{PQ} = \frac{Ng^2}{32\pi^2} F\tilde{F} \tag{15}$$

Now, everything goes through just as in the Weinberg-Salam model. The potential is arranged so that

$$\langle H_U \rangle = \frac{1}{\sqrt{2}} \begin{pmatrix} f_u \\ 0 \end{pmatrix} \qquad \langle H_D \rangle = \frac{1}{\sqrt{2}} \begin{pmatrix} 0 \\ f_d \end{pmatrix} . \tag{16}$$

Then

$$M_W^2 = 4g^2 \left(|f_u|^2 + |f_d|^2 \right) \tag{17}$$

$$m_u = G_u f_u \qquad m_d = G_d f_d \tag{18}$$

etc.

Because QCD explicitly breaks the $U(1)_{PQ}$ symmetry, the relative phases of f_u and f_d are fixed. The important thing which Peccei and Quinn showed is that f_u and f_d would both be real. Thus the masses m_u and m_d would be real, and there would be no strong CP violation in this theory. We can give a brief, heuristic summary of their argument. We first choose phases of the fields so that $\Theta = 0$ (using the $U(1)_{PQ}$ "symmetry"). Then we recall that in QCD it is a good first approximation to set $m_u = m_d = 0$. In this limit, and with $\Theta = 0$, QCD conserves CP, and gives rise to non-zero (and CP conserving) v.e.v.'s for the quantities

$$\langle \bar{u}u \rangle = \langle \bar{d}d \rangle \quad . \tag{19}$$

Now, by a hypercharge transformation, we can make f_u real, $f_d = e^{i\alpha}|f_d|$. To a first approximation, the terms in the potential (vacuum energy) involving α are just given by the expectation value of the Lagrangian, i.e.

$$\begin{aligned} &|m_d|\langle \bar{d}_R d_L \rangle e^{i\alpha} + \text{h.c.} \\ &= |m_d|\langle \bar{d}d \rangle \cos\alpha \quad . \end{aligned} \tag{20}$$

So $\alpha = 0$, and there is no CP violation.

There is, however, another problem.[10] Were it not for the small quark mass terms, the phase, α, would be arbitrary, corresponding to a symmetry. We would simply choose α, breaking the symmetry and giving a Goldstone boson. The quark mass terms we have just considered explicitly break the symmetry. Since the Goldstone field is roughly $|f_d|$ times the phase, α, the axion mass will be of order

$$m_A^2 \sim \frac{m_d \langle \bar{d}d \rangle}{|f_d|^2} \tag{21}$$

To actually compute the mass and other properties of the axion,[10,11] it is helpful to redefine the transformation laws slightly. We require the Lagrangian to be symmetric under

$$\begin{aligned} &\bar{u}_R Q \to e^{-i\alpha X_u}\bar{u}_R Q \qquad H_U \to e^{i\alpha X_u} H_U \\ &\bar{d}_R Q \to e^{-i\alpha X_d}\bar{d}_R Q \qquad H_D \to e^{i\alpha X_d} H_D \end{aligned} \tag{22}$$

We also require $X_u + X_d = 1$ (this is a normalization). The corresponding current is (just this once we include all the quarks)

$$j^{\mu}_{PQ} = X_u H_U^{\dagger} \overleftrightarrow{\partial}_{\mu} H_u + X_d H_D^{\dagger} \overleftrightarrow{\partial}_{\mu} H_D + X_u(\bar{u}\gamma_{\mu}\gamma_5 u + \bar{c}\gamma_{\mu}\gamma_5 c + \bar{t}\gamma_{\mu}\gamma_5 t)$$
$$+ X_d(\bar{d}\gamma_{\mu}\gamma_5 d + \bar{s}\gamma_{\mu}\gamma_5 s + \bar{b}\gamma_{\mu}\gamma_5 b)$$
$$+ X_d(\bar{e}\gamma_{\mu}\gamma_5 e + \bar{\mu}\gamma_{\mu}\gamma_5 \mu + \bar{\tau}\gamma_{\mu}\gamma_5 \tau) \ . \tag{23}$$

Before gauge interactions and instantons are considered, the theory has two Goldstone bosons. These can be thought of as the imaginary parts of the fields H_U^0 and H_D^0. One linear combination of these fields, ϕ, is eaten by the Z boson. Recalling the coupling in the Weinberg-Salam model

$$Z^{\mu}\partial_{\mu}(f_u(H_U - H_U^{\dagger}) - f_d(H_D - H_D^{\dagger})) \tag{24}$$

we see that

$$\phi = \mathrm{Im}\, \frac{f_u H_U - f_d H_D}{f} \tag{25}$$

where

$$f = \sqrt{f_u^2 + f_d^2} \ . \tag{26}$$

The axion field is essentially the orthogonal linear combination,

$$A = \mathrm{Im}\, \frac{f_d H_U + f_u H_D}{f} \tag{27}$$

To find the axion mass and study its properties, it is useful to define the charges, X_u and X_d, so that the current j^{μ}_{PQ} does not create the Higgs, ϕ, from vacuum. In other words, we require

$$\langle 0 | j^{\mu}_{PQ} | \phi \rangle = 0 \ .$$

This gives

$$X_u f_u^2 - X_d f_d^2 = 0 \tag{28}$$

or

$$X_u = \frac{f_d^2}{f^2} \qquad X_d = \frac{f_u^2}{f^2} \ . \tag{29}$$

The decay constant of the axion is defined by

$$\langle 0|j^{\mu}_{PQ}|A\rangle = f_A q^{\mu} \tag{30}$$

so

$$f_A = \frac{f_u f_d}{f} . \tag{31}$$

Now the current, j^{μ}_{PQ} has an anomaly,

$$\partial_{\mu} j^{\mu}_{PQ} = N \frac{g^2}{32\pi^2} F\tilde{F} . \tag{32}$$

From our previous discussion of the U(1) problem, we know that this current is in no sense conserved. There is, however, a current which is nearly conserved and has no anomaly. We can obtain this current by subtracting from j^{μ}_{PQ} a term involving only light quark fields with an equal anomaly (here we are following closely the analysis of Bardeen and Tye[11]). We are working in an approximation where only the u and d quarks are light, so we define

$$J^{\mu}_{x} = j^{\mu}_{PQ} - N\left\{\frac{1}{1+z}\bar{u}\gamma_{\mu}\gamma_5 u + \frac{z}{1+z}\bar{d}\gamma_{\mu}\gamma_5 d\right\} . \tag{33}$$

Obviously, for any z, this current is anomaly free, and its divergence is proportional to light quark masses. This current still creates the axion with strength f_A, and we can do conventional current algebra with it. The choice of z will be dictated by convenience, as we will see below.

One of the basic theorems of current algebra is Dashen's theorem.[12] It states that the mass of a pseudo-Goldstone boson is given by

$$m^2 f^2 = \langle 0|[Q,[Q,H]]|0\rangle \tag{34}$$

Here Q and f are the corresponding charge and decay constant. Let me remind you what this gives for the pion. For the π^0, say,

$$Q^{T_3} = \int d^3x \left\{\bar{u}\gamma^0\gamma_5 u - \bar{d}\gamma^0\gamma_5 d\right\} \tag{35}$$

which gives, plugging in the formula

$$f_{\pi}^2 m_{\pi}^2 = m_u\langle\bar{u}u\rangle + m_d\langle\bar{d}d\rangle . \tag{36}$$

Now, applying this to the axion we must be a bit careful. In particular, for general z,

$$<[Q^{T_3},[Q^x,H]]> \neq 0 \tag{37}$$

and we have to diagonalize a 2 × 2 matrix. Instead, it is simpler and more useful to define z so that (37) vanishes. This gives

$$<0|m_u\bar{u}u - zm_d\bar{d}d|0> = 0 \tag{38}$$

or

$$z = \frac{m_u<\bar{u}u>}{m_d<\bar{d}d>} \cong \frac{m_u}{m_d} \tag{39}$$

With this definition (using again $<\bar{u}u> \cong <\bar{d}d>$) and (Eq. 36)

$$\begin{aligned} f_A^2 m_A^2 &= N^2(1+z)^{-2}\left\{m_u<\bar{u}u> + z^2 m_d<\bar{d}d>\right\} \\ &\cong N^2(1+z)^{-2} z\, m_\pi^2 f_\pi^2 \quad . \end{aligned} \tag{40}$$

To calculate other properties of the axion, such as its couplings and lifetime, we just repeat the usual current algebra analyses, using this current. The two-photon lifetime, for example, is computed from the two-photon anomaly just as for π^0 decays:

$$\tau(A\to 2\gamma) = \tau(\pi^0\to 2\gamma)\left(\frac{m_\pi}{m_A}\right)^5 \frac{N^4 z^3}{(1+z)^4} \, . \tag{41}$$

The couplings of the axion to leptons can simply be read off the Lagrangian. These go as lepton masses over f_A, just like conventional Higgs couplings. For quarks, we have to be a bit more careful, because of strong interaction corrections. Since $<\bar{u}u>$ breaks the non-anomalous X symmetry, there is a significant $\left(\mathcal{O}\left(\frac{m_N}{f_A}\right)\right)$ quark component in the axion. As a result, when one derives Goldberger-Trieman relations for the axion couplings to nucleons, they turn out to be of order m_N/f_A. The main thing to remember is that the axion always couples as $1/f_A$.

As you have probably heard at many meetings, the axion I have described is pretty well ruled out experimentally.[13] If it existed, its mass would be of order 60 KeV, its two-photon lifetime of order 1 sec., and its couplings to matter of order $m\sqrt{G_F}$. Its detailed properties are specified by the ratio $X = f_u/f_d$. It would be produced by nuclear reactions and in beam dump experiments. By now, most of the allowed range of X has been ruled out; the few positive results on axions are now contradicted by other experiments.[14]

This fact left people in a bit of a quandry. If there is no axion, how does the strong CP problem get solved? While pondering some apparently unrelated questions, Willy Fischler, Mark Srednicki, and I stumbled across a solution which in retrospect seems almost obvious.[3] As I remarked before, the couplings of the axion go as m/f_A. In the Peccei-Quinn model, f_A is constrained to be less than $\sqrt{G_F}$. But it is easy to cook up models in which f_A can be arbitrarily large, as I will illustrate below. It is not hard to convince yourself that if f_A is increased by an order of magnitude, no terrestial experiment to date would have seen axions.

However, as one increases f_A, one runs into another problem. Recall that the axion mass also goes as $1/f_A$. As the axion becomes light enough to escape detection, it also becomes light compared to the temperatures in many astrophysical environments, in particular the cores of red giant stars. Under these circumstances, one produces enormous numbers of axions, and, because they interact weakly (their mean free paths are typically larger than the stellar radius) they escape from the stars, carrying off energy.[15] For an axion with a mass of a few KeV, for example, a crude, order of magnitude calculation convinces you that red giants would burn out almost instantly. This is not a problem which will be solved by simply retuning the parameters of someone's stellar model. This process sets a lower limit on f_A, since eventually the axion becomes so weakly coupled that it no longer does any harm. Dicus et al. find a limit[15]

$$f_A \gtrsim 10^9 \text{ GeV}, \tag{42}$$

an enormous scale of energy.

At first, this seems very disturbing. To see why, let me describe a simple model in which f_A is so large.[3] The model is essentially the Peccei-Quinn model, except that we add a field, ϕ, which is an SU(3)×SU(2)×U(1) singlet. Under the $U(1)_{PQ}$ symmetry, we require that the quark, lepton, and Higgs fields transform as before (Eq. 22). The field ϕ transforms as

$$\phi \to e^{-i\frac{\alpha}{2}} \phi . \tag{43}$$

One arranges the potential so that ϕ, H_U, and H_D get v.e.v.'s with

$$\langle\phi\rangle \gg \langle H_U\rangle, \langle H_D\rangle \tag{44}$$

In this case, $f_A \cong \langle\phi\rangle$. For $\langle\phi\rangle$ sufficiently large, our previous analysis shows that the strong CP problem is solved with a harmless axion. The problem appears when we examine the potential. The model must include couplings like

$$m^2|H_U|^2 + m'^2|H_D|^2 + \phi^2 H_U H_D \quad . \tag{45}$$

$\langle\phi\rangle$ is thus like a huge mass for the doublets. We can only obtain the proper value of the Fermi constant if we adjust with enormous precision the parameters of the model so that the effective mass for one doublet is small. So we have, it seems, merely replaced the problem of why Θ is small by the question: why are the parameters of the model so finely tuned?

There are, however, other huge scales in nature. If one believes in grand unification, there is a scale of order 10^{15} GeV. It is, in fact, rather easy to modify standard grand unified theories to accommodate a $U(1)_{PQ}$ symmetry. Simple SU(5) examples were suggested by ourselves[3] and by Georgi, Glashow, and Wise.[16] Examples in more involved (and sometimes better motivated) models have been suggested by many people over the last year. In all of these,

$$f_A \sim m_{GUT} \quad . \tag{46}$$

The fine-tuning described above is still present, but it is the same one which grand unified theories have had to face from the beginning. Thus we truly have one less problem.

Since the unification scale is so large, this axion is completely invisible. Taking $f_A = 10^{15}$ GeV, we find, from our previous analysis (with appropriate modification)[3,16]

$$\begin{aligned} m_A &\sim 10^{-8} \text{ eV} \\ \tau(A\to\gamma\gamma) &\sim 10^{56} \text{ years} \end{aligned} \tag{47}$$

The couplings of the axion to nucleons are of order $10^{-16} g_{\pi NN}$, and purely pseudoscalar. Such couplings do not give rise to long range forces. Actually, once CP violating effects in weak interactions are considered, an extremely small scalar coupling will be induced. However, the corresponding force offers no competition to gravity on macroscopic scales.

There may turn out to be other viable solutions to the mystery of strong CP violation. But at least we have one. The strong CP problem is thus no longer a problem.

REFERENCES

1. G. 't Hooft, Phys. Rev. Lett. 37, 172 (1976).

2. R. D. Peccei and H. R. Quinn, Phys. Rev. Lett. 38, 1440 (1977); Phys. Rev. D16, 1791 (1977).
3. M. Dine, W. Fischler, and M. Srednicki, Phys. Rev. Lett. 104B, 199 (1981). See also J. Kim, Phys. Rev. Lett. 43, 103 (1979).
4. R. Jackiw and C. Rebbi, Phys. Rev. Lett. 37, 172 (1976); C. G. Callan, R. F. Dashen, and D. J. Gross, Phys. Lett. 63B, 334 (1976).
5. S. Adler, Phys. Rev. 177, 2426 (1969); J. S. Bell and R. Jackiw, Nuovo Cimento 60A, 49 (1969); W. A. Bardeen, Phys. Rev. 184, 1848 (1969).
6. S. Weinberg, Phys. Rev. D11, 3583 (1975).
7. V. Baluni, Phys. Rev. D19, 227 (1979); R. Crewther, P. Divecchia, G. Veneziano, and E. Witten, Phys. Lett. 88B, 123 (1979).
8. N. F. Ramsey, Phys. Reports. 43C, 409 (1978).
9. S. Weinberg in Proceedings of Neutrinos - 78, April 28-May 2, 1978, ed. by E. Fowler, Purdue University Press, West Lafayette.
10. S. Weinberg, Phys. Rev. Lett. 40, 223 (1978); F. Wilczek, Phys. Rev. Lett. 40, 279 (1978).
11. W. A. Bardeen and S.-H. H. Tye, Phys. Lett. 74B, 229 (1978); M. Peskin (unpublished); J. Kandaswamy, Per Salomonson, and J. Schechter, Phys. Rev. D17, 3051 (1978).
12. R. Dashen, Phys. Rev. 183, 1245 (1969).
13. R. D. Peccei in Proc. Intern. Conf. Maui (1981), ed. R. J. Cence (Hawaii) to be published.
14. A. Zehnder, K. Gabathuler, J. L. Vuilleumier, Phys. Lett. 110B, 419 (1982); A. Zehnder, Phys. Lett. 104B, 494 (1981).
15. D. A. Dicus, E. W. Kolb, V. L. Teplitz, and R. V. Wagoner, Phys. Rev. D18, 1829 (1978); Phys. Rev. D22, 839 (1980).
16. M. Wise, H. Georgi, and S. L. Glashow, Phys. Rev. Lett. 47, 402 (1981).

BUBBLE CHAMBER EXPERIMENTS ON CHARMED PARTICLE LIFETIMES*

R. C. Field
Stanford Linear Accelerator Center
Stanford University, Stanford, California 94305

ABSTRACT

The three current bubble chamber experiments on charmed particle lifetimes are compared. Their most recently released results are discussed.

Although single charmed decay vertices have been seen in bubble chambers before, this discussion will cover the three current experiments. Two are at CERN and the third at SLAC. All use bubble chambers with relatively high resolution photographic techniques, coupled with downstream detector systems, and can measure lifetimes from decay length distributions.

The interest in such experiments increased after initial comparisons of the $D^{\pm}$ and D^{o} lifetimes. Theory, following the standard model, supposed that charmed particle decays would be dominated by processes involving $\Delta C = \Delta S = 1$ transitions of the charmed quark. Relevant diagrams are given in Fig. 1. An obvious consequence was that $\Gamma(D^{\pm}) = \Gamma(D^{o}) = \Gamma(F^{\pm})$, and that the semileptonic branching ratios should be the same. By comparison with muon decay a charmed lifetime could be obtained, $\tau \sim 5 \times 10^{-13}$ sec, in general agreement with what is found. Hard gluon effects calculated in leading log approximation did not substantially change these expectations.

Reports of differences in the semileptonic branching ratios of $D^{\pm}$ and D^{o} did change the picture. The results from DELCO and Mark II at SPEAR using different techniques, but using $D\bar{D}$ production at the Ψ'' resonance are indicated in Fig. 2. This graph of log likelihood against the ratio of branching fractions — or of lifetimes — originally came from the Mark II publication,[1] but an estimate of the DELCO result,[2] plotted

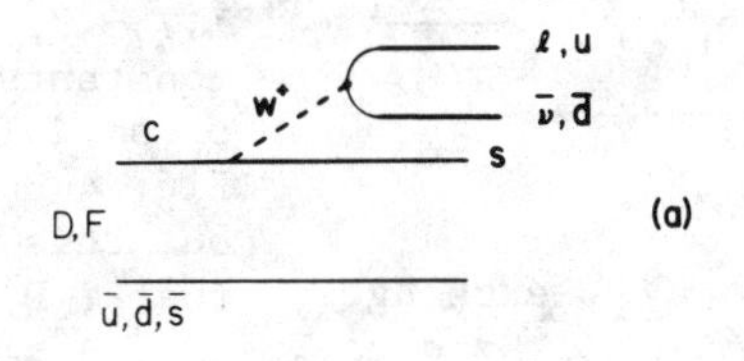

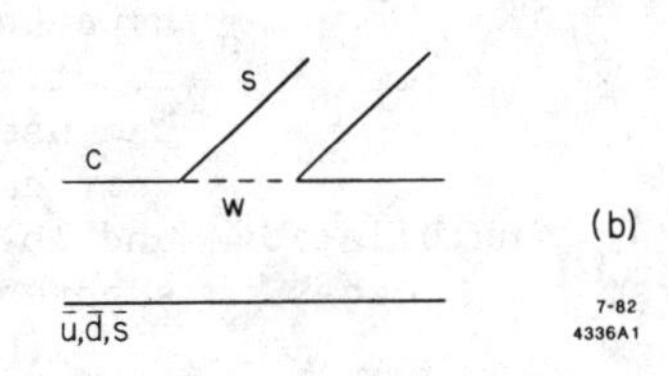

Fig. 1. Quark line diagrams for charmed meson decay — light quark spectator process.

*Work supported by the Department of Energy, under contract DE-AC03-76SF00515.

0094-243X/82/930077-12$3.00

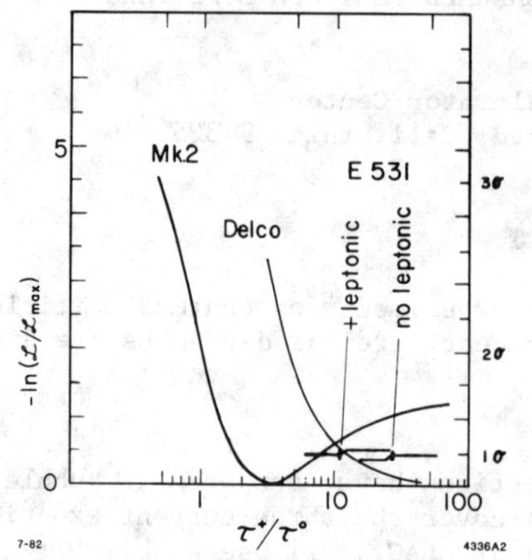

Fig. 2. Plot of -log (likelihood) against charged: neutral D lifetime ratio for Mark II and DELCO experiments. The error bars are for the neutrino emulsion experiment E531, with and without leptonic decays included. The standard deviation equivalent of the vertical scale is also given.

similarly, is shown. In addition, error bars are given for a neutrino-emulsion measurement of the ratio of the lifetimes, FNAL experiment E531.[3] In this case two values are shown. The larger ratio excludes semileptonic events which these experimenters now believe should be considered separately.

The large ratios suggested by these results brought some second thoughts to the theory of the decays. One approach sought a way to enhance the D^o decay rate without affecting the $D^{\pm}$ rate. A diagram available to D^o, but not to $D^{\pm}$ because of the quark content, is the W-exchange diagram of Fig. 3. Helicity mismatch of the light quarks should suppress this, but a new hypothesis[4] was that initial state emission of gluons could eliminate the helicity constraint. An alternative picture[4] considered the gluon component in the initial state wave function. Calculations allowed ratios $\tau(D^{\pm})/\tau(D^o) = 1.7 \sim 7$, depending on quark masses and coupling constants, with a preference for the range 2-3 or so. It should be noticed that the diagram Fig. 4 could similarly enhance the F decay, but this enhancement may be less marked, so that $\tau(D^{\pm}) \approx \tau(F^{\pm})$.

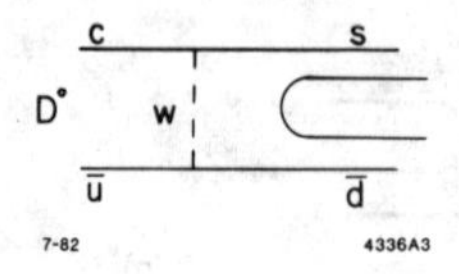

Fig. 3. W-exchange diagram for D^o decay.

A different viewpoint sees the D^+ decay suppressed relative to the D^o and F^+ decays. This is a "sextet dominance" model[5] which hypothesizes an enhancement of diagrams 1b) relative to 1a). 1b) are normally suppressed because of difficulty with color matching. If they can be enhanced, then for D^+, diagram 1a) and 1b) could interfere destructively, thus suppressing the D^+ decays.

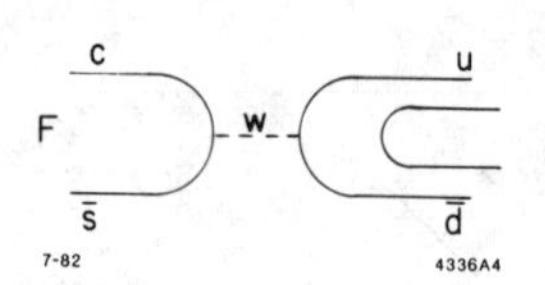

Fig. 4. Annihilation diagram for F decay.

The three bubble chamber experiments are compared in Table 1, and in Figs. 5, 6, and 7. The resolved track widths, which are in the range 30-55 μm, are adequate to observe charm decays with reasonable efficiency. Two of the chambers use liquid hydrogen, the third chamber a Freon.

It can be seen that the small CERN chambers operate with no magnetic field, and the downstream spectrometers allow only ~90% efficiency for finding charged tracks. This has obvious consequences in reconstructing charmed decays. For the data reported, only the SLAC experiment had operating particle identification beyond

Table 1. Comparison of the three bubble chamber experiments.

	LEBC (EHS Collaboration)	SHF (SLAC Collaboration)	BIBC (BERN)
Liquid	H_2	H_2	C_4F_8
Track width	45 μm	55	30
Track bubble density	80 per cm	60	300
B.C. diameter	20 cm	110	6.5
Beam	π^-, p, 360 GeV/c	γ, 20	π^-, 340
Hybrid equipment	2 magnets drift chambers Pb. glass	P.W.C. Cerenkov Pb. glass	Streamer Chamber in magnet
Fraction of tracks momentum analyzed	90% of charm tracks ~50% of decays	100%	90%
π vs. (k,p) identification	small (ISIS consistency check)	~30%	small
Mass res. at D:			
No π^o	12 MeV/c^2	12	~40
With π^o	25	25	--
Mean charged multiplicity in CHARM events:			
At interaction	8	3.8	As LEBC
After decays	11	6.6	plus nuclear effects
Observed π^o	2.5	0.35	0
Observed K^o, Λ	?	0.23	?
Hadronic interactions:			
Total	πp 140K pp 250K	500K	94K
Analyzed	πp 70K pp 80K	200K	75K
Interactions per event in lifetime analysis	πp 5000 pp 9250	8000	5400

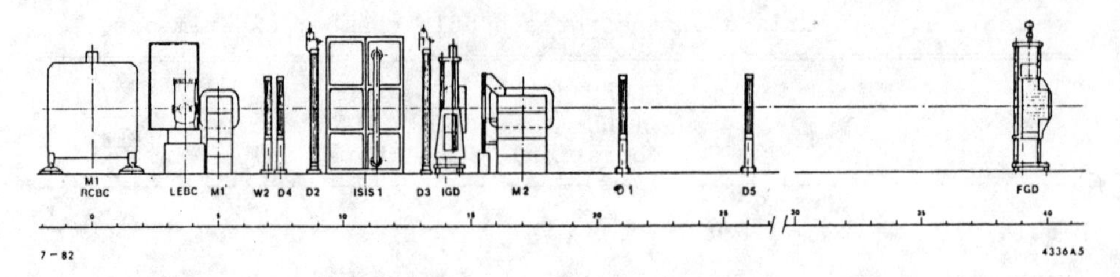

Fig. 5. EHS experiment layout. Behind LEBC are tracking chambers, magnets, ISIS and lead glass detectors.

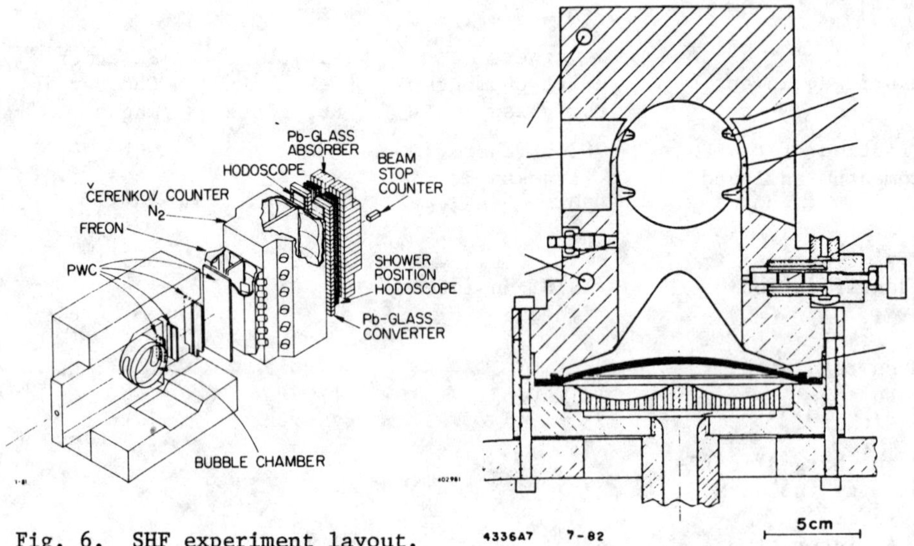

Fig. 6. SHF experiment layout.

Fig. 7. The BIBC freon bubble chamber.

ionization of slow tracks in the chamber liquid. Even in this experiment the charmed decay track identification was limited to ~30% of tracks.

When full reconstruction was possible, mass resolutions were ~10 MeV/c^2 for the hydrogen experiments, but considerably poorer for the heavy liquid experiment which used a streamer chamber for momentum analysis. The resolution was a factor of two worse when π^o's were involved. These were measured in the lead glass arrays behind the two hydrogen experiments. It may be seen that the LEBC experiment had a relatively large number of showers from which to extract its π^o signal, both because of its higher π^o production and larger acceptance solid angle.

The 350 GeV/c momenta of the hadron beams of the CERN experiments also introduce a noteworthy difference in the topology of the events.

The multiplicity at the production vertex is twice as much as at SLAC energies. The track count increases, on average, by three after the charmed decay. The high energy experiments must live with a comparatively sharp forward collimation of their events, Fig. 8.

It is worth noting that the SLAC 20 GeV photon beam, although by no means unique, is somewhat unusual. It is produced by directing ultraviolet photon pulses from a laser against the SLED energy (30 GeV) electron beam. Compton backscattered photons are then collimated into the bubble chamber in a 3 mm diameter, almost monochromatic, pencil beam.

The statistics on the exposures serve to indicate that the LEBC and SLAC collaboration hydrogen experiments are ~40% analyzed, whereas the Bern statistics are almost complete (although their result is still preliminary).

The set of pictures from SLAC, Fig. 9, may be compared with those of the other experiments. The contrast has been slightly enhanced for ease of reproduction. As seen in a) and b), charmed events can be topologically very simple. Indeed the majority of D^+ decays are "kinks" as in a). However, mostly because of the large amount of missing energy, these are presently not used in the lifetime analyses. V^o topologies, like kinks, can be from strange particle decays. In both cases, events whose kinematics show they might be strange decays must be cut from the sample.

In Fig. 9c), a difficulty is illustrated. In a few events it is impossible to be sure which vertex (there are three!) several of the tracks come from. Thus it is possible, in this event, to reconstruct a Λ_c^+ and $\bar{D}^o$ — but not at the same time since the two hypotheses use common tracks. Another problem occurs in the event in Fig. 9d). Although topologically clear, the three-prong decay — which has a π and a K identified — can be reconstructed as a D^+ or F^+ depending on whether the third track is a π or a K. This ambiguity is characteristic of charmed vertex reconstruction. It is ameliorated by improved mass resolution and track identification, and will improve also when one feels confident in the mass of the F, and perhaps with better knowledge of the branching ratios.

The event in Fig. 9e) is a clear example of a charmed pair decaying. It is unfortunately particularly representative of the processes, because in both cases there is substantial missing mass. Neutral particles, π^o, K^o, Λ, n .... even if detected, cannot usually be allocated after reconstruction to a specific decay vertex.

The final example in the figure shows a 4-track neutral decay where two pions are identified and one track is a K^+ (or p). The hypothesis that the fourth track is a π gives a $\bar{D}^o$ mass, and there is no kinematic indication of a missing neutral. It appears to be a clean example of a $\bar{D}^o$. The interesting thing is its lifetime. At 23×10^{-13} sec it would have travelled ~10 lifetimes if certain of the reported D^o lifetime results are correct — a probability of $< 5 \times 10^{-5}$.

In comparing the various analyses there are several subtleties which we must pass over to concentrate on the more general issues.

First the strange particles are removed kinematically. The LEBC experiment believes they may have ~1 event background from a strange V^o. In the other experiments it should be substantially less.

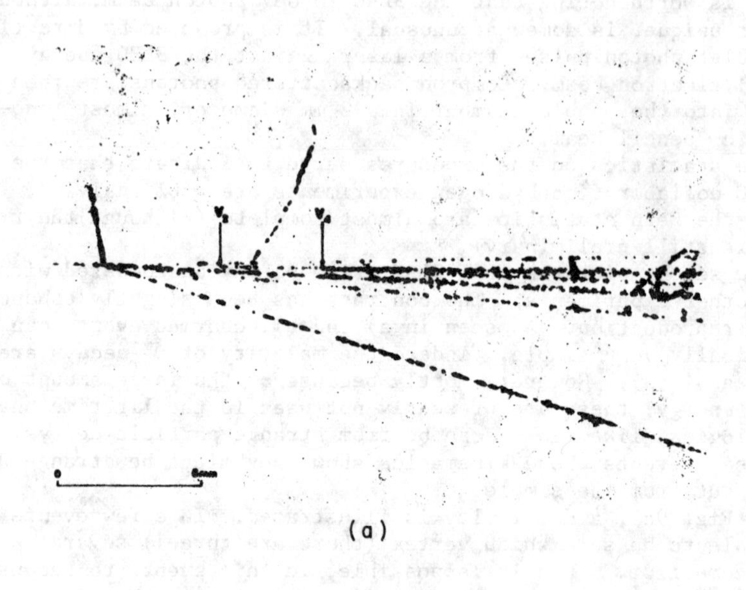

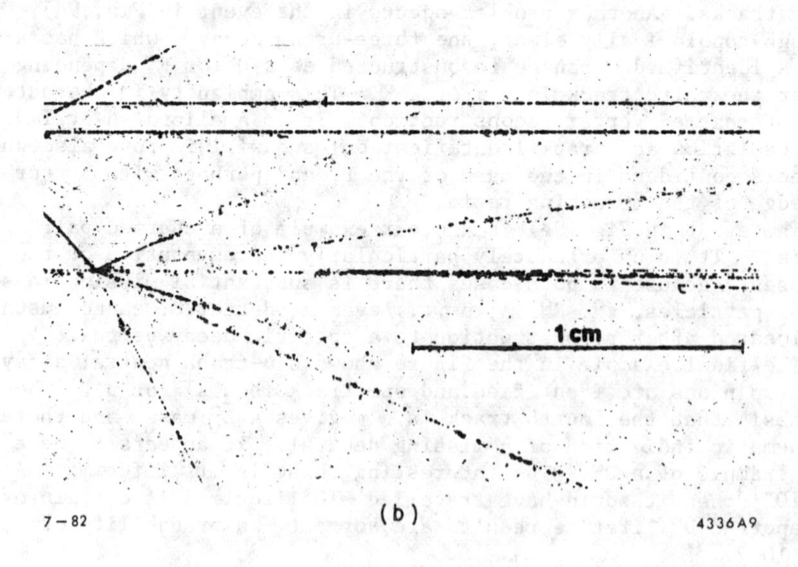

Fig. 8. Examples of events in a) LEBC and b) BIBC.

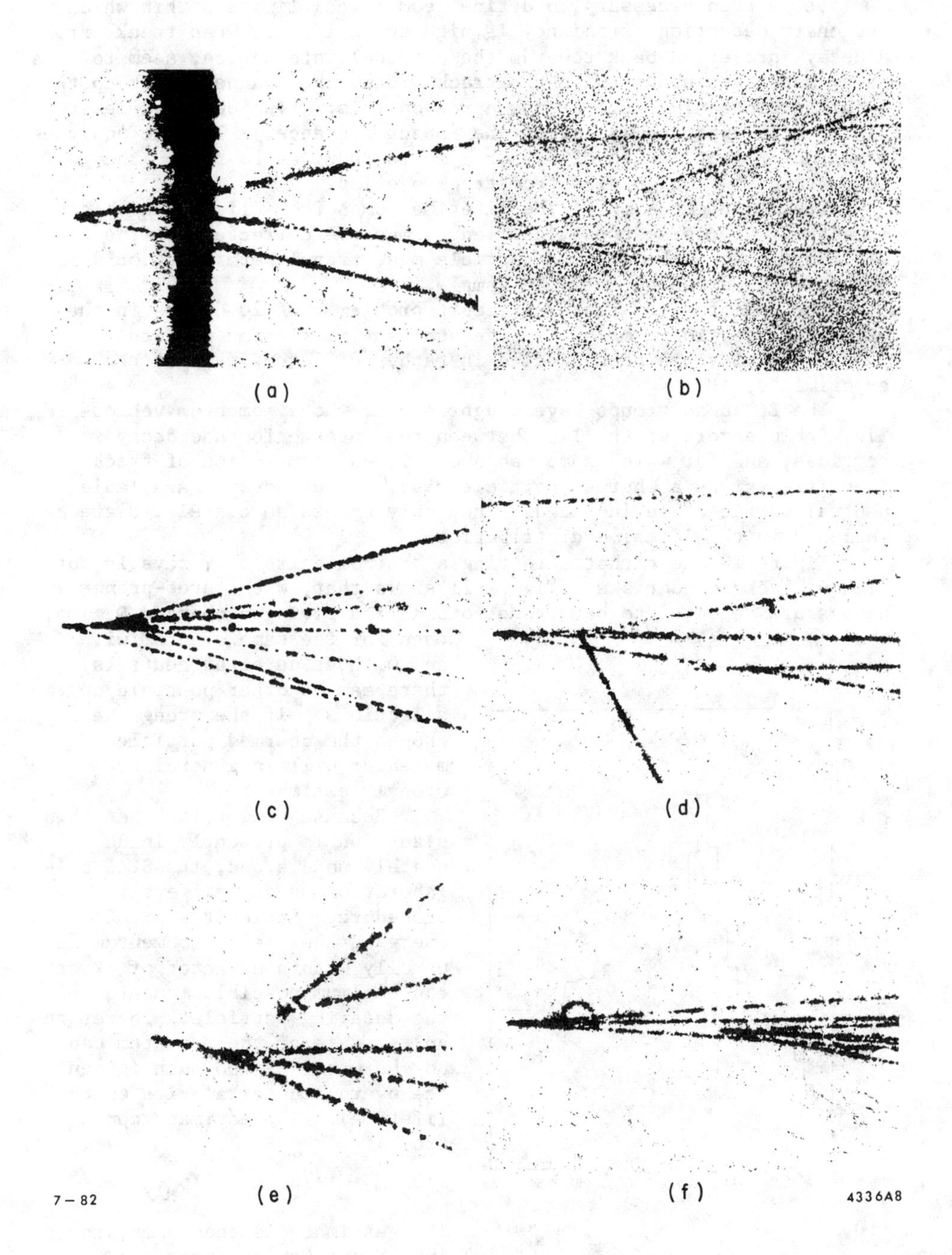

Fig. 9. Events from the SHF data.

It is then necessary to define geometrical limits within which the charm detection efficiency is high and uniform. When tracks from a decay, projected back towards the original interaction, seem to miss it by more than about twice the track width, the scanners pick up the events efficiently. All groups agree on this. (The observed least distance of approach is termed the impact distance.) In some analyses a minimum flight length of 0.5 or 1 mm is required to reduce the reconstruction ambiguities from track overlap.

Given a clean sample, the lifetime comes from flight length and momentum. Of course the flight length must be corrected for the minimum distance the charmed particle must travel before it could be detected and accepted into the sample.

The momentum is a more difficult problem. Of 20 decays in the SLAC collaboration sample, 12 have substantial missing momentum. This difficulty has been handled in rather different ways in the experiments.

The European groups have sought events whose momentum vectors lie within errors of the line between the interaction and decay vertices, and for which some Cabibbo allowed permutation of track identities gives a charmed particle mass. If necessary, available neutral vertices are included. Then only events so classified are included in the lifetime distribution.

Figure 10 shows that selecting a π^0 from pairs of γ rays is not without background risks. Figure 11 shows that, when three-prong decays are selected to have some combination giving a charged D mass, then for the same events, with and without detected neutrals, there may be other possible charm selections. If the wrong one is chosen the charmed particle momentum will in general be wrongly estimated.

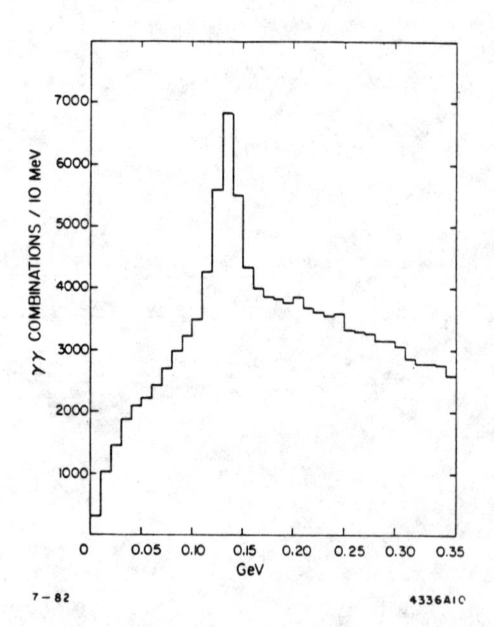

Fig. 10. Mass spectrum of γ pairs from lead glass data (EHS).

Because of worries that this situation is presently inadequately understood, the SLAC collaboration used a different procedure. At least at SLAC energies the visible momentum is usually within a factor of ~2 of the maximum possible momentum of the decaying particle. Therefore an estimate of the momentum cannot be wrong by too much (given the overall uncertainties on the lifetime). The estimate comes from

$$1/p_{EST} = (1/p_{VIS} + 1/p_{MAX})/2.$$

This estimate is then simulated in a Monte Carlo representation of the experiment. In fact, three independent Monte

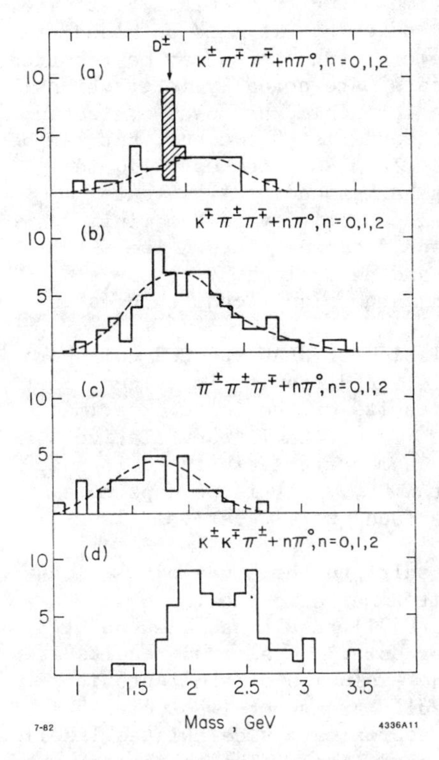

Fig. 11. Reconstructed masses for three-prong plus ($n\ \pi^{o}$) events (EHS). Events are selected to have a three-prong mass at the D mass, and other Cabibbo-allowed masses are plotted.

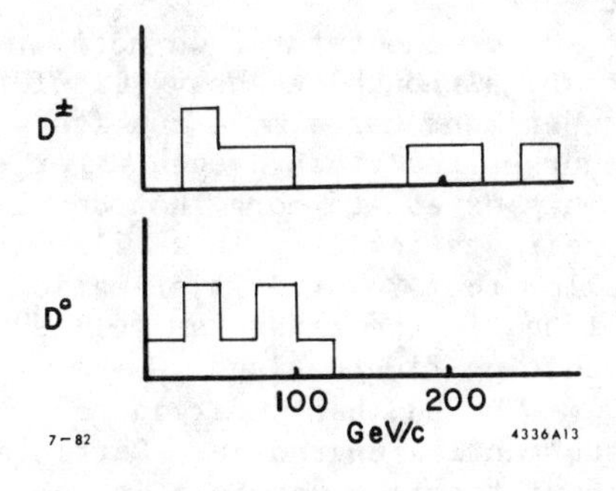

Fig. 13. Spectra of $D^{\pm}$ and D^{o}'s from EHS.

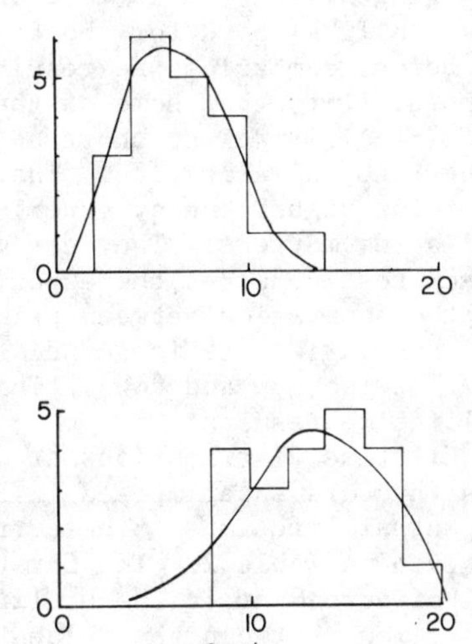

Fig. 12. Spectra of (upper plot) momenta visible at decay vertices and (lower plot) total visible momentum of SHF events. Smooth curves are from the Monte Carlo analysis.

Carlos have been compared with the data and represent it very well. The momentum representation is particularly important and examples of total visible momentum and decay vertex visible momentum plots are given in Fig. 12.

It is interesting to compare this with the broad momentum distributions for the $D^{\pm}$ and D^{o} events from the EHS (LEBC) collaboration (Fig. 13). The reason for, or the significance of, the difference between the charged and neutral spectra from the EHS is obscure to the writer.

Each charmed decay is compared with the Monte Carlo and a maximum likelihood found as a function of lifetime. The measureable quantities compared are: the impact distance (which measures lifetime essentially independent of momentum); the decay length, and the effective lifetime using the momentum estimate. A maximum likelihood for the ensemble of events is also obtained.

Before comparing the results, some useful checks can be reported. The possibility that some of the "decays" are actually interactions with a recoil track so short as to be invisible, has been evaluated by the SLAC collaboration. That background is $\lesssim 0.1$ event, but may be larger for higher energy experiments. In BIBC, the heavy liquid complicates matters. They estimate a background of 1.5 events. Another test concerns the experimental sensitivity to the minimum accepted separation between primary and decay vertices. The SLAC group has varied its impact distance and decay length cuts by a factor of two, up and down. This led to no significant changes in the lifetime results.

Lifetime distributions are available for SLAC and EHS collaborations, and are superimposed in Fig. 14. A disparity is evident in the neutral lifetimes. The current results of the three experiments are given in Table 2. The first thing to notice is the relative level of agreement in the $D^{\pm}$ lifetime numbers. Again there is some disagreement between the values for the D^{0}'s. It is perhaps noteworthy that the few Λ_c and F examples found were relatively long lived.

In an attempt to relate these results to the study with which the talk started, we return to the lifetime ratio plot. In this case, Fig. 15, the neutrino-emulsion (E531) result has been updated with their most recent neutral decay result.[6] Again two results are given: for all decays; and, after the exclusion of semileptonic events. The three bubble chamber results are super-imposed with approximately calculated likelihood lines. The congested figure suggests that a lifetime ratio in the neighborhood of 2.5-3.0 would suit all of the experiments with the exception of DELCO. That one would disagree at a level of perhaps 2.5-3.0 standard deviations. Such a ratio, we recall, would also be well received by present theory.

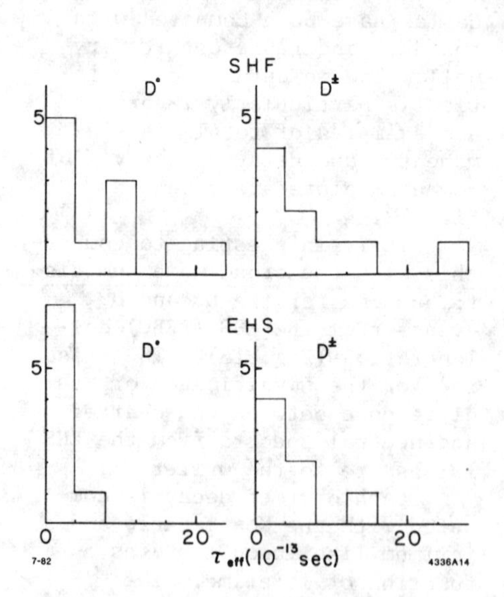

Fig. 14. Effective lifetime plots for $D^{\pm}$ and D^{0} from SHF and EHS.

For the future, we note that, although the heavy liquid experiment has almost run its course, the two hydrogen experiments expect to more than double their statistics. This will allow them to study systematic biases in their data as well as reducing fluctuations.

The EHS has undergone a substantial upgrading. Particle identification has been introduced using aerogel and helium

Table 2. Comparison of results of the three bubble chamber experiments. Units 10^{-13} sec.

	D^{o}	$D^{\pm}$	$D^{\pm}/D^{o}$	$F^{\pm}$	Λ_c^+
LEBC	$2.1\ ^{+1.3}_{-0.7}$ (8 decays)	$6.5\ ^{+4.7}_{-2.1}$ (7 decays)	$3.1\ ^{+2.9}_{-1.4}$	5.5 (1 decay)	7.7 (1 decay)
SHF	$6.7\ ^{+3.5}_{-2.0}$ (11)	$8.2\ ^{+4.5}_{-2.5}$ (9)	$1.2\ ^{+0.9}_{-0.5}$	--	--
BIBC	$3.8\ ^{+2.4}_{-1.2}$ (8)	$6.3\ ^{+6.0}_{-2.5}$ (5)	$1.65\ ^{+1.9}_{-0.9}$	8.3 (1)	--

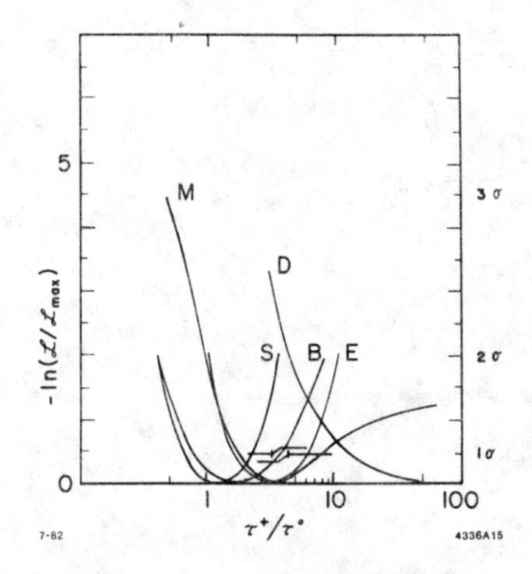

Fig. 15. Figure 2 updated. Experiments are identified by letters: M — Mark II, D — DELCO, S — SHF, B — BIBC, E — EHS. The E531 neutrino emulsion experiment's most recent results are given by the error bars: lower ratio with leptonic decays included, higher ratio without leptonic decays.

Cerenkov counters, as well as a full scale ISIS ionization sampling device, and a transition radiation detector. Upgraded tracking has been included, and perhaps most significant, a new small bubble chamber, HOLEBC, has been made to give resolved track widths of 20 μm with conventional optics. This system has already had its first run.

The SHF collaboration also plans to run again with improved track resolution and particle identification. In two years or so these groups should statistically better their first experiments but with substantially improved systematics.

I wish to thank my various colleagues in the SLAC collaboration, and particularly Dr. G. Kalmus, for useful discussions about this topic.

REFERENCES

[1]R. H. Schindler et al., Phys. Rev. D 24, 78 (1981).
[2]W. Bacino et al., Phys. Rev. Lett. 45, 329 (1980).
[3]N. Ushida et al., Phys. Rev. Lett. 45, 1049 and 1053 (1980).
[4]For example: M. Bander et al., Phys. Rev. Lett. 44, 7 (1980).
H. Fritzsch et al., Phys. Lett. 90B, 455 (1980).
[5]S. P. Rosen, Phys. Rev. Lett. 44, 4 (1980).
[6]N. Ushida et al., Phys. Rev. Lett. 48, 844 (1982).

MEASUREMENT OF THE LIFETIMES OF THE CHARMED $D^{\pm}$, D^{o} AND F^{+} MESONS AND Λ_c^+ CHARMED BARYON

The Fermilab Experiment 531 Collaboration

N. Ushida, T. Kondo, G. Fujioka, H. Fukushima, Y. Takahashi, S. Tatsumi, C. Yokoyama, Y. Homma, Y. Tsuzuki, S. Bahk, C. Kim, J. Park, J. Song, D. Bailey, S. Conetti, J. Fischer, J. Trischuk, H. Fuchi, K. Hoshino, M. Miyanishi, K. Niu, K. Niwa, H. Shibuya, Y. Yanagisawa, S. Errede, M. Gutzwiller, S. Kuramata, N. W. Reay, K. Reibel, T. A. Romanowski, R. Sidwell, N. R. Stanton, K. Moriyama, H. Shibata, T. Hara, O. Kusumoto, Y. Noguchi, M. Teranaka, H. Okabe, J. Yokota, J. Harnois, C. Hebert, J. Hebert, S. Lokanathan, B. McLeod, S. Tasaka, P. Davis, J. Martin, D. Pitman, J. D. Prentice, P. Sinervo, T. S. Yoon, H. Kimura, and Y. Maeda

Aichi University of Education, Kariya, Japan, and Fermi National Accelerator Laboratory, Batavia, Illinois 60510, and Physics Department, Kobe University, Rokkodai, Nada, Kobe 657, Japan, and Faculty of Liberal Arts, Kobe University, Tsurukabuto, Nada, Kobe 657, Japan, and Department of Physics, Korea University, Seoul 132, Korea, and Department of Physics, McGill University, Montreal, Quebec H3A 2T8, Canada, and Department of Physics, Nagoya University, Furo-Cho, Chikusa-Ku, Nagoya 464, Japan, and Physics Department, Ohio State University, Columbus, Ohio 43210, and Physics Department, Okayama University, Okayama, Japan, and Physics Department, Osaka City University, Sugimoto-Cho, Sumiyoshi-Ku, Osaka 558, Japan, and Science Education Institute of Osaka Prefecture, Karito-Cho 4, Sumiyoshi-Ku, Osaka 558, Japan, and Department of Physics, University of Ottawa, Ottawa 2, Ontario, Canada, and Institute for Cosmic Ray Research, University of Tokyo, Tokyo, Japan, and Physics Department, University of Toronto, Toronto, Ontario M5S 1A7, Canada, and Faculty of Education, Yokohama National University, Hodogaya-Ku, Yokohama, Japan

Presented by S.Errede *
The Ohio State University, Columbus, Ohio, 43210

ABSTRACT

Recent results from a hybrid emulsion experiment in the Fermilab wide-band neutrino beam of the measurement of the lifetimes of the charmed $D^{\pm}$, D^{o} and F^{+} mesons and Λ_c^+ charmed baryon are presented.

* Now at the University of Michigan, Ann Arbor, Mich. 48109

0094-243X/82/930089-16$3.00

Fermilab Experiment 531 uses a hybrid emulsion spectrometer to measure the lifetimes of short-lived particles produced in neutrino interactions. The lifetime results presented here[1] are from the first exposure, based on a fitted sample of 11 D^{+}, 19 D^{o}, 3 F^{+} and 8 Λ_c^{+} decays. A second larger exposure with an improved spectrometer is presently being analyzed and will eventually increase our sample by an additional factor of 2.5.

The elevation view of the hybrid emulsion spectrometer is shown in Figure 1. The target consists of 23 liters of Fuji ET-7B emulsion divided into equal amounts of two types of modules containing sheets of emulsion oriented either edge-on, or normal to the beam direction. The 5cm thick emulsion target was exposed to neutrinos from the Fermilab single horn wide-band beam, with a total of $7.2x10^{18}$ 350 GeV protons on the neutrino target.

The large aperture spectrometer is capable of locating events in the emulsion with high accuracy and efficiency, in addition to providing good magnetic momentum analysis and particle identification from time-of-flight (TOF), necessary for kinematic fitting and identification of charmed particle species. The energies of photons and electrons are measured in a lead glass array. The decays of K_s^o and Λ^o are observable in the drift chambers, and the energies of long-lived neutral hadrons are measured in the iron-scintillator calorimeter. Muons are identified by penetration through 4 GeV of iron absorber. The essential parameters of the spectrometer and emulsion are summarized in Table I.

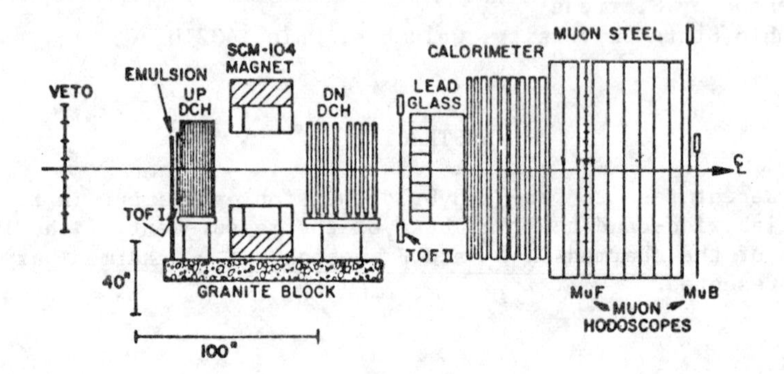

Fig. 1 Elevation view of the hybrid emulsion spectrometer.

One of the major stumbling blocks of previous hybrid emulsion experiments was the inability to locate events in the emulsion. In order to make such a hybrid system work, it is essential to have precise electronic tagging and accurate, stable mounting and registration of all critical components. Thus, the emulsion target and drift chambers were mounted on a granite optical bench, and the relative positions were known of each of the emulsion modules, fiducial sheet

TABLE I.

E-531 SPECTROMETER PARAMETERS

Drift Chambers: Multi-hit capability (up to 15 hits per wire)

12 Upstream: σ_x=125μm, σ_θ=0.6mrad, $\sigma_{2\text{-track}}$=1.8mm.

8 Downstream: σ_x=175μm, σ_θ=0.8mrad.

Charged Particle Momentum Resolution

$\sigma_P=\{(0.013P)^2 \oplus (0.005P^2)\}^{1/2}$ (up-down tracks)

$\sigma_P=0.35P^2$ (upstream only tracks in fringe field)

Time-of-Flight: $\sigma_{TOF\ II}$=100psec (narrow counters)

$\sigma_{TOF\ II}$=150psec (wide counters)

π/K separation to 3.2 GeV/c (1 σ)
K/p separation to 5.5 GeV.c (1 σ) narrow counters

Lead Glass Array: $\sigma_{PbG}=0.15\sqrt{E}$ ($\delta E/E=0.15/\sqrt{E}$)

Calorimeter: $\sigma_{Cal}=1.1\sqrt{E}$ ($\delta E/E=1.1/\sqrt{E}$) (five planes)

Muon Identifier: σ_x=15cm, threshold: P_μ> 4GeV/c (two planes)

Muon detection efficiency $\varepsilon_\mu=0.90 \times 0.82 = 0.74 \pm 0.05$
(2 planes x geometry)

EMULSION PARAMETERS

30 grains/100μm/minimum ionizing track

Angular Resolution:

Horizontal: $0.5\sigma_x = \sigma_y = 0.0033 + 0.02\Theta$ (rad)
(edge-on)

Vertical: $\sigma_x = \sigma_y = 0.00]5 + 0.01\Theta$ (rad)
(normal-to)

Momentum Resolution: $\sigma_P=0.25P^2$ for $P\beta$ < 700 MeV/c (multiple scattering)

Particle Identification below $P\beta \sim$ 800 MeV/c

(see below), drift chambers and mounting stands were known from optical survey(s), continuous electronic monitoring via position sensors and continuous calibration with background (beam associated) muons to an accuracy of ± 50μm or better at all times. But having a well-engineered experiment is only one of the "necessary" conditions for success - equally important is the great skill, care and patience of the hard-working emulsion people in our collaboration, who were able to locate 1248 of the predicted 1829 neutrino interactions within the emulsion fiducial volume, for a net finding efficiency of 69%. The difference between the found and predicted coordinates was better than 0.4mm in the transverse (x-y) directions, and 1.4mm in the longitudinal (z) direction.

The vertices predicted by the spectrometer were searched for by volume scan and track follow-back techniques[2]. In the former method, the emulsion is scanned at low power (100x magnification) in a volume $4x4x20mm^3$ centered around the predicted vertex. Location of events by this method with few or no black tracks (from nuclear break-up) is quite difficult. With the follow-back technique, most useful for vertical emulsion (emulsion sheets mounted normal to the beam), drift chamber tracks are first extrapolated with a typical accuracy of ±300μm to an emulsion fiducial sheet mounted on the downstream end of the emulsion target. The fiducial sheet consists of an 800μm thick lucite sheet coated on both sides with 330μm of emulsion. The fiducial sheet was changed every few days during the exposure, and served as a high resolution detector, coupling drift chamber tracks to the downstream end of the emulsion stack with an accuracy of ± 50μm. Once found in the downstream end of the main body of emulsion, tracks were followed back to their point of origin with an overall follow-back efficiency of 90 ± 2%. The event finding efficiency vs. depth in emulsion is shown in Figure 2, for the two types of emulsion used in the experiment. The distributions show that the ability to locate events is largely independent of the depth in emulsion.

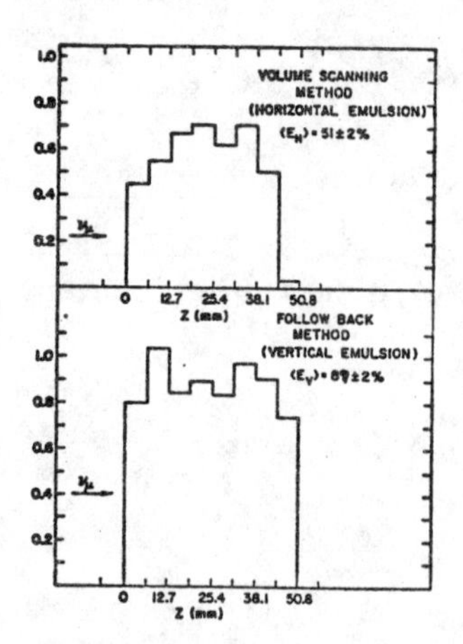

Figure 2. Event Finding Efficiencies in Emulsion.

After location of a neutrino interaction in the emulsion, a sytematic scan for the decays of short-lived particles was performed, using three methods. Charged decays were searched for by following all tracks out from the primary vertex within $\Theta = 0.3$ radians of the beam direction for a minimum distance of 6mm, or until exiting the emulsion. Neutral decays are sought by volume scanning a cylinder 300μm in radius x 1000μm long downstream of the primary vertex with an efficiency of 60%. A third, and very powerful method for the detection of short-lived particles is the scanback method[2], whereby spectrometer tracks with no matching emulsion tracks at the primary vertex with P > 700 MeV/c and pass within 2mm of the primary vertex are followed into the emulsion via the fiducial sheet. The scanback technique, most compatible with the normally-mounted emulsion is equally sensitive to charged and neutral decays of short-lived particles, gamma conversions and secondary nuclear interactions at all distances from the primary vertex, and also allows cross-calibration of the scanning efficiencies for the methods used in observing these processes. The scanback technique is very similar to that of track follow-back, with a track scanback efficiency of 90 ± 2%.

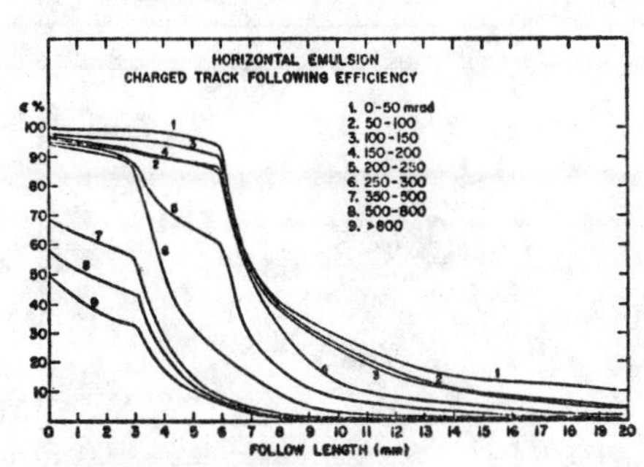

Figure 3. Horizontal Emulsion Charged Track Following Efficiency.

The charged track following efficiency vs. distance is shown in Figure 3. The charged track following efficiency for the vertical emulsion is similar. The drop in efficiency at large distances merely reflects the number of tracks followed out past that distance.

The efficiency for finding single prong or "kink" decays[3,6] as a function of kink angle Θ_k is shown in Figure 4. The kink detection efficiency drops sharply below 40 mrad due to the increasing difficulty in the detection of small-angle deflections along a track. The kink detection efficiency is reasonable good with respect to the decays of charmed particles produced in neutrino interactions, as the observed single prong charged charm decays have kink angles well in excess of 40 mrad. This is particularly true for the charmed baryons, which have a lower momentum spectrum and larger Q than the charged charmed mesons.

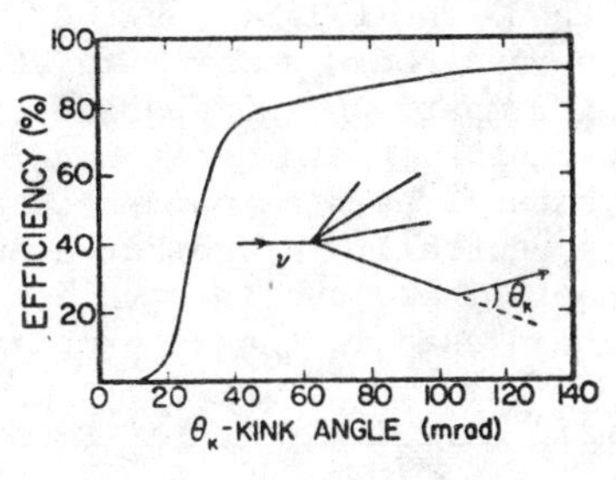

Figure 4. Kink Detection Efficiency.

The angular distribution of found charm decays is shown in Figure 5. The distribution falls to zero before the limit of the angular scanning criteria is reached, i.e. $\Theta = 300$ mrad.

The scanback efficiency for finding e^+e^- pairs from gamma conversion in the emulsion is shown in Figure 6a. The efficiency is high and also relatively flat in $Z' = Z_{\gamma\to e^+e^-} - Z_{production}$. The number of found e^+e^- pairs per unit length, dN/dZ' vs. distance from the primary vertex is shown in Figure 6b. The predicted number of e^+e^- pairs (solid curve) for a conversion length of 4.1cm in emulsion is in good agreement with the observed distribution of found pairs. The measured conversion length in our emulsion is 4.2^{+2}_{-1} cm.

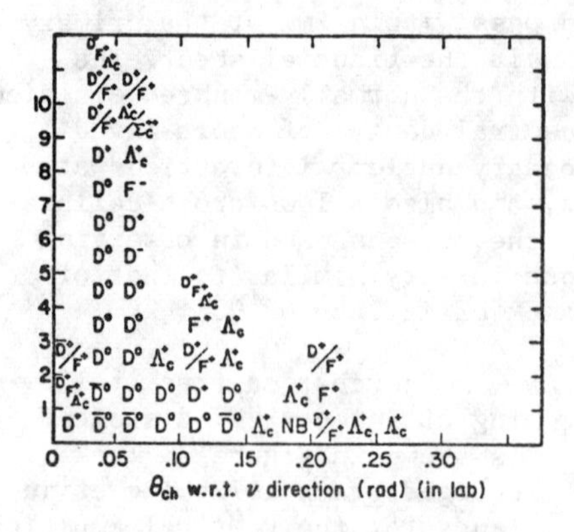

Figure 5. Angular Distribution of Charmed Particles Produced in Neutrino Interactions.

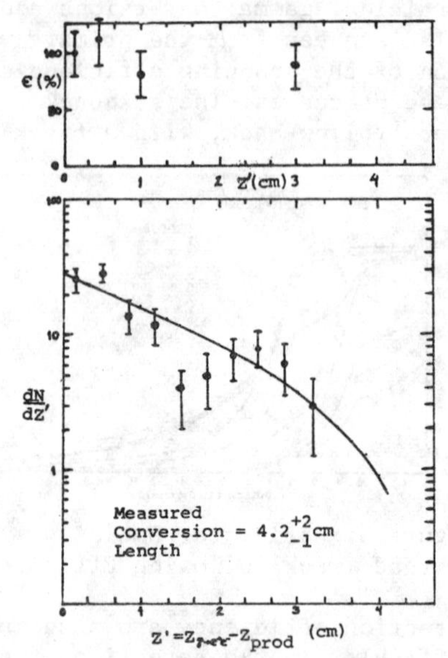

Figure 6a. Scanback Efficiency for e^+e^- Pairs in Emulsion.

Figure 6b. Number of e^+e^- Pairs Found by Scanback vs. Z'

The decay finding efficiency for charged and neutral multiprong decays is shown in Figure 7. The drop in efficiency below 10μm is due to shadowing near the primary vertex, but remains high at large distances because of the scanback method. The flight lengths of observed charm decays are shown for comparison. 25 charged and 20 neutral charm decay candidates were found. In addition to these events, 2 events found occurred with the magnet off (one charged, one neutral). In another event, the decay occurred in the fiducial sheet. These events have been excluded from the sample of charmed decays used in the determination of charmed particle lifetimes.

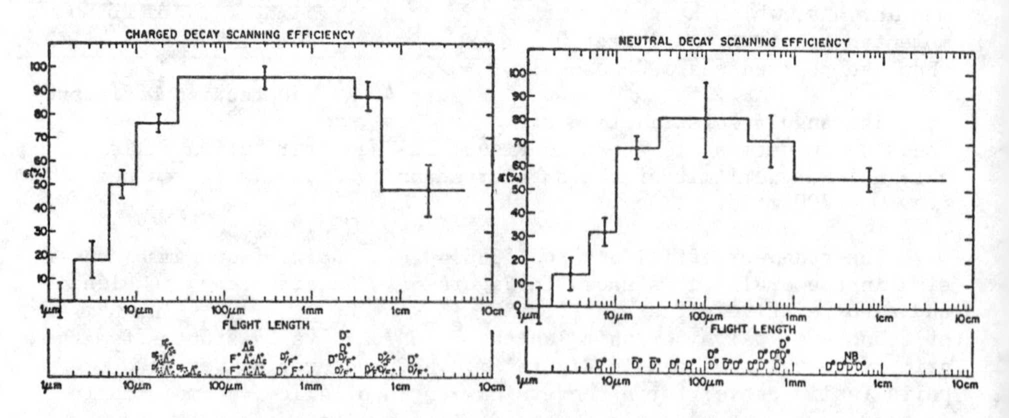

Figure 7. Charged and Neutral Charm Decay Scanning Efficiencies.

Two K^o_s and two Λ^o decays were found in the emulsion. 194 gamma conversions (including 138 found by scanback) were observed.

With regard to the question of background within our sample of charmed events, for the multi-prong candidates none are consistent with strange particle decay. Secondary hadronic interactions in the emulsion can mimick charm decay if and only if there are no observed nuclear fragments (black tracks) and charge appears to be conserved. From the characteristics of 89 hadronic interactions observed in our experiment (85 charged, 4 neutral) we calculate[6] the multiprong hadronic background over all incident particle momenta to be $0.8^{+1.1}_{-0.6}$ events for charged decays, and $0.05^{+0.10}_{-0.03}$ events for neutral decays. However, $\geq$ 80% of all secondary hadrons from neutrino interactions in our experiment have momentum $P \leq 4$ GeV/c, while less than 10% of all charm decay candidates have momentum $P \leq 4$ GeV/c. The impact of the hadronic background is largest for the charmed baryon candidates, and is essentially negligible for the charmed mesons.

The mean absorbtion length and interaction length in emulsion as measured by this experiment are $\lambda_{abs} = 39$ cm, $\lambda_{int} = 25$ cm in good agreement with the Particle Data Book.

In order to measure the lifetimes of charmed particles, one must identify the charmed species in each decay. Identification of particle type, measurement of angles and momenta of all particles emitted in the decay is important. In addition to spectrometer techniques for momentum measurement and particle identification in our experiment, the momenta of soft charged particles which miss the magnet can be determined to $\pm$ 25% from multiple scattering measurements in the emulsion, and their identity can be determined among π, K and p by ionization from grain counting.

A direct measurement is made in the emulsion of the decaying particle's direction, which is of great importance for constraints used in kinematic fitting of decays, as it allows the use of transverse momentum balance to determine the presence of neutral particles emitted in the decay, and which neutral particles observed in the spectrometer (e.g. π^o, K^o, Λ^o...) come from the decay vertex.

Kinematic fits to each decay are performed.[4,5] 3-C fits are obtained for those decays with no unseen neutrals, with an assumed parent mass. All parent masses are cycled over, consistent with the identified secondaries. For unidentified tracks, all allowed particle i.d.'s are cycled over. All decay hypotheses are kept with 3-C confidence levels $\geq$ 1%. 2-C fits are similar, except with the mass used as a parameter. For decays with unseen neutrals, e.g. semi-leptonic decays, 0-C fits are performed, yielding (usually) two momentum solutions. Only Cabibbo-favored decay hypotheses are considered in 0-C fits, unless kinematically forbidden.

The fitted charged charm decays are summarized in Table II. The lifetimes for each species of charmed particle is determined by the maximum likelihood method,[5] taking scanning efficiencies into account. An underlined decay product in the Table signifies identification of that particle by TOF or emulsion methods to better than 90% confidence level, while parentheses indicate unseen neutrals.

The three $F^{\pm}$ decays are very clean, in that the kinematic $D^{\pm}$ background is very small (the Λ_c^+ background is negligible). Specifically, the confidence levels for the D-hypotheses for each of these events is 1.3%, 3.0% and 0.01% respectively. Note that all of the D-hypotheses would also be Cabibbo-unfoavored decays. Furthermore, if the TOF information for the K^+ identification is ignored in the last two events, i.e. the K^+ are treated as π^+, the summed 3-C confidence level for the D^+ hypotheses is less than 1.8% for both events. The measured F^+ lifetime and F^+ mass are:

$$\tau_{F^+} = 2.0^{+1.8}_{-0.8} \times 10^{-13} \text{ sec.} \quad M_{F^+} = 2042 \pm 30 \text{ MeV/c}^2$$

TABLE II.

CHARGED CHARM DECAY CANDIDATES

SUMMARY OF F^+ DECAY CANDIDATES

EVENT NUMBER	MUON	DECAY MODE HYPOTHESIS	D.L. (μm)	P_C (GeV/c)	MASS (MeV/c²)	DECAY TIME (x10⁻¹³ sec)
527-3682	–	$F^+ \to \underline{\pi}^+\underline{\pi}^-\pi^-\underline{\pi}^0$	670.0±4.0	12.2±0.3	2026±56	3.70±0.08
597-1851	*	$F^+ \to \underline{K}^-\underline{\pi}^+\underline{\pi}^+K_L^0$	130.0±1.0	9.3±0.4	2057±110	0.97±0.09
638-9417	-	$F^+ \to \underline{K}^+K^-\pi^+\underline{\pi}^0$	153.0±8.0	6.0±0.1	2050±45	1.72±0.09

* No muon observed

SUMMARY OF Λ_c^+ DECAY CANDIDATES

EVENT NUMBER	MUON	DECAY MODE HYPOTHESIS	D.L. (μm)	P_C (GeV/c)	MASS (MeV/c²)	DECAY TIME (x10⁻¹³ sec)
476-4449	-	$\Lambda_c^+ \to \underline{p}\underline{\pi}^+\underline{\pi}^-(K_L^0)$	27.2±1.0	2.7±0.1 4.8±0.1	0-C	0.79±0.04 0.44±0.04
498-4985	-	$\Lambda_c^+ \to \underline{\Lambda}^0\underline{\pi}^+\underline{\pi}^-\underline{\pi}^+$	180.0±5.0	8.4±0.1	2274±41	1.63±0.05
499-4713	-	$\Lambda_c^+ \to \underline{\Sigma}^0\pi^+$	366.0±6.0	4.2±0.1	2269±17	6.60±0.14
549-4068	-	$\Lambda_c^+ \to \underline{p}K^-\underline{\pi}^+(\pi^0)$	20.6±2.0	1.9±0.1 2.5±0.1	0-C	0.77±0.07 0.63±0.07
567-2596	-	$\Lambda_c^+ \to \underline{p}\ K_L^0$	175.0±5.0	5.8±0.1	2204±207	2.30±0.08
602-2032	-	$\Lambda_c^+ \to \underline{p}\underline{\pi}^+\underline{\pi}^-(K_S^0)$	282.5±5.0	6.3±0.1	0-C	3.40±0.10
610-4088	-	$\Lambda_c^+ \to \underline{\Lambda}^0\underline{\pi}^+\underline{\pi}^-\underline{\pi}^+$	221.0±4.0	4.7±0.2	2374±62	3.60±0.19
650-6003	-	$\Lambda_c^+ \to \underline{\Lambda}^0\underline{\pi}^+\pi^-\underline{\pi}^+$	40.6±2.0	5.7±0.1	2131±63	0.54±0.03

TABLE II. (CONT.)

SUMMARY OF D^+ DECAY CANDIDATES

EVENT NUMBER	MUON	DECAY MODE HYPOTHESIS	D.L. (μm)	P_C (GeV/c)	MASS (MeV/c²)	DECAY TIME ($\times 10^{-13}$ sec)
512-5761	-	$D^+ \to K^- \pi^+ \underline{\pi}^+ \pi^0$	457±5	10.4±0.1	1829±35	2.77±0.05
		$F^+ \to K^- K^+ \underline{\pi}^+ \pi^0$		10.3±0.1	2011±33	3.00±0.05
546-1339	-	$D^+ \to K^- \pi^+ \underline{\mu}^+ (\nu_\mu)$	2150±50	16.6±0.2	0-C	8.06±0.22
		$F^+ \to \pi^- \pi^+ \underline{\mu}^+ (\mu_\nu)$		13.3±0.2	0-C	10.94±0.16
				36.8±0.2		2.95±0.02
580-4508	+	$D^- \to \pi^- \underline{K}^+ e^- (\bar{\nu}_e)$	2307±50	9.46±0.15	0-C	15.20±0.40
				10.01±0.13		14.37±0.36
598-1759	-	$D^+ \to K^- K^+ \underline{\pi}^+ \underline{\pi}^0$	1802±15	17.4±0.3	1862±25	6.44±0.13
663-7758	-	$D^+ \to K^- \pi^+ e^+ (\nu_e)$	13000±50	114.3±7.6	0-C	7.08±0.44
		$F^+ \to \pi^- \pi^+ \underline{e}^+ (\nu_e)$		96.8±7.6	0-C	8.36±0.47
493-1235	-	$D^+ \to \pi^+ \pi^+ \pi^- K_L^0$	2203±10	11.9±1.3	2061±156	11.5±1.2
		$F^+ \to \pi^+ \pi^+ K^- K_L^0$		11.7±0.7	2246±166	12.7±0.8
		$\Lambda_c^+ \to \pi^+ \pi^+ K^- n$		13.3±2.0	2330±123	12.6±1.9
656-2631	-	$D^+ \to \pi^+ K^- \pi^+ \pi^0$	570.0±11.0	32.6±1.2	1933±73	1.09±0.05
		$F^+ \to K^+ K^- \pi^+ \pi^0$		32.4±1.1	2099±73	1.19±0.05
		$\Lambda_c^+ \to p K^- \pi^+$		31.7±1.2	2317±76	1.36±0.06
522-2107	-	$D^+ \to \pi^+ \underline{\pi}^- K^- (\pi^0)$	13600±100	23.5±1.3	0-C	36.0±1.8
				31.7±1.3		26.7±1.3
		$F^+ \to K^+ \underline{\pi}^+ K^- (\pi^0)$		22.5±1.3	0-C	40.9±1.9
				32.7±1.3		28.1±1.1
		$\Lambda_c^+ \to p \underline{\pi}^+ K^- (\pi^0)$		22.5±1.3	0-C	46.0±2.5
				31.5±1.3		32.9±1.3
529-271	-	$F^+ \to K^+ (K_L^0)$	2547±30	43.1±0.2	0-C	4.0±0.1
		$D^+ \to \pi^+ \pi^0 (K_L^0)$		55.4±0.5	0-C	2.9±0.1
		$F^+ \to K^+ \pi^0 (K_L^0)$		38.4±0.5	0-C	4.5±0.1
533-7152	-	$F^+ \to K^+ (K_L^0)$	5246±50	34.8±0.2	0-C	10.2±0.1
		$D^+ \to \pi^+ \pi^0 (K_L^0)$		40.1±0.4	0-C	8.1±0.1
547-3192	-	$F^+ \to \pi^+ \pi^- \pi^+ (\pi^0)$	185±10	9.6±1.2	0-C	1.3±0.2
		$D^+ \to \pi^+ \pi^- \pi^+ (\pi^0)$		9.6±1.2	0-C	1.2±0.2

The eight Λ_c^+ decays all have a well-identified baryon in the final state (e.g. p or Λ^o). Thus the D^+ and F^+ contamination within the Λ_c^+ sample is negligible. The background due to secondary nuclear interactions is also small, as the mean charmed baryon momentum is of order 5 GeV/c. Note that this is substantially lower than for charged and neutral charmed mesons. The measured Λ_c^+ lifetime and Λ_c^+ mass are:

$$\tau_{\Lambda_c^+} = 2.3^{+1.0}_{-0.6} \times 10^{-13} \text{ sec.} \qquad M_{\Lambda_c^+} = 2265 \pm 30 \text{ MeV/c}^2$$

The eleven $D^\pm$ decays are not as clean as the F^+ and Λ_c^+ decay samples, having typical 3-C confidence levels of 10% for F^+ hypotheses. Three of the eleven events are semi-leptonic decays and several of the events are kinematically ambiguous due to lack of charged particle identification and/or unobserved neutrals. In order to determine the degree of F^+ / Λ_c^+ contamination within this sample of events, a two-parameter maximum likelihood fit to the D^+ lifetime was performed[7] in which the D^+ lifetime and the fraction of D^+ within the samples were treated as independent variables, with a contaminant F^+/Λ_c^+ lifetime of $2.0 \pm 0.8 \times 10^{-13}$ sec. The results of this test yielded the same D^+ lifetime as with the 1-parameter maximum likelihood method, with the fraction of D^+ within the sample of eleven events $f_{D^+} = 0.99^{+0.01}_{-0.29}$. The 1, 2, 3σ... curves for the two parameter fit to the D^+ lifetime and D^+ fraction are shown in Figure 8. The measured D^+ lifetime and D^+ mass are:

$$\tau_{D^+} = 11.4^{+6.6}_{-4.4} \times 10^{-13} \text{ sec.} \qquad M_{D^+} = 1851 \pm 20 \text{ MeV/c}^2$$

Note that the D^+ lifetime is substantially longer than the F^+ and Λ_c^+ lifetimes.

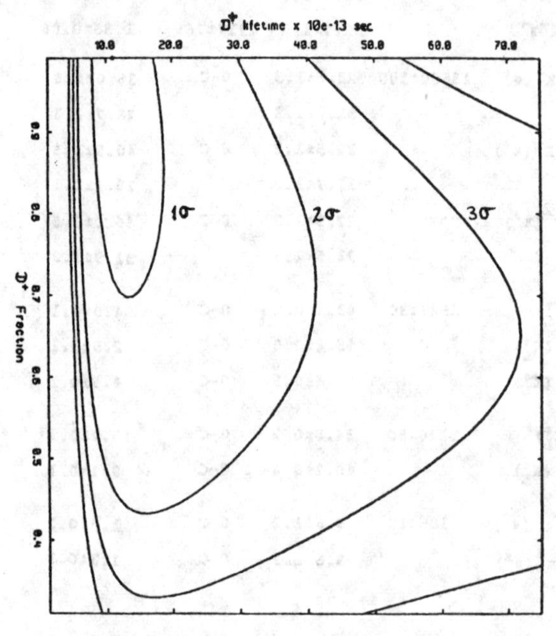

Figure 8. Two-Parameter Likelihood Curves for the Determination of τ_{D^+} and the Fraction f_{D^+} within the D^+ sample.

The 19 neutral decays consistent with a D^0 interpretation have a well-identified sub-sample of six $D^{*\pm} \to D^0 \pi^\pm$, with a D^* mass of M_{D^*} = 2008.7±0.9 MeV/c^2, as shown in Figure 9. The lifetime of the D^0 as determined from this sub-sample of events is $1.9^{+1.2}_{-0.6} \times 10^{-13}$ sec.

If all 19 events within the D^0 sample are used, the lifetime of the D^0 is $3.2^{+1.0}_{-0.7} \times 10^{-13}$ sec. The last three events in the D^0 sample (see Table III.) are semi-leptonic decays, and while consistent in all kinematic aspects as the decays of D^0 mesons, and given the increased uncertainty in their decay times, due to the two-fold momentum ambiguity for each decay, these three events appear to be substantially longer-lived than the majority of the rest of the sample. The probability that these events have the same lifetime as the rest of the D^0 sample is less than 2%. Whether or not this is due to small statistics or is an indication of some other phenomenon or physical process

TABLE III.

NEUTRAL CHARM DECAY CANDIDATES

SUMMARY OF D^0 DECAY CANDIDATES

EVENT NUMBER	MUON	DECAY MODE HYPOTHESIS	D.L. (μm)	P_C (GeV/c)	MASS (MeV/c^2)	DECAY TIME ($\times 10^{-13}$ sec)	Comments
478-2638[a]	−	$D^0 \to \underline{\pi}^- \pi^- \underline{\pi}^+ \underline{\pi}^+ K^- \pi^+ \pi^0$	126	7.5	1897±26	1.03±.08	
493-177	−	$D^0 \to K^0 \pi^+ \underline{\pi}^-$	324	11.3	1819±80	1.77±.13	
513-8010	+	$\bar{D}^0 \to K^+ \pi^+ \underline{\pi}^- \underline{\pi}^- \pi^0$	27	9.2	1766±48	0.18±.02	
518-4935	−	$D^0 \to \pi^+ K^- \pi^0 \pi^0$	116	30.1	1955±132	0.24±.02	
529-3013[b]	ns	$D^0 \to K^- \pi^+ \pi^0$	626	12.9	1856±79	3.02±.18	
529-3013[b]	ns	$\bar{D}^0 \to K^+ K^-$	3307	47.7	1832±124	4.59±.19	
547-2197	−	$D^0 \to \pi^+ \pi^- \underline{\pi}^+ K^- \pi^0$	4056	23.6	1861±39	10.7±.34	
547-3705	−	$D^0 \to K^- \pi^+ \underline{\pi}^+ \pi^-$	748	13.5	1947±99	3.44±.21	D^{*+}
556-152	−	$D^0 \to \pi^- K^- \underline{\pi}^+ \underline{\pi}^+ \pi^0$	41	15.4	1855±43	0.17±.01	D^{*+}
638-5640	+	$\bar{D}^0 \to \pi^- K^+ \pi^0 \pi^0$	183	22.4	1825±68	0.51±.05	
654-3711	−	$D^0 \to \pi^+ \underline{\pi}^+ K^- \pi^- \underline{\pi}^- \underline{\pi}^+$	6.5	19.2	1923±46	.02±.003	
661-2729	−	$D^0 \to \underline{K}^0_s \underline{\pi}^+ \underline{\pi}^- \pi^0$	734	12.4	1835±41	3.66±.19	D^{*+}
666-5294	−	$D^0 \to K^- \pi^- \underline{\pi}^+ \pi^+$	653	55.2	1865±101	0.73±.04	
486-6857	−	$D^0 \to K^- \pi^- \underline{\pi}^+ \pi^+ (\pi^0)$	256	12.9		1.24±.09	D^{*+}
577-5409	−	$D^0 \to \pi^+ \underline{\pi}^- (\bar{K}^0)$	67	11.3		0.37±.04	D^{*+}
670-7870	+	$\bar{D}^0 \to K^+ \underline{\pi}^- (\pi^0)$	187	6.8		1.71±.12	D^{*-}
522-3061	−	$D^0 \to \pi^- \pi^+ \underline{\mu}^+ K^- (\nu)$	5479	35.5		9.58±.36	
				58.5		5.72±.22	
597-6914	−	$D^0 \to \underline{e}^+ K^- (\nu)$	4374	29.8		9.13±.44	
				62.7		4.33±.21	
661-6517	−	$D^0 \to K^- \underline{\mu}^+ (\nu)$	2647	22.8		7.20±.37	
				38.7		4.24±.22	

[a] 1C fit.
[b] Events are an associated pair from the same interaction.

Event 635-4949

Neutral Baryon Candidate

Decay Mode	Mass (MeV/c^2)	P(GeV/c)	τ ($\times 10^{-13}$ sec)	χ^2	C.L.
NB → p $\pi^- K^0_s$	2450±15	4.64±0.51	77.2±0.9	0.13	0.94
NB → p $K^- K^0_s$	2647±11	4.64±0.51	83.4±0.9	0.13	0.94

is uncertain at this time. More statistics are required. While the data from the entire D^0 sample are consistent with one lifetime (50% confidence level), it is felt that due to the uncertainty in the ability to positively determine the identity of the three semi-leptonic candidates, as well as the added uncertainty in knowing which momentum solution is the correct one in each case and the disparity in lifetimes between these three events and the rest of the D^0 sample, we exclude these events from the lifetime sample for the time being and determine the D^0 lifetime from only the fully-constrained sample of the 16 remaining events. The D^0 lifetime and D^0 mass are thus:

$$\tau_{D^0} = 2.3^{+0.8}_{-0.5} \times 10^{-13} \text{ sec.}$$

$$M_{D^0} = 1865 \pm 14 \text{ MeV}/c^2$$

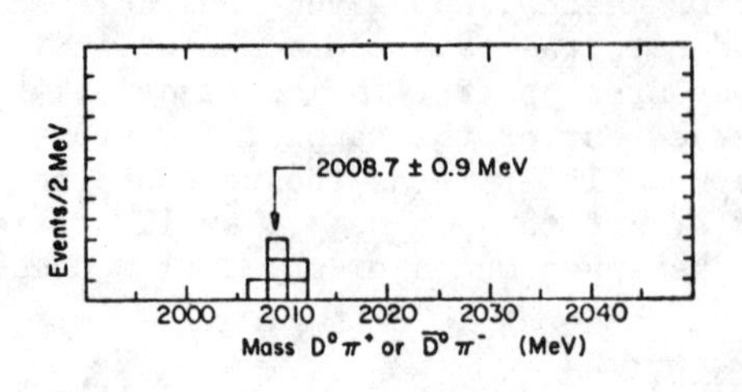

Figure 9. D^0 $\pi^\pm$ Invariant Masses.

The log-likelihood curves for the D^0 lifetime from the 16 constrained events, 6 D^* events and 3 semi-leptonic events are shown in Figure 10. The differential decay time spectrum for the 16 constrained events is shown in Figure 11.

With the forthcoming results from the second exposure of E-531, the situation for the D^0 lifetime should become much more clear.

The ratio of D^+ to D^0 lifetimes using the 11 D^+ and 16 D^0 events is[7]

$$\tau_{D^+}/\tau_{D^0} = 5.0^{+3.0}_{-2.0}$$

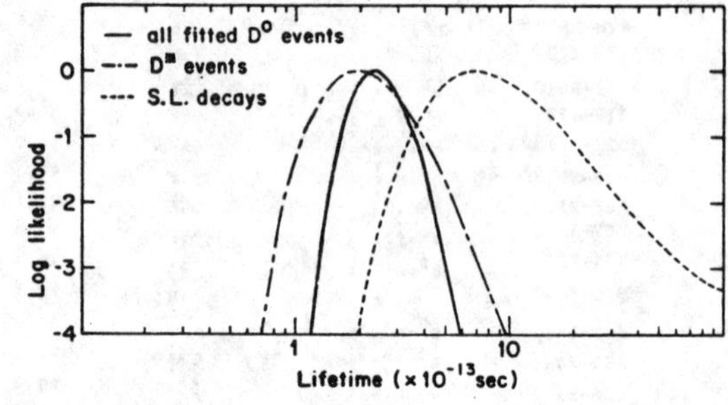

Figure 10. Likelihood Curves for the D^0 Lifetime. Solid curve - all events. Dashed - D^* events. Dotted - Semi-leptonic.

which differs from unity by about three standard devations. Use of all 19 D^0 candidates does not change this ratio within the quoted errors.

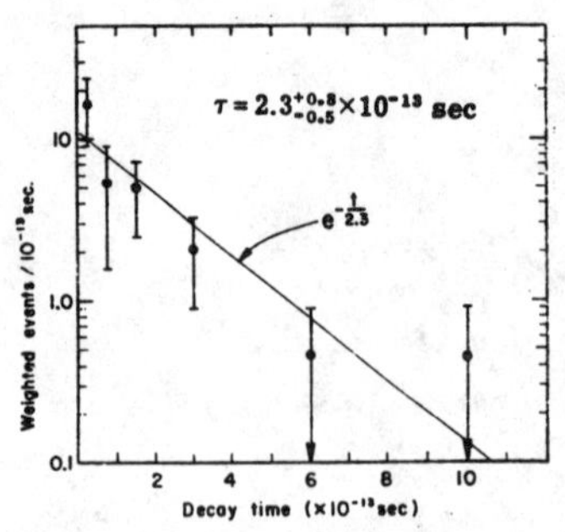

Figure 11. Differential Decay Spectrum for Constrained Events.

The events are weighted by the reciprocal of the scanning efficiency for neutral charmed decays. The curve is from a maximum likelihood fit to the data.

The last event in the neutral decay candidate sample is a curious one, in that the event (635-4949) appears to be the decay of a high-mass long-lived baryon.[4] The proton is well identified by TOF and comes from a clean two prong vertex (found by scanback of the proton track) having no recoil stubs, 4.4mm downstream of the primary vertex. The second track from this vertex is π^-/K^- ambiguous. 0.66 GeV/c of missing transverse momentum at the second vertex is neatly balanced by a K^o_s observed to decay in the drift chambers. The second vertex cannot be due to Λ^o decay, by kinematical analysis[4] and also because the second vertex does not point to the primary vertex. The event is not consistent with a decay of a D^o meson, because of the identified proton and also because of its mass (assuming the proton to be otherwise a π^+ or a K^+). Although we expect less than 0.05 decay-faking K^o interactions within our data sample, this event is an unlikely interaction candidate, due to its relatively high momentum (4.6 GeV/c) and transverse momentum (900 MeV/c) of the neutral particle producing the second vertex. The one surprizing aspect associated with this event is its decay time $\sim 80 \times 10^{-13}$ sec., approximately 40x longer than the lifetime of the Λ_c^+. Although one event is no existence proof for a new particle, should it be real, more will certainly be seen in the second run.

With the sample of charged and neutral charm decay candidates, we have determined the charged-current neutrino production cross - section for charm[6,8]. Figure 12 shows the relative neutrino charged current single charm production cross section. The solid and dashed curves are from Reference 9 and 10, and are in good agreement with the experimental data. Measurement of the neutrino production cross section for charm as determined from opposite-sign dimuon data[11] is also in agreement with our results. The total relative production cross section for single charm production by neutrinos above $E_\nu \geq 10$ GeV is measured to be:

$$\frac{\sigma_{charm}}{\sigma^{\nu}_{c.c.}\mu(E_\nu > 10\text{GeV})} = 6.5^{+1.8}_{-1.5}\%$$

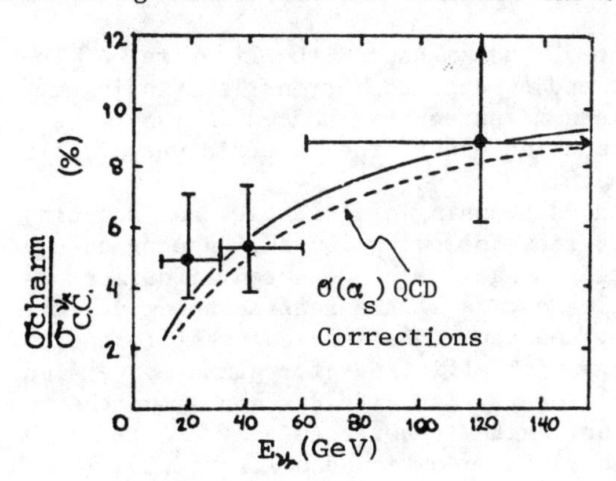

Figure 12. Relative Charm Production Cross Section for Charged Current Neutrino Interactions. Theoretical curves are from References 9 and 10.

As mentioned earlier, analysis of the second exposure of E-531 is presently under way. Many improvements were made for the second run, both to the detector and the beam. The energy of the protons incident on the neutrino target was raised to 400 GeV; the shielding in the berm was substantially upgraded, such that the beam associated flux of muons was considerably reduced, relative to the first run. The emulsion target thickness was increased by a factor of 1.4. More drift chambers and TOF counters were installed. A particle identification dE/dX chamber was installed in the magnet gap. The lead glass array was augmented with a gamma converter and three planes of proportional tubes. The hadron calorimeter was also upgraded with proportional tubes in the first three gaps in the calorimeter. The muon identifier and upstream veto were also upgraded.

- All of the improvements to the detector will result in a large improvement in the ability to reconstruct and identify charmed particles. The momentum resolution and mass resolution for charmed particles will improve substantially due to the improvements in charged track momentum resolution and an overall improvement on the detection and measurement of angles and energies of neutral particles.
- Improvements to the beam have resulted in a significant increase in the ability to find events in the emulsion, due to the reduction of background tracks.

To date 5200 events have been reconstructed with a vertex in the target, and 4200 events are within the useful emulsion volume. 1600 neutrino interactions have been found so far, with a finding efficiency in excess of 85% for both types of emulsion. 30 multi-prong charm candidates (11 charged and 19 neutrals) have been found. Approximately 150 charm decays are expected from the second run.

To summarize, Experiment-531 has measured the lifetimes of the charmed D^{+}, D^{o} and F^{+} mesons and Λ_c^+ charmed baryon. The finding efficiency for charm decays has been measured and checked in many ways:

1.) The measured neutrino event finding efficiency is flat with target depth.
2.) The measured charm scanning efficiencies and scanning criteria (e.g. track following length, angular cuts,etc.) are good. Little change in the measured lifetimes of charmed particles occurs if the real scanning efficiencies are replaced by constant (flat) scanning efficiencies.
3.) The measured scanback efficiency for gamma conversions within the emulsion is flat with distance, and the measured conversion length is correct.
4.) The measured hadronic absobtion and interaction lengths in emulsion are correct. The obsreved number of interactions is in good agreement with expectations.
5.) The measured single charm production cross section from charged-current neutrino interactions is in good agreement with theoretical predictions and other experimental data.

The charmed particle lifetimes and masses as determined from the first exposure of Experiment-531 are;

$$\tau_{D^+} = 11.4^{+6.6}_{-4.4} \times 10^{-13}\text{sec.} \quad M_{D^+} = 1851 \pm 20\ \text{MeV}/c^2$$
(11 Events)

$$\tau_{D^o} = 2.3^{+0.8}_{-0.5} \times 10^{-13}\text{sec.} \quad M_{D^o} = 1865 \pm 14\ \text{MeV}/c^2$$
(16 Events)

$$\tau_{F^+} = 2.0^{+1.8}_{-0.8} \times 10^{-13}\text{sec.} \quad M_{F^+} = 2042 \pm 30\ \text{MeV}/c^2$$
(3 Events)

$$\tau_{\Lambda_c^+} = 2.3^{+1.0}_{-0.6} \times 10^{-13}\text{sec.} \quad M_{\Lambda_c^+} = 2265 \pm 30\ \text{MeV}/c^2$$
(8 Events)

The ratio of D^+ to D^o lifetimes is measured to be:

$$\tau_{D^+}/\tau_{D^o} = 5.0^{+3.0}_{-2.0}$$; More than three standard deviations from unity.

Although the central values of the charmed particle lifetimes and D^+/ D^o lifetime ratio measured in this experiment may differ from those obtained in other experiments[12,13,14,15] , the errors on the central values are such that the individual experiments are all within one to two standard deviations of each other.

The results from the second exposure of Experiment-531 should be forthcoming in the immediate future. The improvements made in the second run to the detector and the beam will result in a corresponding improvement in the quality and the quantity of events, which will be reflected in a significant reduction in the lifetime errors.

This work has been supported in part by the U.S. Department of Energy, the U.S. National Science Foundation, the Natural Sciences and Engineering Research Council of Canada, the Department of Education of the Province of Quebec, Canada, the Ashigara Research Laboratories of the Fuji Film Co., Ltd., the Mitsubishi Foundation, the Ministry of Education of Japan, the Japan Society for the Promotion of Science and the Japan - U.S. Cooperative Research Program. Many people have contributed to the success of this experiment. In particular, the technical personnel of Ohio State University, McGill University, Toronto University and the Science Workshop at Carleton University, and the very patient and careful scanning staffs at Kobe University, Nagoya University, Osaka University and the

University of Ottawa have been vital to our success. We are grateful to the accelerator and neutrino area crews at Fermilab for our very smooth run(s) and their cooperation.

REFERENCES

1. Previous published lifetime results from this experiment include
N.Ushida, et al., Phys.Rev.Lett.45, 1049 (1980).
N.Ushida, et al., Phys.Rev.Lett.45, 1053 (1980).
N.Ushida, et al., Phys.Rev.Lett.48, 844 (1982).

2. See for example:
K.Niu in "Proceedings of the XX International Conference on High Energy Physics", L.Durand and L.G.Pondrom, eds. (Madison, 1980), p.352.
N.R.Stanton in "Proceedings of the 1981 International Conference on Neutrino Physics and Astrophysics", R.J.Cence, E.Ma, A.Roberts, eds. (Maui,Hawaii,1981), p.491.

3. N.Ushida, et al., Phys.Rev.Lett.47, 1694 (1981).

4. S.Errede, Ph.D. Thesis, Ohio State University, (unpublished), (1981).

5. M.Gutzwiller, Ph.D. Thesis, Ohio State University, (unpublished), (1981).

6. D.Bailey, Ph.D. Thesis, McGill University, (unpublished), (1982).

7. D.Pitman, Ph.D. Thesis, University of Toronto, (unpublished), (1982).

8. N.Ushida, et al., "Cross Sections for Charm Production by Neutrinos", to be submitted to Physics Letters.
N.Ushida, et al., "Characteristics of Charmed Hadrons Produced by Neutrino Interactions", to be submitted to Physics Letters.
N.Ushida, et al., "New Result for the $D^{\pm}$, F^+ and Λ_c^+ Lifetimes", to be submitted to Physical Review Letters.

9. B.Campbell, Ph.D. Thesis, McGill University, (unpublished),(1979).
Y.Afek, et al., Z.Phys.C6, 251 (1980).

10. T.Gottschalk, Phys.Rev.D23, 56 (1981).

11. See for example,
M.Jonker, et al., Phys. Lett.107B, 241 (1981).
H.C.Ballagh, et al., University of Hawaii preprint UH-511-429-81, (1981).

12. R.C.Field, talk given at this conference.

13. G.Bellini, talk given at this conference.

14. L.Foà in "Proceedings of the 1981 Symposium on Lepton and Photon Interactions at High energies", (Bonn,1981).

15. R.H.Schindler, Ph.D. Thesis, Stanford University, (unpublished), (1979). and W.Bacino, et al., Phys.Rev.Lett.45, 329 (1980).

MEASUREMENTS OF CHARMED PARTICLE LIFETIME WITH ELECTRONIC TECHNIQUES.

G. Bellini
Istituto di Fisica dell'Università di Milano
Istituto Nazionale di Fisica Nucleare - Sezione di Milano

ABSTRACT

Live target techniques used to search on charmed particles and to measure their lifetime are described in this paper. Using this technique, clear signals of $D^{\pm}$, $D^{*\pm}$ and $F^{\pm}$ have been obtained, together with the first measurements of the $D^{\pm}$ lifetime.

Recent developments in this domain concern improvements of the target granularity and of the signal to noise ratio.

1. INTRODUCTION

The measurement of charmed particle lifetimes with electronic devices has been based, until now, on the live target[1] technique. In the lifetime measurements carried on at CERN SPS (NA1 experiment) a telescope of 40 Silicon detectors, 300 µm thick, 150 µm apart, has been used.

The live target has to be used in association with a forward multiparticle spectrometer, which selects events by recostructing the masses and identifying the topology. The candidates to the charmed particle production are analyzed in the telescope target to identify the interaction and the decay detectors, which are signaled by steps in ionization.

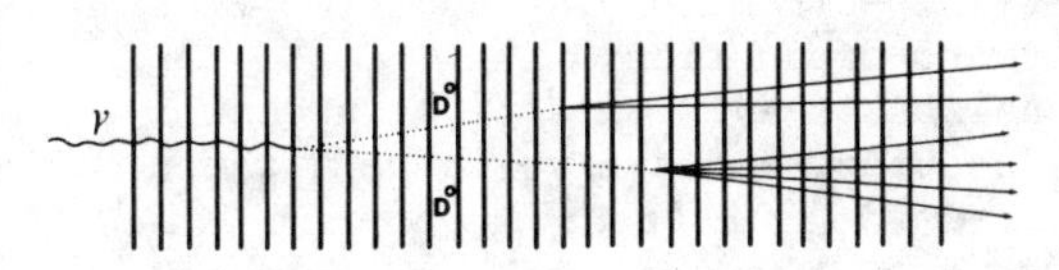

Fig. 1. Pictorial sketch, of $D^{\pm}$ and D^{o} associated productions, drawn within the live target.

This technique is more efficient if the events we search for are restricted to coherent production of charmed particles, because changes in ionization are easier to be recognized if the total multiplicity is lower. In addition in the coherent production, the interaction detector is well labelled, in many cases, by the nuclear recoil.

In the incoherent production most of the products of the nuclear break cross more than one detector; the rejection of events showing spikes in two or more adjacent layers provides a purified sample of coherent interactions.

A single large spike signalling a nuclear recoil is the unique label for the production detector, when a pair of neutral charmed particles is produced. In fig. 1, pictorial sketchs of the coherent production of charmed and neutral D's within the target may help in understanding the method.

The resolution in lifetime measurements is strictly limited by the spatial resolution of the live target. Due to the Landau fluctuations and the electronic noise, which introduce uncertainties in the multiplicity identification, more than one adjacent layer are required to give a response consistent with the same multiplicity in order to define a step in ionization. In the NA1 experiment, at least four detectors have been required to define an ionization level: the minimum detectable lenght was actually $\sim$1.8 mm.

The Landau fluctuations and the electronic noise increase, when the junction thickness is decreased, due to smaller energy loss and larger capacitance, respectively. Big efforts are so devoted to keep the overall capacitance of each Silicon junction below very small values and to use solid state detectors with higher Z and larger density (as Germanium) with respect to the Silicon.

The role of the live targets in the search on charmed particles is not limited to the lifetime measurements, but their use is a strong tool to purify the sample of charmed events. In the NA1 experiment the search of ionization steps corresponding to decay processes within the target was a powerful filter, which guarantees a right selection of the charmed particle events.

2. SEARCH OF CHARMED PARTICLES

In the NA1 experiment , the coherent photoproduction of pairs of charmed particles off Silicon nuclei has been studied. The restriction of the event sample to coherent interactions not only simplifies the analysis of the live target spectrum, but offers the following further advantages:

i) the pair of charmed particles is produced without additional pions, reducing the number of combinations;

ii) the photon energy flows into the charmed mesons giving a high stretching γ-factor;

iii) the acceptance of the magnetic spectrometer is full, for F and D pairs, due to the very forward production.

The experimental set-up consists of a multiparticle spectrometer, for charged particles and photons, and a live target, which detects the lifetime of the produced particles decaying within it.

The multiparticle spectrometer is sketched in fig. 2. Sets of drift chambers, interspaced with four magnets, provide the measurement of the charged particles with a resolution $0.5 < \Delta p/p < 1\%$. This resolution is constant in the momentum range 1-150 GeV/c, because faster is the particle, longer is the track detected in the spectrometer. The geometrical acceptance is $\Delta\Omega \simeq 0.8 \times 10^{-13}$ srs.

Five photon detectors, positioned in front of the frame of

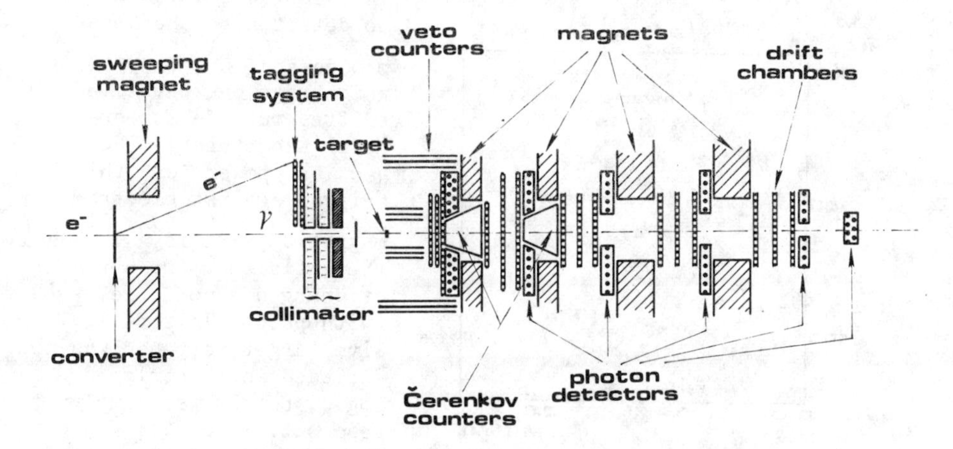

Fig. 2. Top view of the NA1 spectrometer.

each magnet, detect the photons in a $\Delta\Omega \simeq 0.25$ srs. The first two counters, the largest ones, are lead sandwhiches while the following three are lead glass counter matrices. The resolutions are $\Delta E_\gamma/E_\gamma = 0.4/\sqrt{E_\gamma}$ and $\simeq 0.14/\sqrt{E_\gamma}$, respectively.

Two Cerenkov counters, filled one with CO_2 and the second one with air, are placed inside the first two magnets.

The photon beam, with 150 GeV top energy, was tagged and was obtained by an electron beam of $\sim 2\times10^6$ electrons/burst.

Fig. 3 and fig. 4 show the resolution in recostructing K^o's ($\pi^+\pi^-$ invariant mass) and π^o's ($\gamma\gamma$ spectrum).

The trigger was organized with the main goal to identify hadronic events from the electromagnetic background, without dead region in the horizontal plane. It consisted of the following requirements:

i) tagged photon energy >40 GeV;
ii) at least two hadrons or one hadron and one photon in the spectrometer;
iii) at most one charged particle outside the spectrometer.

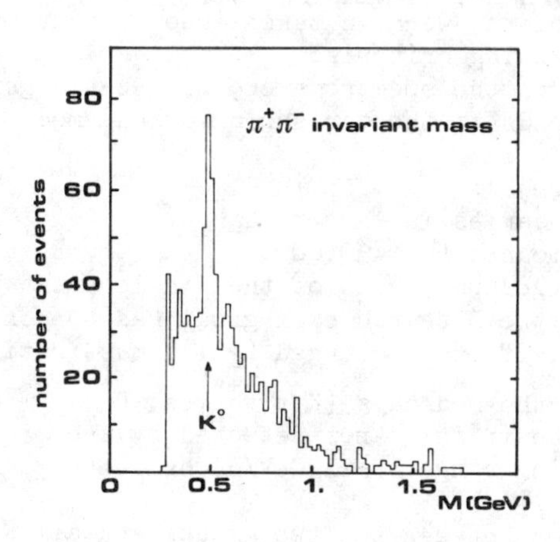

Fig. 3. Spectrum of $\pi^+\pi^-$ invariant masses, recostructed in the spectrometer.

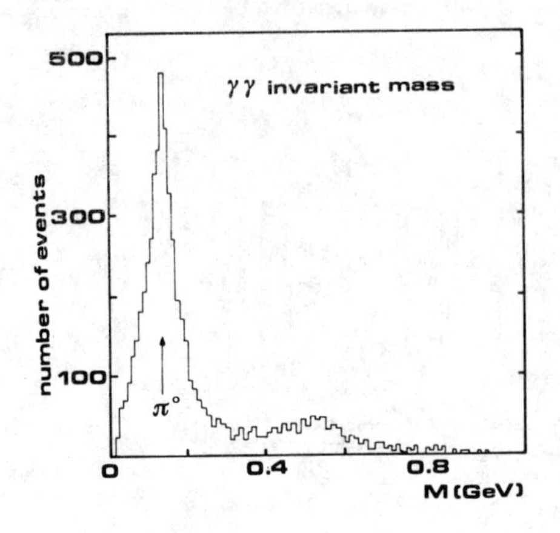

Fig. 4. Spectrum of γγ invariant masses, recostructed in the spectrometer.

This last condition allows to detect also the decay products of D^*, the 30% of which have one particle (π) emitted outside the spectrometer acceptance. The rejection power of this trigger against electromagnetic pairs is $\geq 10^{-5}$ and the efficiency for a number of charged secondaries ≥ 4 (with photons) is full. The live target response was not included in the trigger logic, but it was simply read out.

With these conditions about one million of hadronic events have been collected in the period January - February 1980.

2.1 EVENT SELECTION FOR D SEARCH

The analysis procedure for D and F search followed a multistep approach.

First selections of the events have been carried on using only the spectrometer response and searching on charmed particles both in inclusive and exclusive way. A further analysis based on the target spectra provides strongly selected samples (see §3).

In order to search on charm, we have taken into account only the events which fulfil the following criteria:

i) the total energy, measured in the spectrometer, must be larger than 70 GeV, which corresponds to the threshold for charmed particle production;

ii) events with electrons are rejected;

iii) the number of charged secondaries is larger than 4.

The number of events survived amounts to ∿ 30.000.

The jet of neutral and charged particles of the single event has been divided in two groups. For the D search each group has to contain at least one kaon candidate and, if it is charged (± 1), must satisfy the Cabibbo selection rules for the D-decays ($K^-\pi^+\pi^+$ yes; $K^+\pi^-\pi^+$ no). A particle is defined K candidate if it is not detected by the Cerenkov counters in their working range (5-21 GeV/c) or it has a momentum <5 or >21 GeV/c.

Particles are then exchanged between the two groups and all K candidates are considered in turn, in order to build up all possible configurations. In fig. 5a and 5b, the invariant mass of all possible configurations for one of these particle groups (M_2) is plotted,

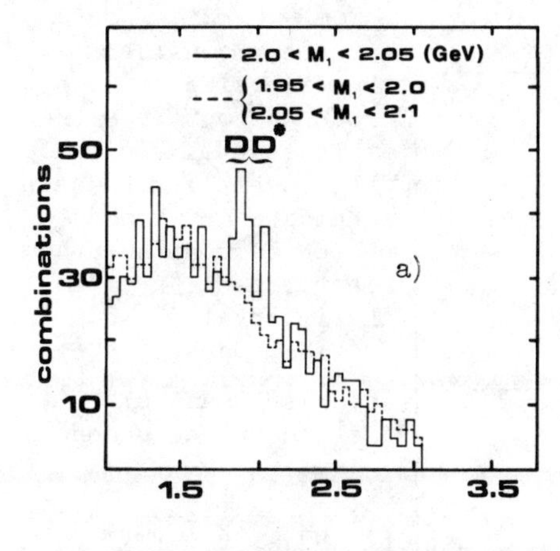

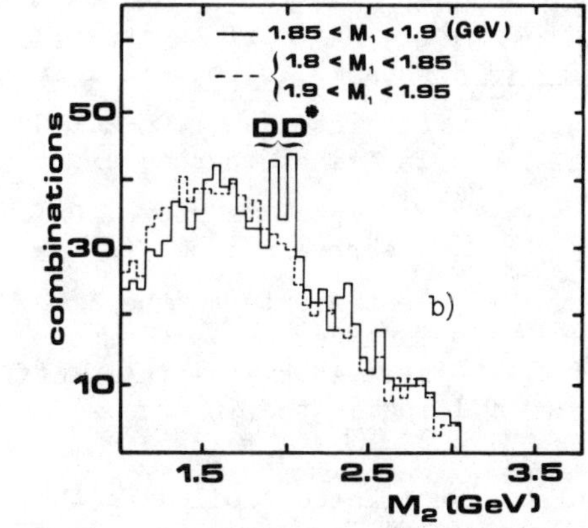

Fig. 5a, 5b. M_2 inclusive mass spectra; for the selection of M_1, all the channels are taken into account (see text).

when the invariant mass of the other group (M_1) falls in the D^* and in the D regions, respectively. The histograms (full line) show clear peaks at the D and D^* masses and are compared with the M_2 background plot (dashed line) obtained by selecting M_1 in the first two side bins excluding just D and D^*.

In the background histograms no normalization factors are introduced and the M_2 values, simply averaged on the two M_1 selected bins, are plotted.

The same analysis procedure has been applied choosing exclusive decay channels for M_2 and both for M_1 and M_2. In fig. 6 all the combinations for the M_1 invariant mass are plotted, when the invariant mass $M_2 \to K^-\pi^+\pi^+\pi^o$ falls in the D^* region. Fig. 7a and 7b show the invariant mass distributions (M_2) for the channels $K^-\pi^+\pi^+(n\pi^o)$ when M_1 corresponds to $D^- \to K^+\pi^-\pi^-\pi^o$ and $\overline{D^o} \to K^+\pi^-\pi^o$.

In all cases clear peaks in the D region are exhibited, proving that D's are produced in our event sample[2].

2.2 SEARCH OF F's PARTICLES

In the search of $F^\pm$, the same analysis procedure as in the case of D's has been followed. All possible configurations were built up assigning in turn the particles to the M_1-M_2 groups, but the criteria used to form the groups were different. Taking into account the spectator quark model, the decay channels:

$$F \to KK\ (n\pi)$$
$$F \to \eta\ (n\pi)$$

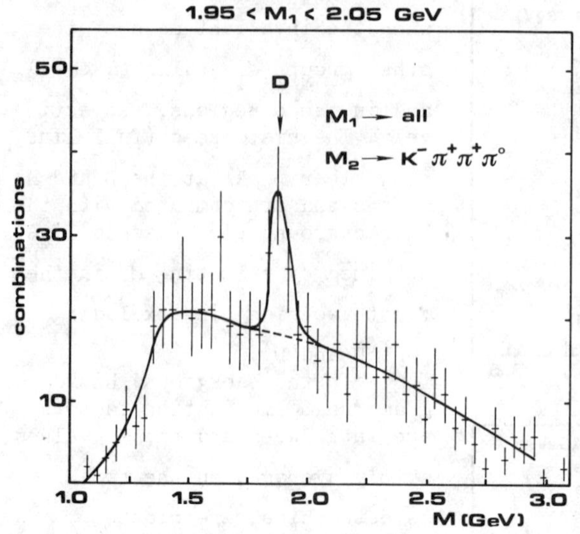

Fig. 6. M_1 inclusive mass spectrum; (for M_2 only the channel $K^-\pi^+\pi^+\pi^o$ is considered).

have been considered thus imposing to each group to contain either at least two kaon candidates or at least one η candidate defined as a photon pair with invariant mass falling in the interval 500-600 MeV. In figs. 8a,b,c, the M_2 invariant masses are plotted in the channels: $\eta\pi^+\pi^-\pi^{\pm}\pi^o$; $K^+K^-\pi^{\pm}\pi^o$; $\eta\pi^{\pm}(n\pi)^o$, $K^+K^-\pi^{\pm}(n\pi)^o$, $\eta\pi^{\pm}(n\pi)^o$ all together; when M_1 is contained in the interval 1.98-2.1 GeV. In all histograms a clear $F^{\pm}$ signal is shown, which disappears if M_1 is selected outside the F mass region, as demonstrated by the full line plot of fig. 8b, which corresponds to M_1 falling in the intervals 1.86-1.98 and 2.10-2.22 GeV.

The channel $F \rightarrow (n\pi)$ has been **also** investigated, following the same procedure, but the bump in the F region is wider than the experimental resolution. Therefore the evidence of F production is not so clear as in the previous channels.

I estimate that the proof of the existence of a charmed-strange particle as the F meson obtained in NA1 experiment is the clearest evidence produced until now; its mass is 2.05 ± 0.020 GeV.

3. LIFETIME MEASUREMENTS

The lifetime measurements have been performed with the live target placed in front of the magnetic spectrometer. The target was a telescope array consisting of forty silicon junctions, 300 μm thick, 150 μm apart, 14 mm in diameter (useful area). Each detector is associated to an analogue processor including a head amplifier and a filter.

The electron pairs created by the conversion of the primary photons in the target ($\sim$ 0.15 radiation lenght) give rise to a counting rate growing along the telescope and approaching $\sim 5\times10^5 - 10^6$ pulses/s in the last junctions, when the beam flux is $\sim 2\times10^6$ p/s. To cope with this non-uniform rate, the second group of twenty layers

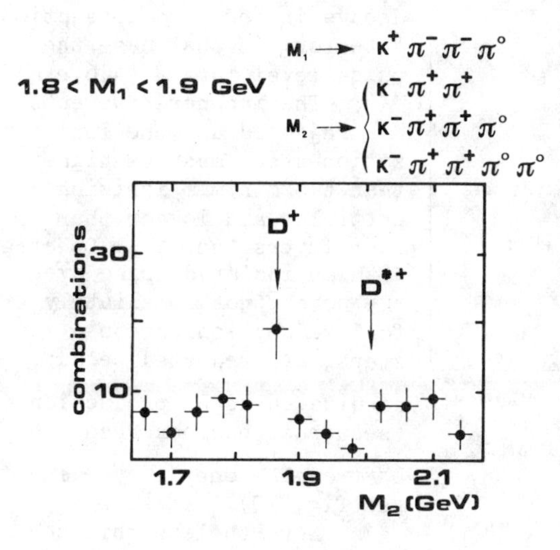

a)

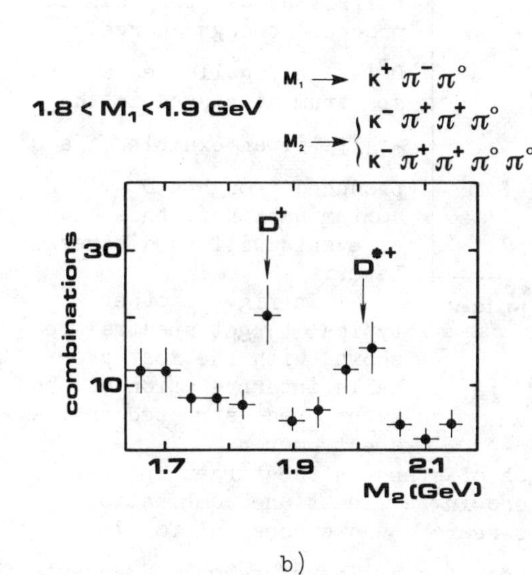

b)

Fig. 7a, 7b. M_2 exclusive mass spectra (M_1 is selected only in the channels $K^+\pi^-\pi^-\pi^o$ (7a) and $K^+\pi^-\pi^o$ (7b)).

was associated to triangular shapers, where the pulse width (∿ 0.1 ns) was shortened with respect to the trapezoidal filters (∿ 0.5 ns) used for the first twenty detectors[3]. The resolution of the acquisition channels is shown in fig. 9, where the Landau distributions of one, two, four and six minimum ionizing particles are displayed for one detector connected to the slower and to the faster processors.

3.1 LIFETIME MEASUREMENTS OF $D^\pm$ MESONS

The sample of events taken into account to measure the charged D lifetime, fulfilled to the following criteria:

i) the number of charged secondary must be ≥ 6. This requirement favours the charged D's production and tends to cut down the D^o pairs;
ii) at least one combination has to fall in the D or D^* mass region.

The target pulse spectra of this sample have been analyzed looking for sequences: production (with a possible nuclear recoil higher than the noise level) - $D^\pm$ path - one decay or two decays (the events with one D decaying outside the target are also accepted). The ionization level corresponding to two minimum ionizing particles which signals the $D^\pm$ travel before they decay, may be missing because one $D^\pm$

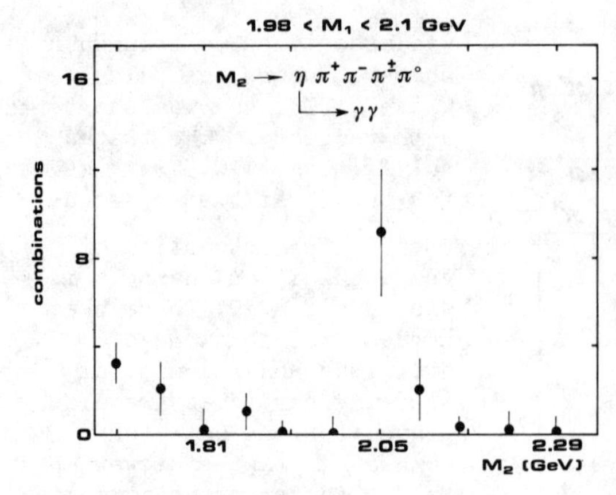

Fig. 8a. $M_2 \rightarrow \eta\pi^+\pi^-\pi^\pm\pi^o$ invariant mass spectrum.

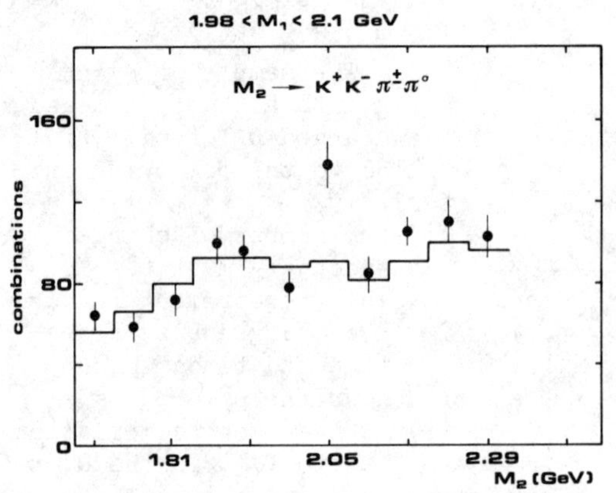

Fig. 8b. $M_2 \rightarrow K^+K^-\pi^\pm\pi^o$ invariant mass spectrum.

decays in the production detector. In that case the pulse level is > 2 (4,6,etc).

The incoherent events are rejected and the ionization steps must be higher than two minimum ionizing particles and longer than four layers (see §1). Spectra with an isolated spike over the noise, not immediately followed by ionization steps, are rejected because candidates to D° production (sequence: nuclear recoil - D° travel - one decay - etc. see fig. 10).

Nevertheless this selection is not enough to reject interactions with D° pairs, because they can be produced trough charged D^*'s. In fig.11 the target spectrum displayed in the middle is an example of a D° produced trough a D^{*-}. The contamination of this class of events will be discussed later.

In fig. 11 other typical target spectra are shown, with the most probable interpretation of the event reconstructed in the spectrometer.

The search of good sequences of steps in the target spectra have been applied to the events showing at least one combination with $3.65 < M_1 + M_2 < 4.2$ GeV. This search was successful for 76 events, in which 98 decays have been found. Therefore only 23 events display two decays within the target.

The $M_1 + M_2$ histograms for this final sample are shown in fig. 12, where all the combinations are plotted.

The 98 decay paths measured in the target have been corrected to take into account the following sources of bias:

i) the finite lenght of the target (18 mm);

ii) the minimum detectable path (1.8 mm);

iii) the loss of the second step masked by the first. This bias is

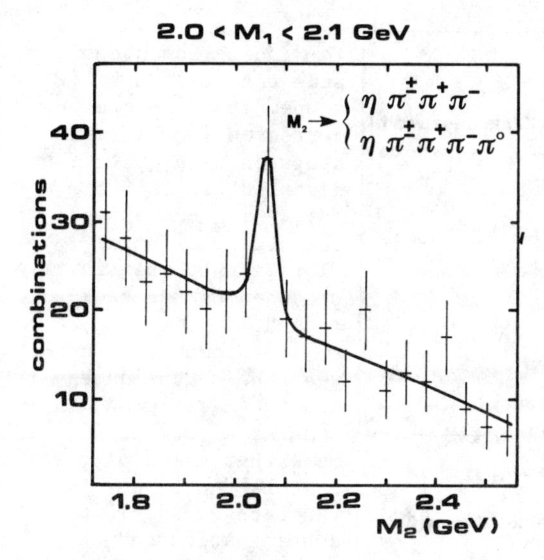

Fig. 8c. M_2 invariant mass spectrum: the channels $\eta\pi^{\pm}(n\pi)^{\circ}$, $K^+K^-\pi^{\pm}(n\pi)^{\circ}$ $\eta\pi^{\pm}(n\pi)^{\circ}$ are plotted all together.

not negligible due to the definition of steps which have to consist of four layers at least: for $\tau \simeq 10^{-13}$ s this correction is $\sim$10%.

The lifetimes corresponding to the 98 decays measured in the target are plotted in fig. 13. Due to the ambiguity in associating the paths detected in the target with the D momenta measured in the spectrometer(°), the γ factor has been assumed:

$$\gamma_D = \gamma_{\bar{D}} = \frac{E_{tot}}{M_D + M_{\bar{D}}}$$

where E_{tot} is the total energy of the $D\bar{D}$ pair.

A fit with only one exponential gives for the lifetime of the charged D's:

$$\tau_{D^{\pm}} = 8.9^{+2.9}_{-1.7} \times 10^{-13} \text{ s.}$$

The possible sources of background in this measurement are:

i) fake ionization steps in the live target due to the photon conversion and secondary interactions;

ii) D° contamination.

The background steps due to electron conversions are cut in large part by rejecting events with secondary tracks recognized as electrons. Most of the secondary interactions can also be avoided because they produce spikes at the beginning of step due to the recoil of the nucleus or to nuclear break products. In addition part of the incoherent interactions are simply rejected by the trigger requirements, which accept at most one charged particle outside the spectrometer acceptance.

Nevertheless in order to evaluate the effect of the remaining background on the $D^{\pm}$ lifetime, a sample of events having the same structure as the $D^{\pm}$ candidates, but falling outside the selected mass ranges, have been studied analyzing the target spectra and looking

(°) In principle this ambiguity can be solved when the two charmed particles decay into different numbers of charged secondaries. Moreover, when the particle combinations corresponding to charmed particles are not unique, the ambiguity remains in any case.

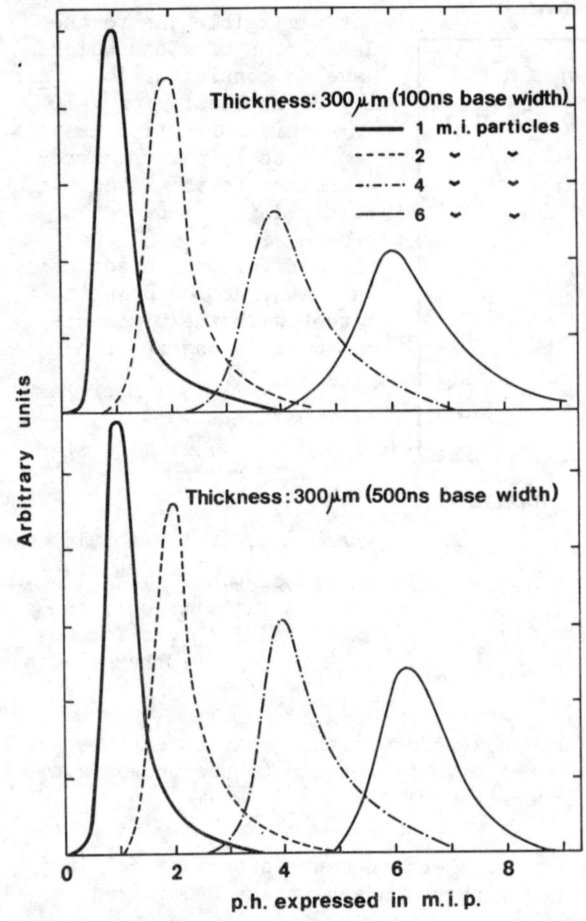

Fig. 9. Landau distributions for one, two, four and six minimum ionizing particles crossing a 300 μm thick detector, obtained with the 0.5 ns processor and with the 0.1 ns processor.

for ionization steps inside the Silicon telescope. The time spectrum corresponding to fake steps shows a quite flat distribution (see fig. 14), which guarantees that the D lifetime slope is not artificially produced by the background.

A comparison between the D-candidates and the background samples allows to estimate that a 10% of the D steps can be due to the background. This contamination reduces the charged D lifetime to 8.2×10^{-13} s, but does not influence the errors.

The D^o-contamination, has been estimated by applying to simulated events the same analysis procedure as in selecting our $D^{\pm}$ sample. I recall that the D^o contamination which escapes our selection is mostly due to $D^{*\pm} \to D^o \pi^{\pm}$ decay.

If the branching ratios available in the literature are used and a D^o lifetime of 2.5×10^{-13} s is assumed, the contribution of D^o's in the $D^{\pm}$ sample is $\simeq 20\%$. This percentage is smaller when D^o lifetime is shorter. A fit on the 98 decay times (fig. 13), which takes into account the D^o contamination, gives for the charged D a lifetime:

$$\tau_{D^{\pm}} \simeq (9.5^{+3.1}_{-1.9}) \times 10^{-13} \text{ s.}$$

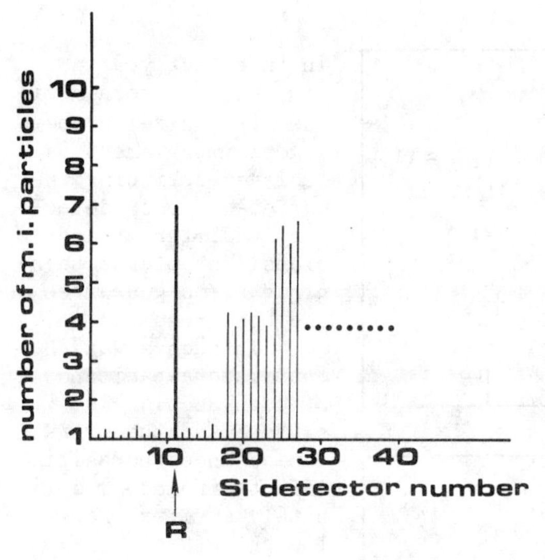

Fig. 10. Sketch of a possible target pulse spectrum corresponding to a D^{o} production event.

3.2 F LIFETIME MEASUREMENT

To measure the F lifetime, we have analyzed the target spectra of the events falling in the bin corresponding to the F meson in the channels: $\eta\pi^{+}\pi^{-}\pi^{\pm}\pi^{o}$ (fig. 8a) and $K^{+}K^{-}\pi^{\pm}\pi^{o}$ (fig. 8b). Only for 8 events the pulse spectrum shows clear steps. These data have been treated with the maximum likelihood method, once corrected for the finite lenght of the target, for the minimum detectable path and for the shadow effect of a step on the following one.

An approximate value of the F lifetime has so been estimated. It results to be:

$$\tau_F = 5^{+5}_{-2.5} \times 10^{-13} \text{ s.}$$

4. RECENT DEVELOPMENTS IN THE DOMAIN OF THE LIVE TARGETS

Live targets with finer granularity and smaller capacitance and processors with shorter pulse width and lower noise have been developed

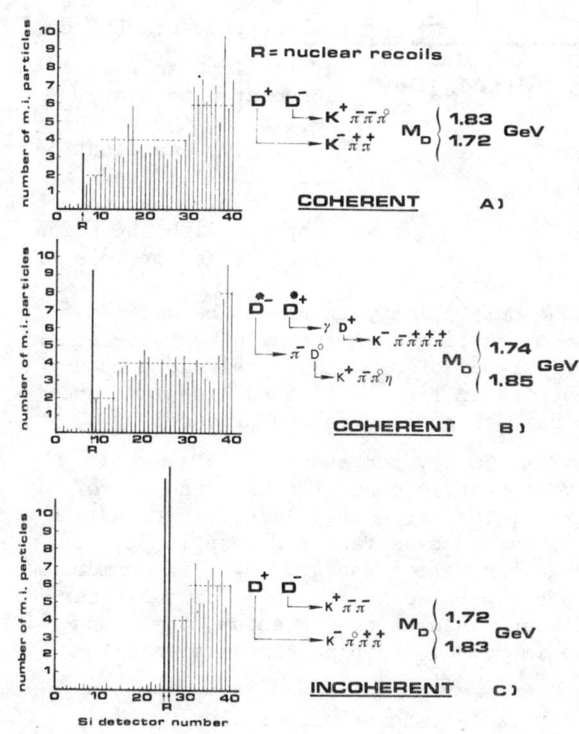

Fig. 11. Example of three different events detected by the target and compared with the spectrometer outcome. The signal amplitudes from the Silicon detectors are expressed in units of single minimum ionizing particles.

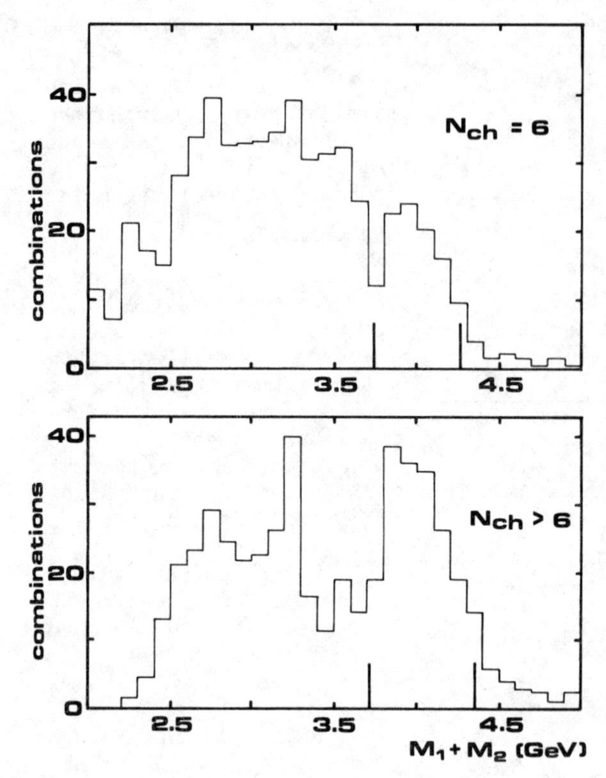

Fig. 12. $M_1 + M_2$ distributions for the events selected using the target spectra (see text).

in this last year.

In the domain of the Silicon telescopes, a good improvement is achieved splitting the surface of each detector in different sections. The planar technique seems guarantee the best result.

In such a way the intrinsic capacitance of each section is < 55 pF and the overall capacitance (parasitic effects included) definitely smaller than 100 pF.

A new telescope target has so been developed,consisting of Silicon detectors, 200 μm thick, 40 μm apart.

A new very fast pulse processor (60 ns of total pulse width)[4] associated to every section of the detectors, gives an electronic noise ≃ 17 KeV (FWHM) which has to be compared with the noise level of 40 KeV or more of the NA1 experiment target.

With this resolution the multiplicity level can be defined by only 3 layers instead of 4, so reducing the minimum detectable lenght to 0.72 mm. The detection efficiency of a 50 layer telescope manufactured in this way is presented in fig. 15 (dashed line) as function of the lifetime,in the case of charmed mesons produced by a bremsstrahlung photon beam with 150 GeV top energy[5].

A second important development is connected with the use of Germanium junctions. For fixed paths the energy loss by a relativistic particle in Germanium is larger by a factor 2 than in Silicon. In addition,due to the average ionization energy, lower in Germanium than in Silicon, the corresponding pulse height is ∿ 2.5 times larger in Germanium than in Silicon. Then using Germanium, the granularity along the beam can be improved, keeping a good signal-to-noise ratio[6].

Germanium target prototypes have been built following the

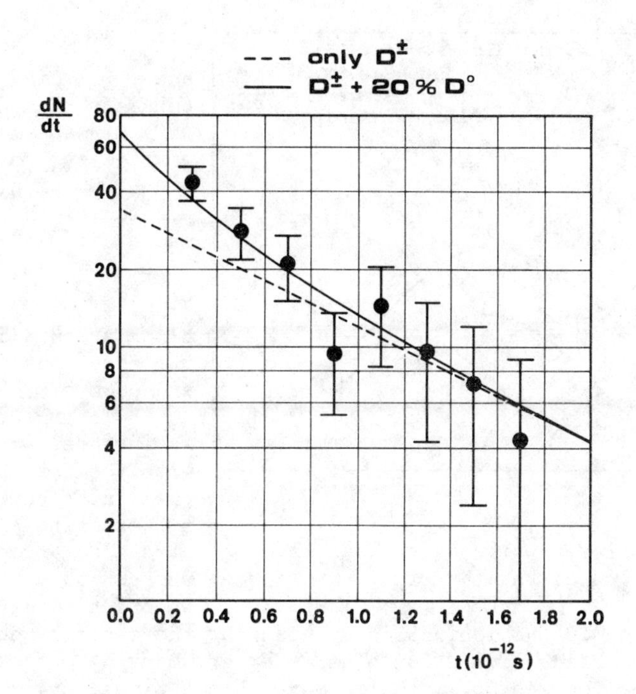

Fig. 13. Plot of the lifetime of 98 $D^{\pm}$'s decaying in the target. The black points are the experimental data. The dashed line corresponds to the fit with only one exponential. The full line takes into account a further 20% contribution from a lifetime = 2.5×10^{-13}s (D^{o})

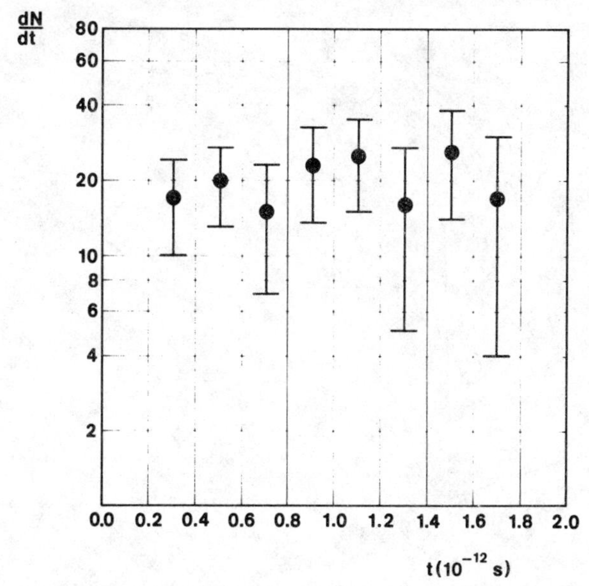

Fig. 14. Time distribution of the background events.

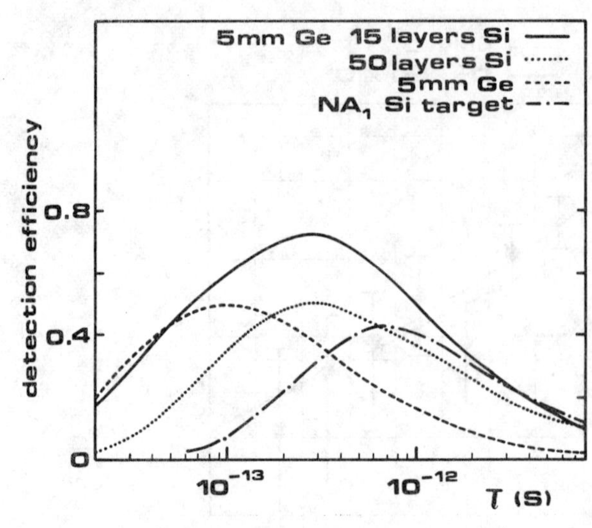

Fig. 15. Efficiency for given mean lifetimes detected by the targets.

scheme of fig. 16. It consists of a high purity Germanium block with 40 strips (50 μm wide, 100 μm pitch) deposited on the top face and a diffused electrode on the bottom. The overall dimensions of the block are 4 × 50 mm^2 for the electrode surfaces and 5 mm of thickness.

The bulk detector is intended to be used with the beam entering the region between the two opposite electrodes at a zero angle with respect to their planes, in order to give a spatial sampling of the ionization by means of the strip signals.

An important limitation of the Germanium as live target is due to its radiation lenght, which is more than 4 times larger than the Silicon radiation lenght. Then the total lenght of a Germanium monolithic target has to be restricted to few millimeters (2 - 5), limiting to a small lifetime range the application of this type of

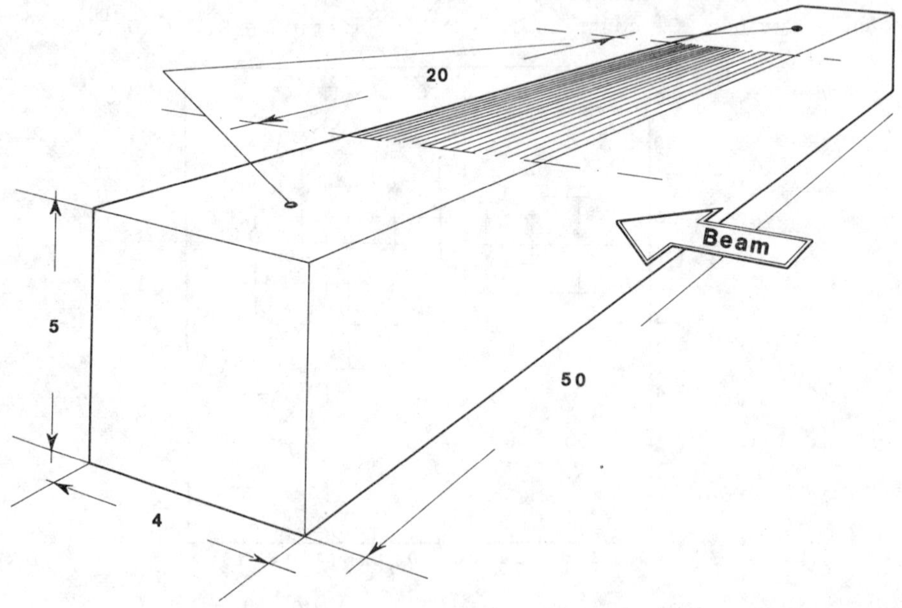

Fig. 16. The sketch of a Germanium bulk detector used as live target.

target (fig. 15).

Moreover a good solution can be obtained associating a Germanium bulk detector with a Silicon telescope. In fig. 15 the detection efficiency of a Germanium crystal, 5 mm long, followed by a telescope of 15 Silicon detectors, 200 μm thick, is displayed. In this exercise, the first 8 Silicon junctions are spaced 200 μm and the remaining 7 are spaced 400 μm. Due to the smaller collision lenght in Germanium, 85% of all the production takes place in it.

This solution provides probably a good tool to study shorter lifetimes(D^o, $F^{\pm}$, Λc) at the energies available at the SPS.

REFERENCES

1 G.Bellini, Live targets, invited talk at the EPS International Conference on High-Energy Physics, Lisbon, 1981; Proceedings in press.

2 E.Albini et al., Phys. Lett. 110B (1982) 339.

3 G.Bellini et al., Nucl. Instrum. and Meth. 196 (1982) 351.

4 P.d'Angelo et al., Nucl. Instrum. and Meth. 193 (1982) 533.

5 Frascati - Milano - Pisa - Torino - Trieste Collaboration - Addendum to Proposal P170 CERN/SPSC/82-33 SPSC/P170/Add. 1 (L982).

6 G.Bellini et al., Physics Reports 83 (1982) 1.

A COMPOSITE MODEL OF THE WEAK INTERACTIONS*

L. F. ABBOTT**
Physics Department
Brandeis University
Waltham, MA 02254

ABSTRACT

A model is presented in which quarks and leptons are composite, with the weak interactions resulting from residual interactions between their constituents. The model can account for low-energy data while making high-energy predictions different than those of the standard model. Some modifications and extensions of the original model are also discussed.

I. INTRODUCTION

In this talk, I will describe a model in which the weak interactions are a result of the composite nature of quarks and leptons. The model is a composite model which correctly describes the known fermions. In addition, it may account for the low-energy weak interactions while making high-energy predictions which differ from those of the standard model. This work was done in collaboration with E. Farhi.

One motivation for this model is the notorious gauge hierarchy problem. In the standard $SU(3) \times SU(2)_L \times U(1)$ model, the mass scales of the strong and weak interactions are set in very different ways. The scale of the strong interactions, which is something like 100 MeV - 1 GeV, is determined by the energy Λ_3 at which the running QCD coupling constant, α_3, gets large. This is a very satisfactory way for a mass scale to be set. We now believe that the fundamental mass scale in elementary particle physics is something very large like $M_{GUT} \approx 10^{15}$ GeV or $M_{Planck} \approx 10^{19}$ GeV. We must then explain why the strong and weak interaction scales are so much smaller than this fundamental scale. For QCD this is easy. At the large, fundamental scale α_3 is something like 1/50. The strong interaction scale is much smaller than this scale because α_3 only changes logarithmically with energy so it runs a long way before it gets big. On the other hand, the scale of the weak interactions, $G_F^{-1/2} \approx$ 300 GeV, is determined by the vacuum expectation value of a scalar field and has nothing to do with Λ_2, the energy at which the $SU(2)_L$ coupling constant would get large if $SU(2)_L$ were not spontaneously broken. In a theory with a fundamental scale around 10^{19}GeV, there is no way (except for supersymmetry - but these models have their

*Invited talk given at 5th International Conference on Particle Physics at Vanderbilt University, May 1982.

**Supported in part by DOE contract DE-AC03-76ER03230-A005.

own problems) to explain why a scalar vacuum expectation value is only about 300 GeV. This is the gauge hierarchy problem.

In the model I will describe,[1] the scale of the weak interactions is not determined by a scalar vacuum expectation value but rather by Λ_2, the energy at which the $SU(2)_L$ coupling constant gets big, in exactly the same way as the strong interaction scale is determined by the QCD Λ_3. This is done by taking the standard model and changing the scalar potential so no spontaneous symmetry breaking occurs and then increasing the value of the $SU(2)_L$ coupling constant until its Λ parameter, Λ_2, gets to be of order $G_F^{-1/2} \approx 300$ GeV. In addition, since no symmetry breaking occurs the U(1) of the standard $SU(3) \times SU(2)_L \times U(1)$ model is identical to the U(1) of electromagnetism so its coupling constant must be set equal to the electromagnetic coupling constant. Aside from these changes the model described here is identical in its particle content and quantum number assignments to the standard model. However, these changes make the model behave quite differently. Since the $SU(2)_L$ gauge theory is now unbroken and strong-coupling at the weak interaction scale all the particles in the standard model with non-trivial $SU(2)_L$ transformation properties are confined at a scale of $G_F^{1/2}$. They are thus preons and the observed fermions must be constructed from bound states. In the next section, I show how quarks and leptons can be constructed in this way. These bound state fermions will interact through residual gauge interactions between their constituents in exactly the same way as hadrons interact through the nuclear force. In section II, I also show how these interactions can give rise to the observed weak interactions.

Although the model presented here does account for the huge discrepancy between the large, fundamental mass scale and the weak interaction scale in a natural way, it unfortunately contains a cheat. The model uses a light scalar particle as a constituent and it is unnatural for such a light scalar to exist. To fix this, one must build models without fundamental scalar preons. In section III, I describe a model of this type and also discuss some other models. The most interesting phenomenological feature of this alternate approach to weak interactions is the new physics which appears at energies near $G_F^{-1/2}$. This is discussed in the final section.

II. THE MODEL

Our composite weak interaction model is just the standard $SU(3) \times SU(2)_L \times U(1)$ model with the $SU(2)_L$ theory considered in an unbroken, confining mode. Just as in the standard model, there are N left-handed $SU(2)_L$ doublets ψ_L^a [a = 1, 2, ... N] with N = 4 × number of generations, one complex scalar doublet ϕ and right-handed $SU(2)_L$ singlets ψ_R^b [b = 1, 2 ...]. The U(1) quantum number assignments are just as in the standard model: -1/2 or 1/6 for the left-handed doublets, -1/2 for the scalar doublet and for the right-handed singlets a U(1) hypercharge equal to the electric charge, -1, -1/3, or 2/3. Recall that for us this U(1) is equal to electromagnetism so these quantum numbers are just the charges of these particles. Since the right-handed fermions are singlets of $SU(2)_L$ they are not confined and are just the right-handed quarks and leptons. The $SU(2)_L$ doublets $\psi_L{}^a$ and ϕ, however, are confined by the unbroken $SU(2)_L$ gauge theory at a scale of $\Lambda_2^{-1} \approx G_F{}^{1/2}$. They are thus the preons from which we must construct the observed left-handed fermions.

Like QCD, the confining $SU(2)_L$ gauge theory is expected to produce a rich spectrum of particles. Most of these particles will have masses around $\Lambda_2 \approx 300$ GeV and therefore will be irrelevant to low-energy physics. However, some of the bound states of the $SU(2)_L$ theory can have very light masses because of chiral symmetries in the model. These light states are quarks and leptons.

Because the $SU(2)_L$ theory is strongly interacting in this model, its dynamics are difficult to analyze. I will base most of the discussions on global symmetries and try to make as few dynamical assumptions as possible. Therefore, it is essential to identify the global symmetries of the model. At a scale of $G_F{}^{-1/2}$, QCD, electromagnetic and Yukawa interactions are all much weaker than the strong $SU(2)_L$ interaction. Let us then begin by ignoring them. The $SU(2)_L$ sector then has N identical fermion doublets $\psi_L{}^a$ resulting in an SU(N) symmetry. The U(1) symmetry associated with these doublets has $SU(2)_L$ anomalies and is therefore not a good symmetry. In addition, the scalar sector has an overall O(4) symmetry. This O(4) can be expressed as two SU(2)'s. One of these is the gauged $SU(2)_L$ but the other is a global SU(2) symmetry. Thus, the total global symmetry group of the model is SU(N) × SU(2).

The quarks and leptons are constructed out of bound states of the left-hand fermions $\psi_L{}^a$ and the scalar ϕ in the following way

$$\left.\begin{aligned} (\nu, u, c, \ldots) &= \phi_i{}^* \, \psi_{Li}^{\ a} \\ (e, d, s, \ldots) &= \varepsilon^{ij} \phi_i \psi_{Lj}^{\ a} \end{aligned}\right\} a = 1, 2, \ldots \tag{2.1}$$

These states are massless because the SU(N) × SU(2) global symmetry forbids them from having any mass terms. Their charges are just the sums of the charges of the constituents which can easily be checked to give the correct quark and lepton charges.

The states constructed above have the correct charges to be the quarks and leptons and, as promised, they are much lighter than $G_F{}^{-1/2}$ - at this point they are in fact massless. Recall, however, that we have up to now been ignoring Yukawa couplings. These have the form

$$\lambda \bar{\psi}_R \, \phi_i{}^* \, \psi_{Li} \text{ and } \lambda \bar{\psi}_R \, \phi_i \, \psi_{Lj} \, \varepsilon^{ij} \tag{2.2}$$

In the standard model, they produce left-right transitions when ϕ develops a vacuum expectation value and thus give the fermions a mass. Here they do the same thing for a different reason. The combinations $\phi_i^* \psi_{Li}$ and $\phi_i \psi_{Lj} \varepsilon^{ij}$ appearing in these Yukawa terms are just the bound states we have identified with the observed fermions so these terms produce transitions between the fundamental right-handed fermions and the composite left-hand fermions giving the quarks and leptons their masses. On dimensional grounds the mass produced will be of order $\lambda\Lambda_2 \approx \lambda G_F^{-1/2}$ so for Yukawa couplings of the same order as those of the standard model, the correct fermion masses can be produced.

We have thus been able to construct the known quarks and leptons as bound states in our model. We must now account for the weak interactions. To do this we write down a Lagrangian for the low-energy ($E < G_F^{-1/2}$) interactions produced between composite fermions by residual $SU(2)_L$ interactions. This is a four-Fermi interaction with the scale of the interaction set by Λ_2. This parameter is of course adjusted so that the scale for these interactions comes out to be exactly G_F. The four-Fermi interaction can only involve left-handed fermions since only these feel the $SU(2)_L$ force and thus it is automatically a V-A interaction. The most general such interaction consistent with the $SU(N) \times SU(2)$ global symmetry of the model is

$$\mathcal{L} = \frac{4G_F}{\sqrt{2}} \left[J_{\mu L}^+ J_{\mu L}^- + (J_{\mu L}^3)^2 \right] + \frac{4G_F}{\sqrt{2}} \xi \, (J_{\mu L}^{\circ})^2 \tag{2.3}$$

where $\vec{J}_{\mu L}$ and $\overset{\circ}{J}_{\mu L}$ are the usual weak isovector and isoscalar currents. The first term in this Lagrangian can be recognized as part of the weak interaction Lagrangian of the standard model. The second term, however, is not present in the data. The parameter ξ is known to be less than about 0.05. The additional J_μ^{EM} term in the standard weak interactions arises in this model from the fact that the left-handed bound-state fermions have a non-trivial electromagnetic form factor. If a vector meson dominance model is used to estimate this form factor the correct J_μ^{EM} term is induced into the weak interaction Lagrangian[1,2] as was first shown by Bjorken and by Hung and Sakurai.[2] The result of all this is that the observed weak interaction Lagrangian,

$$\mathcal{L} = \frac{4G_F}{\sqrt{2}} \left[J_{\mu L}^+ J_{\mu L}^- + (J_{\mu L}^3 - \sin^2\theta_W J_\mu^{EM})^2 \right] \tag{2.4}$$

can be produced in this model provided that the following dynamical assumptions are made:

1) The parameter ξ must be small ($\xi < 0.05$). We do not know why ξ should be so small in our model but we do have reason to believe that the isovector and isoscalar terms in the general interaction (2.3) should have different strengths. This is because they arise from exchanges of different sorts of particles. The isovector term comes from exchanges of bound states of two scalars with much the same properties as the W's and Z of the standard model. The isoscalar term arises from exchanges of bound states of two fermions. Thus, the assumption that ξ is small is equivalent to the assumption that scalar exchanges dominate. Perhaps this is due to a large scalar self-coupling.

2) In this model, the parameter $\sin^2\theta_W$ has nothing to do with a ratio of the $SU(2)_L$ and U(1) coupling constants as it does in the standard model. Rather it is related to the amplitude for a photon to mix with one of the vector mesons of the $SU(2)_L$ theory. In order to fit the data this parameter must be 0.23. The analagous parameter in QCD, the photon-rho meson mixing parameter is only about 0.02. Thus, we must assume that there is large photon-vector meson mixing in the model.

Aside from these assumptions, the structure of the weak interactions arises from symmetries of the model. There are in addition a few phenomenological features of the model of interest:

1) There are no problems with strangeness - changing neutral currents. The GIM mechanism operates as in the standard model.

2) In order to account for the small value of the muon magnetic moment we must assume a factor 2-3 suppression in the magnetic form factor. This in itself is not a problem but it is worrisome that we need a suppression here and an enhancement in the electric form factor which enters into the neutral current as discussed in #2 above.

3) Universality of the weak interactions is assured by the SU(N) symmetry of the model. However, there are QCD corrections which do not respect this symmetry. Thus, we expect corrections to lepton-quark universality of order $\alpha_3(G_F^{-1/2})$ or $\alpha_3(G_F^{-1/2})/4\pi$. Corrections of order $\alpha_3(G_F^{-1/2})$ are too big while corrections of order $\alpha_3(G_F^{-1/2})/4\pi$ are perfectly acceptable so this is marginal but probably okay.

Finally, there are two interesting features of the model I would like to discuss. The first concerns the possibility of spontaneous breaking of the SU(N) × SU(2) symmetry. Confining gauge theories often spontaneously break their chiral symmetries by forming condensates. QCD is an example of this. If a dynamical breaking of the chiral SU(N) × SU(2) symmetry in the weak interaction model occurs, this would be a disaster since all the arguments presented are based on the validity of this symmetry. 'tHooft has given a consistency condition[3] for checking whether or not chiral symmetry must be broken. The model presented here satisfies the 'tHooft conditions and thus it is consistent to assume that the global SU(N) × SU(2) symmetry is not broken. This check also guarantees that the light bound states have been correctly identified.

A second interesting feature of the model is baryon number violation. As in the standard model, baryon number is not conserved in our model due to anomalies. In the standard model, baryon number violation from instantons is suppressed by a factor $\exp(-8\pi/g_2^2)$ which gives it an extremely small rate.[4] However, for us, the $SU(2)_L$ coupling constant is big and no such suppression occurs. Instead baryon number violation is suppressed because there are three (or more) generations of fermions. In the model with three generations $\Delta B = 3$ process like $p + p \rightarrow \bar{p}$ + leptons are allowed. They are produced by an operator O which is a determinant involving all 12 fermion doublet fields in the three generation model.

On dimensional grounds then, the effective Hamiltonian for $\Delta B = 3$ process is something like

$$H = G_F{}^7 O. \tag{2.5}$$

We must now take the matrix element of this Hamiltonian between initial and final states in the process $p + p \rightarrow \bar{p}$ + leptons. All the mass scales relevant to this matrix element and to the integrations over final state momenta are of order the proton mass m_p. Furthermore, since the determinant O involves fields of all three generations and the p's and $\bar{p}$ consist of only first generation quarks there are Cabbibo suppression factors. The resulting rate for this process is then

$$\Gamma \approx G_F{}^{14}\, m_p{}^{29}\, (\sin\theta_c)^{18} \approx 10^{-82}\ \text{GeV}.$$

This is many orders of magnitide slower than present limits on proton decay so $\Delta B = 3$ process do not occur at an observable rate in the model.

III. FURTHER DEVELOPMENTS

There are several models which are related to the work I described in section II and some more recent developments of it. H. Fritzsch and G. Mandelbaum[5] have also considered the idea that the weak interactions arise from composite structure. R. Casalbuoni and R. Gatto[6] have constructed composite models with scalar preons and noted that they satisfy the 'tHooft anomaly conditions. A left-right symmetric version of the composite model was created by R. Barbieri, R. N. Mohapatra and A. Masiero[7] and recently Y.-P. Kuang and S.-H. Tye[8] developed an interesting grand unified model which incorporates the composite weak interaction idea.

The development I will describe here is the construction of a model without fundamental scalar preons. Such a model was constructed by E. Farhi, A. Schwimmer, and me[9] and also, independently, by F. Bordi, R. Casalbuoni, D. Dominici, and R. Gatto.[10] In this model the quarks and leptons are bound states of three spin 1/2 preons. The binding force which serves also to induce the weak interactions is due to two gauge theories, an SU(M) technicolor and $SU(2)_L$. Both of these theories have Λ parameters of order $G_F^{-1/2}$. The preons in the model are: Ψ_L, an M of SU(M) technicolor and a doublet of $SU(2)_L$; A_R and B_R which are M's of SU(M) and $SU(2)_L$ singlets; and $\psi_L{}^a$ which are N singlets of SU(M) and doublets of $SU(2)_L$. The model has the global symmetry SU(N) × SU(2) × U(1). Note the presence of the SU(N) and SU(2) global symmetries needed to get the correct low-energy weak interactions. The SU(M) × $SU(2)_L$ confining force produces massless composites

$$\bar{A}_R \Psi_L \psi_L{}^a \text{ and } \bar{B}_R \Psi_L \psi_L{}^a. \tag{3.1}$$

In these bound states the A_R and B_R are bound to the Ψ_L by SU(M) technicolor and then this combination (which plays the role of the fundamental scalar in the model of section II) is bound to $\psi_L{}^a$ by the $SU(2)_L$ force. The model satisfies all of the 'tHooft anomaly conditions needed to assure that these bound states are indeed massless.

The three-fermion bound states described above cannot get masses unless the model is extended. If we take the SU(M) of technicolor to be SU(3) then SU(3) technicolor and $SU(2)_L$ can be unified into an SU(5) (not the usual GUT SU(5)). If we spontaneously break SU(5) down then the X and Y bosons of the SU(5) model mediate four-Fermi interactions which induce bound-state fermion masses. This extended model is much like extended technicolor. It has the same successes and unfortunately the same failures as standard extended technicolor.

IV. CONCLUSIONS

One of the best features of the weak interaction model I have described is that we will soon know whether or not it is correct. If a narrow Z boson resonance is discovered in $\bar{p}$ - p collisions and it has the mass and width predicted in the standard model, then the model presented here is undoubtedly wrong. If such a resonance does not appear what should we expect to see? In the composite model there are of course heavy vector bosons which will be produced but since they decay through the strong $SU(2)_L$ force they should be broad. We expect their masses to be in the range of 100 - 180 GeV and their widths 20 - 30 GeV. The most characteristic feature of their decays is the appearance of anomalously large numbers of leptons. This is because the strong $SU(2)_L$ force does not differentiate between quarks and leptons and so should produce them in equal numbers. Of course, there are three colors of quarks for each lepton and furthermore extra hadrons are produced when the quarks get confined. However, anomalous numbers of leptons should still appear as a fairly clear signal.[11] If such a signal does appear, then a rich new strong interaction spectrum - all the states of the confining $SU(2)_L$ theory - will also await discovery.

ACKNOWLEDGEMENTS

I thank my collaborators E. Farhi and A. Schwimmer. In addition, I am grateful to R. Panvini and E. Berger for their organizational work at the Vanderbilt conference.

REFERENCES

1. L. F. Abbott and E. Farhi, Phys. Lett. 101B, 69 (1981); Nucl. Phys. B189, 547 (1981).
2. J. D. Bjorken, Phys. Rev. D19, 335 (1979); P. Q. Hung and J. J. Sakurai, Nucl. Phys. B143, 81 (1978).
3. G. 'tHooft, Cargèse Summer Institute Lectures (1979), Plenum Press, N. Y. (1980).
4. G. 'tHooft, Phys. Rev. Lett. 37, 8 (1976).
5. H. Fritzsch and G. Mandelbaum, Phys. Lett. B102, 319 (1981).
6. R. Casalbuoni and R. Gatto, Phys. Lett. 103B, 113 (1981).
7. R. Barbieri, R. N. Mohapatra and A. Masiero, Phys. Lett. 105B, 369 (1981).
8. Y.-P. Kuang and S.-H. H. Tye, Cornell University Preprint CLNS 82/529.
9. L. F. Abbott, E. Farhi and A. Schwimmer, Nucl. Phys. (to be published).
10. F. Bordi, R. Casalbuoni, D. Dominici and R. Gatto, University of Geneva preprint UGVA-DPT 1982/04-346.
11. A. DeRujula, Phys. Lett. 96B, 279 (1980).

RECENT RESULTS FROM CLEO

Paul Avery
Laboratory of Nuclear Studies
Cornell University
Ithaca, N.Y. 14853

ABSTRACT

I review recent results obtained by the CLEO collaboration in the last year, focussing principally on charged multiplicity in B decay, kaon and baryon production, vector meson production, fractional momentum distributions, the $\Upsilon(3S)\rightarrow\pi^{+}\pi^{-}\Upsilon(1S)$ decay and inclusive $D^{*\pm}$ production.

I. INTRODUCTION

The CLEO detector[1] shown in Figure 1 is one of two experiments operating at the Cornell Electron Storage Ring (CESR). Charged particle tracking is provided by a 3 layer proportional chamber and a 17 layer drift chamber located inside a 1m radius solenoid magnet operating at 1.0 T. The momentum resolution attained by this system can be parametrized at high momentum by $\Delta p/p = 0.012p$, with p in GeV/c. Eight identical octants together covering one half of the total solid angle provide most of the particle identification for the experiment. Hadrons are identified either by pressurized specific ionization (DE/DX) chambers ($\sigma/E = 0.06$) or by the system of 96 time of flight counters (σ = 400 psec). Electrons and photons are identified by their characteristic interactions in lead proportional tube shower counters which surround the time of flight counters and enclose the central detector with two endcaps. Muons must penetrate iron and leave hits in a two dimensional array of planar drift chambers which enclose the rest of the detector in a giant cube.

0094-243X/82/930129-26$3.00

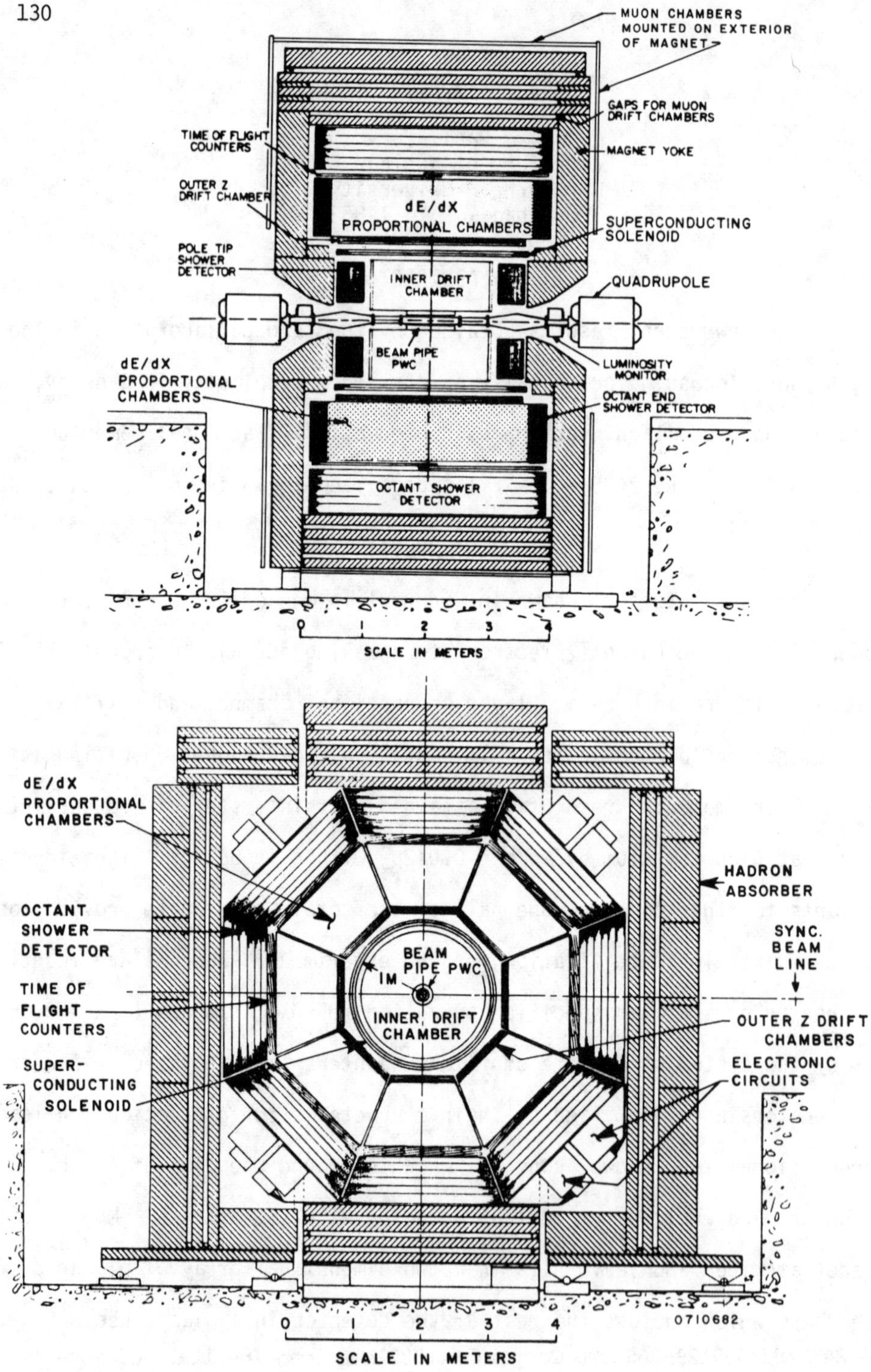

Fig. 1. The CLEO Detector

A mini-β system installed in the fall of 1981 has provided CLEO with a factor of 4 - 5 increased luminosity. We typically accumulate about 350 nb^{-1} per day near the $\Upsilon(4S)$ resonance. The increased flow of data, coupled with the factor of 3 improvement in momentum resolution brought about by the installation of a superconducting coil (which raised our field from 0.42 T to 1.0 T), has enabled us to pursue topics such as resonance production more vigorously.

The table below summarizes the data taken to date.

Table I: Summary of Luminosity (nb^{-1}) taken up to July 15, 1982

	0.42 T	1.0 T
$\Upsilon(1S)$	411	753
$\Upsilon(2S)$	1237	0
$\Upsilon(3S)$	3290	8080
$\Upsilon(4S)$	5550	13800
Continuum	3200	9240

I will discuss the following topics in this talk:

- Charged particle multiplicity from the continuum and B decay
- Heavy particle production ($K^{\pm}$, K_S, $\overline{P}$, Λ) on the $\Upsilon(1S)$, $\Upsilon(3S)$, $\Upsilon(4S)$ and continuum.
- Fractional Momentum Distributions.
- Vector meson production ($K^{*\pm}$, K^{*o}, ϕ) from the $\Upsilon(1S)$, $\Upsilon(3S)$, $\Upsilon(4S)$ and continuum.
- Study of the process $\Upsilon(3S) \rightarrow \pi^+\pi^-\Upsilon(1S)$
- Measurement of inclusive $D^{*\pm}$ production.

Note: In the remainder of this talk I will occasionally label $\Upsilon(4S)$ quantities as "$B\overline{B}$" since it is known that the $\Upsilon(4S)$ decays 100% of the time into B mesons.

II. CHARGED PARTICLE MULTIPLICITY

The data for this particular study[2] was obtained before the 1.0 T superconducting coil was installed and consists of 5502 nb^{-1} of integrated luminosity taken on the $\Upsilon(4S)$ ($10.538 < W < 10.558$ GeV) and 3204 nb^{-1} taken below ($10.378 < W < 10.528$ GeV). After imposing conditions on the data to select hadronic events we are left with 16562 events on resonance and 6684 off resonance, yielding a direct $B\bar{B}$ sample size of 5082±190 events. Visual scanning of approximately 1300 events verified that the data sample suffered negligible contamination from beam-gas, beam-wall, two photon, $\tau^+\tau^-$ or radiative bhabha processes.

Figure 2 shows the observed $B\bar{B}$ and continuum charged multiplicity distributions normalized to the $B\bar{B}$ sample size. The means of these distributions are 8.00±0.03 and 9.69±0.10 for the continuum and $B\bar{B}$ respectively. To determine the true distributions we apply a standard matrix unfolding algorithm using a Monte Carlo simulation.[3] This procedure yields the following multiplicity values:

B decay	$\langle N \rangle = 5.75 \pm 0.1 \pm 0.2$	$D/\langle N \rangle = 0.40 \pm 0.03$
Continuum	$\langle N \rangle = 8.10 \pm 0.1 \pm 0.3$	$D/\langle N \rangle = 0.34 \pm 0.03$

where $D = \sqrt{\langle N^2 \rangle - \langle N \rangle^2}$ is the dispersion. The continuum multiplicity and dispersion agree well with measurements made at nearby center of mass energies.[4]

The above results can be combined with our inclusive lepton measurements[5] to calculate separately the semileptonic and nonleptonic charged multiplicities in B decay. To see this, let us express the mean charged multiplicity as a weighted sum of semileptonic and nonleptonic processes,

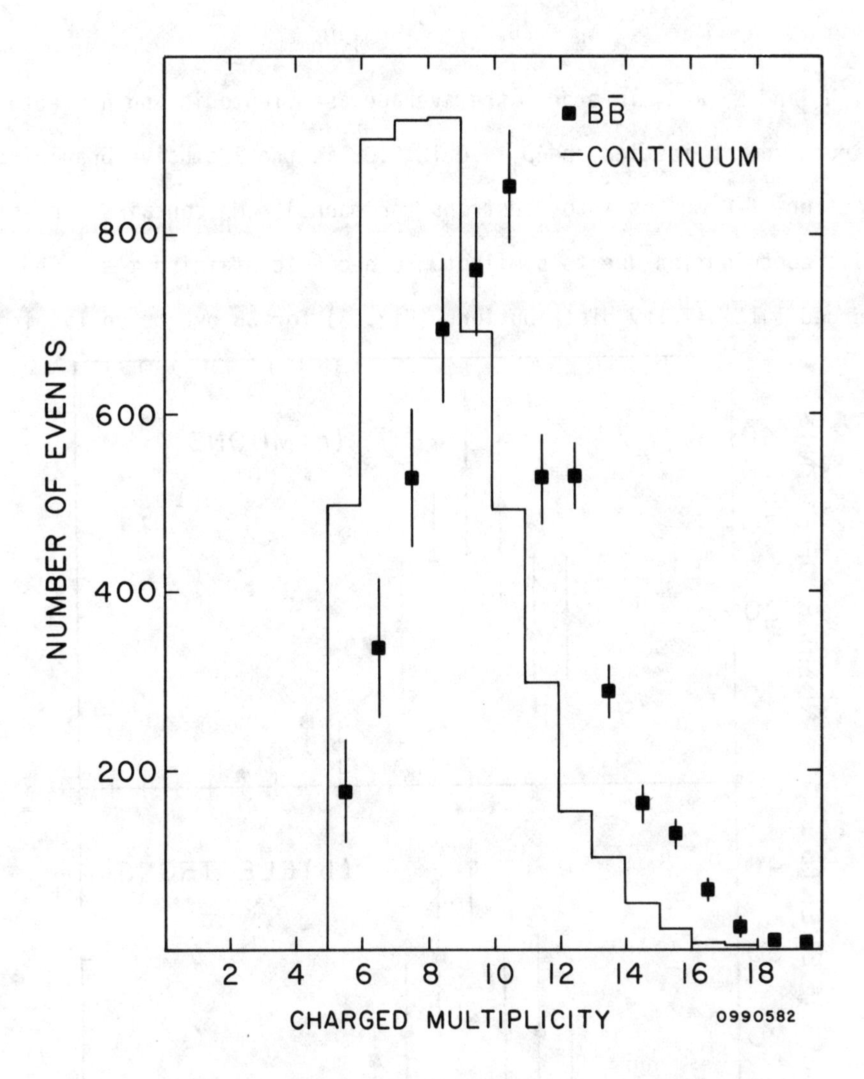

Fig. 2. Charged multiplicity distributions for $B\bar{B}$ events and continuum events. The continuum data has been normalized to the $B\bar{B}$ sample size.

viz.

$$N_B = 2B_1 N_s + (1-2B_1)N_h$$

N_B, N_s and N_h are, in order, the average, semileptonic and nonleptonic charged multiplicities and $B_1 = 0.12 \pm 0.02$ is the inclusive branching fraction of B mesons into electrons (or muons). N_h contains a possible small contribution due to semileptonic decays containing τ's. The charged multiplicity distributions (Fig. 3) for $B\bar{B}$ events containing

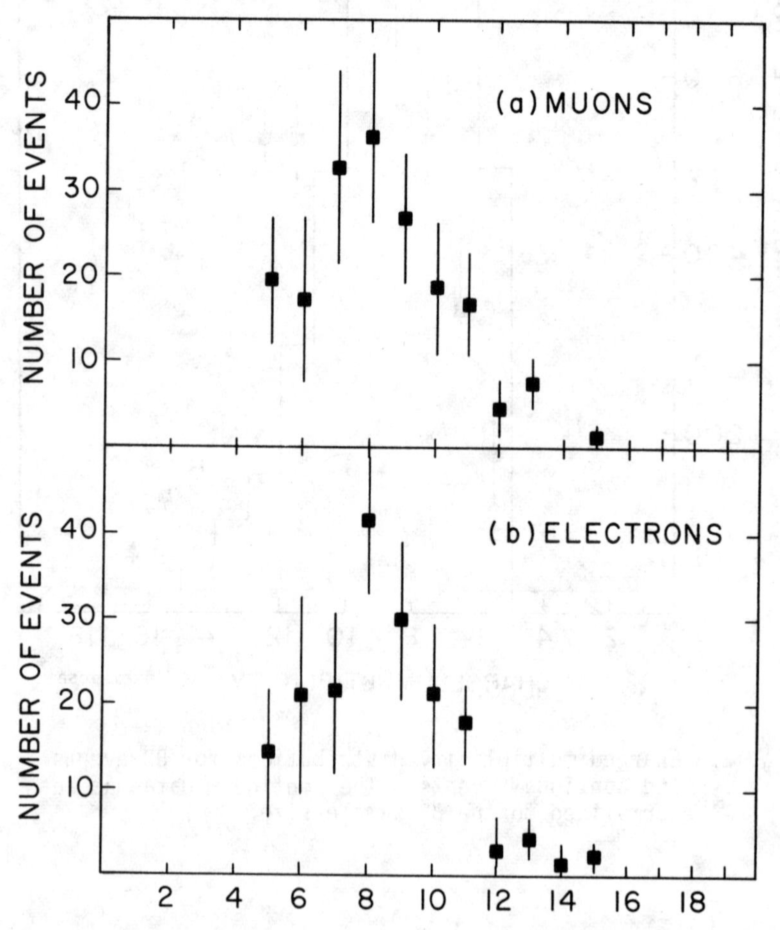

Fig. 3. Charged multiplicity distributions for $B\bar{B}$ events containing (a) muons and (b) electrons.

electrons or muons were obtained using lepton selection criteria described in previous reports.[5] They have been fully corrected for hadron contamination and secondary leptons resulting from τ, D and F decay. The mean multiplicities obtained from these distributions are $\langle N_{lep} \rangle = 8.30 \pm 0.27$ and $\langle N_{lep} \rangle = 8.39 \pm 0.30$ for muons and electrons respectively.

Combining these numbers and using the matrix procedure mentioned above, we find the following values for semileptonic charged multiplicity N_s and the nonleptonic charged multiplicity N_h in B decay:

$$N_s = 4.1 \pm 0.35 \pm 0.2$$
$$N_h = 6.3 \pm 0.20 \pm 0.2 \ .$$

SPEAR data[6] show that a mixture of D and D^* mesons has a mean charged multiplicity of 2.5 ± 0.1. If we assume for the moment that $b \rightarrow c$ dominance holds and that every B decay produces a D meson (F mesons should not change the answer), then an "average" B decay would behave like one of the following decays:

Semileptonic: $B \rightarrow D + 0.5\ \pi^{\pm} + 0.25\ \pi^{0} + e\nu$
Nonleptonic: $B \rightarrow D + 4\ \pi^{\pm} + 2\ \pi^{0}$

The presence of 1/2 - 1 "extra" particles in B semileptonic decay agrees with our observation[5] that the electron spectrum observed in B decay can be fit to a V-A process of the type $B \rightarrow Xe\nu$ where the system "X" has a mass of approximately 2.0 - 2.1 GeV/c^2.

The high average multiplicity observed in B nonleptonic decay indicates that B reconstruction will be very difficult without large statistics.

III. KAON AND BARYON PRODUCTION IN THE Υ REGION

Charged kaons and antiprotons (protons are not used because of a contamination due to beam-gas interactions) are identified by either time of flight counters or the DE/DX system. Identification takes place over a solid angle covering approximately one half of 4π in a momentum interval that depends on both the particle type and device as shown below.

	$K^{\pm}$	$\overline{P}$
TOF	$0.50 < p < 1.0$	$0.85 < p < 1.45$
DE/DX	$0.45 < p < 0.85$	$0.65 < p < 1.45$

K^0's and Λ's are identified by their secondary decays within the drift chamber in the momentum region $p > 0.3$ GeV/c.

To calculate the number of each of these particles per event we extrapolated the momentum spectrum using a modified version of the LUND Monte Carlo.[3] Because the baryon momentum spectrum is not as well understood as that of lighter mesons, we assign a generous systematic uncertainty to the calculated number of baryons per event.

The estimated numbers of particles per event for the four particle species are shown in table II. The values for the narrow resonances have had the 1γ decay contribution removed by subtraction.

TABLE II: Kaon and Baryon Production

The errors shown below are statistical only. Systematic errors are approximately 7% for charged kaons and protons, 12% for K^0's and 15% for Λ's.

	Υ(1S)	Υ(3S)	Υ(4S)	Continuum
K^0/Event	1.02 ± 0.07	0.84 ± 0.09	1.24 ± 0.21	0.89 ± 0.4
$(K^+ + K^-)$/Event	0.95 ± 0.04	0.93 ± 0.05	1.58 ± 0.13	0.90 ± 0.04
Λ/Event	0.25 ± 0.03	0.15 ± 0.03	0.006 ± 0.06	0.08 ± 0.01
$2\overline{P}$/Event	0.55 ± 0.05	0.45 ± 0.02	0.21 ± 0.15	0.27 ± 0.02

Several conclusions can be drawn from this table. First, the numbers of kaons per event are approximately the same for narrow resonances and the continuum. This rough equality is presumably fortuitous since we expect a significant fraction of K's from the continuum to arise from D decay. Since charm production on the narrow resonances is expected to be highly suppressed due to phase space considerations, some other mechanism must be at work to "make up" the extra kaons.

Second, a significant enhancement in kaon production can be seen at the $\Upsilon(4S)$, where the weak quark decay sequence $b \to c \to s$ is expected to produce extra K's. In fact, one can use the $\Upsilon(4S)$ kaon enhancement to make a model dependent estimate of the relative contribution of $b \to c$ and $b \to u$ processes to B decay. If we assume each charm quark produces a single K and use our narrow resonance and continuum data to estimate the number of kaons resulting from normal hadronic fragmentation, then we obtain a value of 74±18±9% for the $b \to c$ fraction in B decay.[7] A somewhat more model dependent procedure[7] using the lepton spectrum from B decay yields a lower limit for this fraction of 80-90%. This technique relies on the assumption that u quark fragmentation produces low mass hadronic systems recoiling against the electron (see Fig. 4).

Third, the narrow resonances seem to be about a factor of 2 or so more efficient at producing baryons than the continuum (no statistically significant baryon signal can be seen in B decay). There appears to be a discrepancy between the $\Upsilon(1S)$ and the $\Upsilon(3S)$ in terms of baryon production, with the $\Upsilon(1S)$ being the more copious source. The difference is seen in both $\overline{P}$ and Λ production, although the effect is somewhat larger for the former than for the latter. The resolution of this

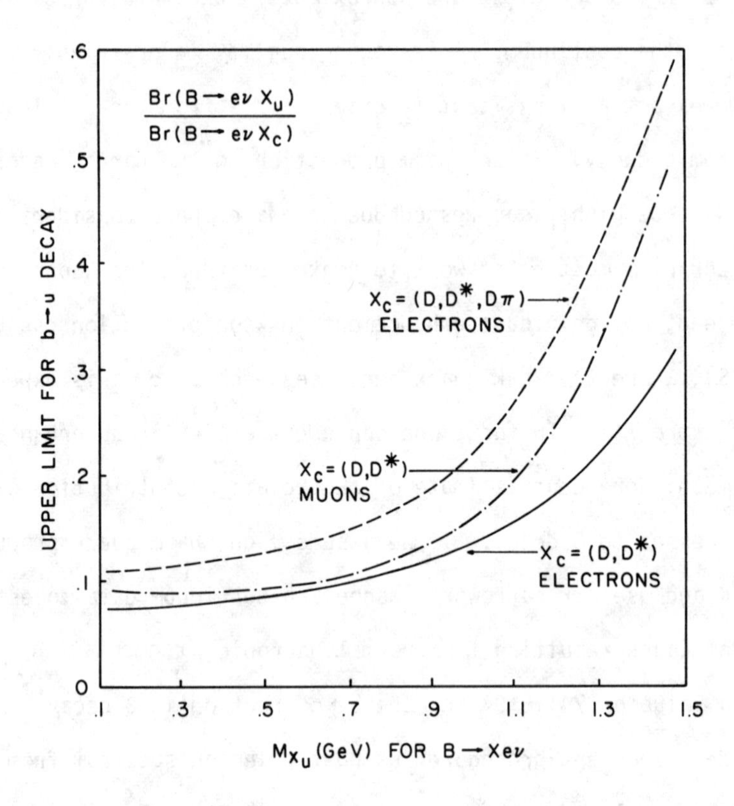

Fig. 4. Upper limit for $B(b \to u)/B(b \to c)$ as a function of M_{Xu}, the mass of the hadronic system in semileptonic B meson decay.

discrepancy might ultimately rely on the fact that a significant fraction (25 - 30%) of $\Upsilon(3S)$ decays[7] involve transitions to x states, which are expected to decay approximately 70% of the time into two gluons. However, a plausible mechanism seems difficult to imagine.

IV. FRACTIONAL MOMENTUM DISTRIBUTIONS

Figure 5 shows the inclusive sdσ/dx (x = 2P/W) distributions of

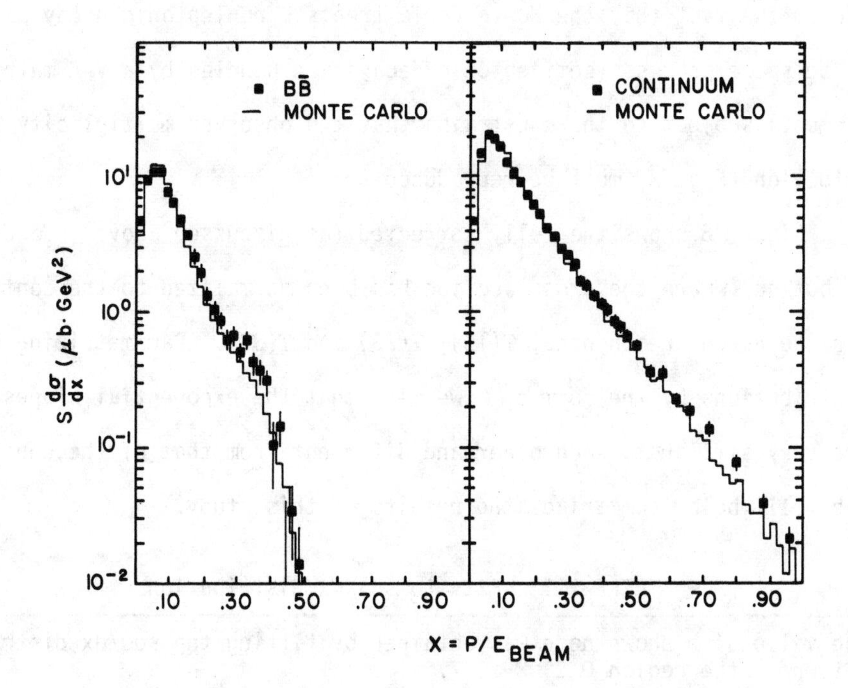

Fig. 5. Distribution of sdσ/dx for $B\bar{B}$ and continuum events.

all charged particles in $B\bar{B}$ events and continuum events respectively. These distributions have been fully corrected for acceptance, resolution photon conversions into e^+e^- pairs in the beam pipe, $\tau^+\tau^-$ production, two photon events, beam wall and beam gas collisions and radiative bhabha processes.

The continuum distribution appears to be well described by the Monte Carlo, in which the primary $q\bar{q}$ pair hadronizes according to the Feynman-Field prescription, with the exception that four momenta is conserved at every vertex and gluons are emitted (and allowed to fragment) according to QCD. The $B\overline{B}$ spectrum, however, appears to be somewhat "harder" than the Monte Carlo prediction. This discrepancy might be due to the fact that the Monte Carlo treats B nonleptonic decay as a phase space process (semileptonic decays are handled by a V-A matrix element) subject to the constraint that the observed multiplicity distribution (Fig. 2) must be reproduced.

Figure 6 shows the fully corrected (as discussed above) $sd\sigma/dx$ distribution (where the cross section has been normalized to the continuum) for the narrow resonances, $\Upsilon(1S)$, $\Upsilon(2S)$ and $\Upsilon(3S)$. Parametrizing these distributions by the form e^{-bx} we find that the exponential slopes, b, are very similar to each other and different from that of the continuum. Table III below summarizes the results of this study.

TABLE III: Fits to $sd\sigma/dx$ Distributions

The value of b shown here was obtained by fitting the $sd\sigma/dx$ distribution in the region $0.2 < x < 0.7$.

Region	b
$\Upsilon(1S)$	11.3 ± 0.1
$\Upsilon(2S)$	11.8 ± 0.1
$\Upsilon(3S)$	10.9 ± 0.1
Continuum	8.3 ± 0.1

The fractional x distributions for charged and neutral kaons, antiprotons and lambdas are displayed in Figures 7 and 8. For these distributions, the variable x is defined in terms of energy, i.e. $x = 2E/W$.

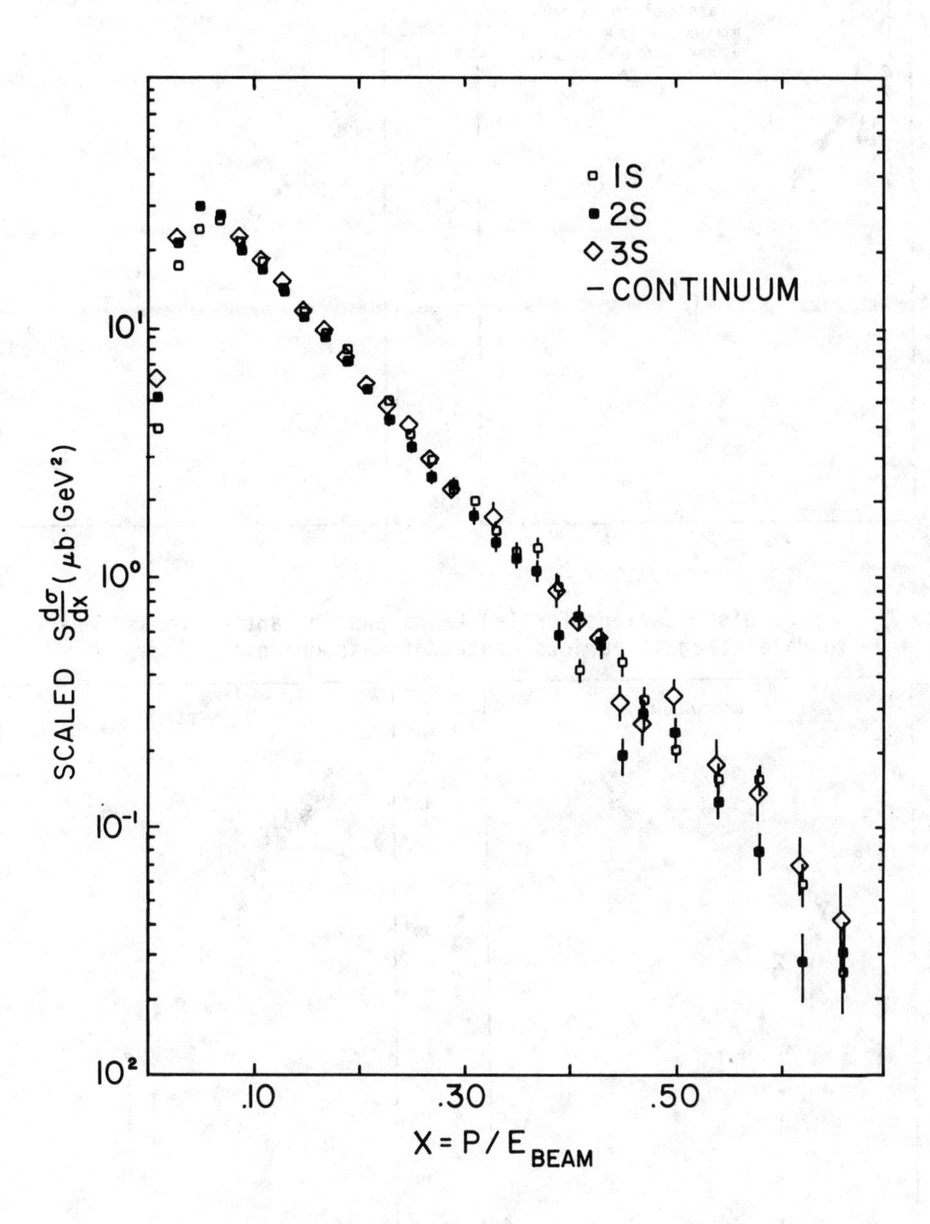

Fig. 6. Distribution of sdσ/dx for the narrow resonances. The cross section has been normalized to the continuum cross section.

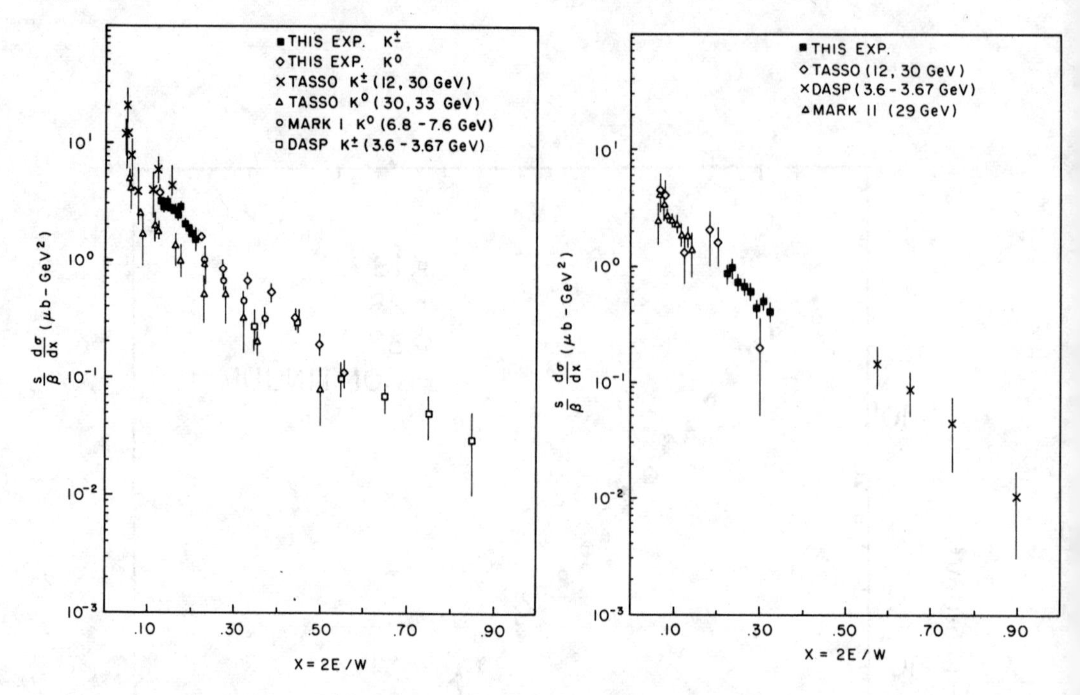

Fig. 7. $sd\sigma/dx$ distributions for (a) kaons and (b) antiprotons compared to data taken at various center of mass energies.

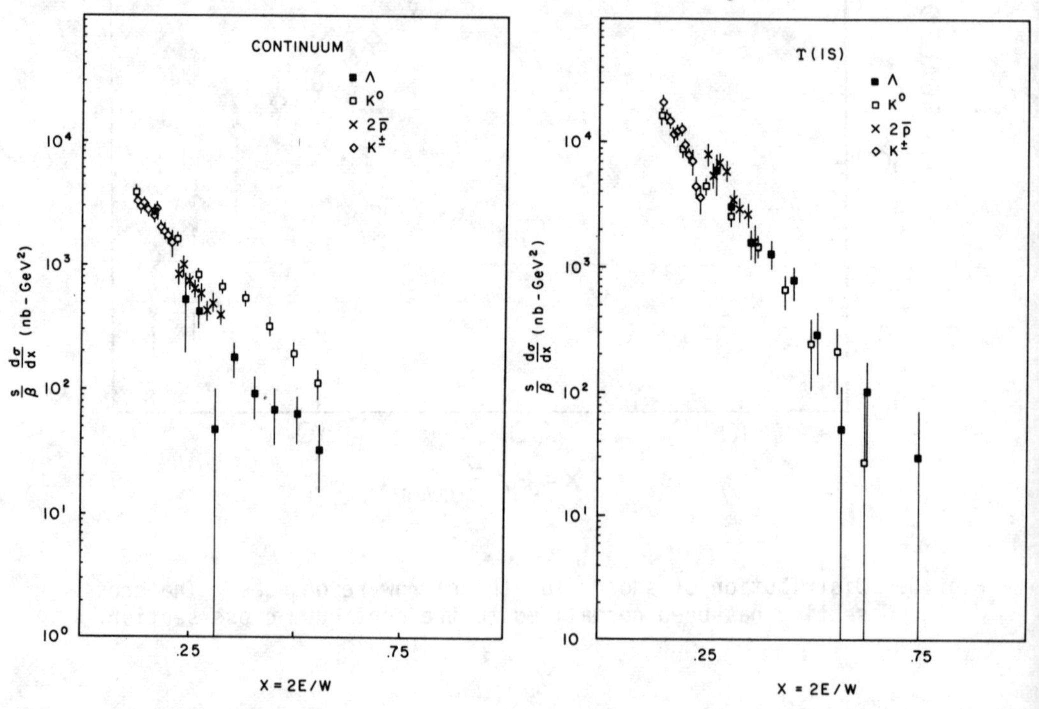

Fig. 8. $sd\sigma/dx$ distributions for kaons and baryons for (a) continuum and (b) $\Upsilon(1S)$ data.

V. VECTOR MESON PRODUCTION IN THE Υ REGION

We have measured the inclusive production of K^{*o}, $K^{*\pm}$ and ϕ mesons at the $\Upsilon(1S)$, $\Upsilon(3S)$, $\Upsilon(4S)$ and continuum using identified charged and neutral kaons. Sample mass distributions for $K^{\pm}\pi^{\pm}$ combinations are shown in Figure 9. Table IV summarizes the results of resonance fits to these distributions. Within the large statistical and systematic errors we see that the number of each of these particles per event does not depend significantly on the origin, whether continuum, 3 gluon or weak decay of B mesons. Approximately 0.1 ϕ's and 1 K^*'s per event are observed.

TABLE IV: Vector Meson Production

Note: the values shown below for the various resonances do not have the continuum contribution subtracted. Systematic errors resulting from uncertainties in the acceptance calculation and the fitting process amount to about 30%.

	$\Upsilon(1S)$	$\Upsilon(3S)$	$\Upsilon(4S)$	Continuum
K^{*+}/Event	0.66 ± 0.26	0.55 ± 0.11	0.56 ± 0.11	0.55 ± 0.15
K^{*o}/Event	0.35 ± 0.16	0.27 ± 0.07	0.38 ± 0.13	0.43 ± 0.13
ϕ/Event	0.12 ± 0.05	0.08 ± 0.02	0.09 ± 0.04	0.11 ± 0.04

The K^* number, when compared with our measurement of total K production, measures the relative amount of vector and pseudoscalar particle production from quark and gluon fragmentation. Even without a direct calculation it is obvious that K^* mesons are produced at rates comparable to those of K mesons. We obtain for the ratio of vector (V) to pseudoscalar (P) K production a value of $V/P = 1.3^{1.5}_{-0.6}$ on the $\Upsilon(1S)$. The corresponding value for the continuum is difficult to calculate reliably because of contribution to K production of D decay,

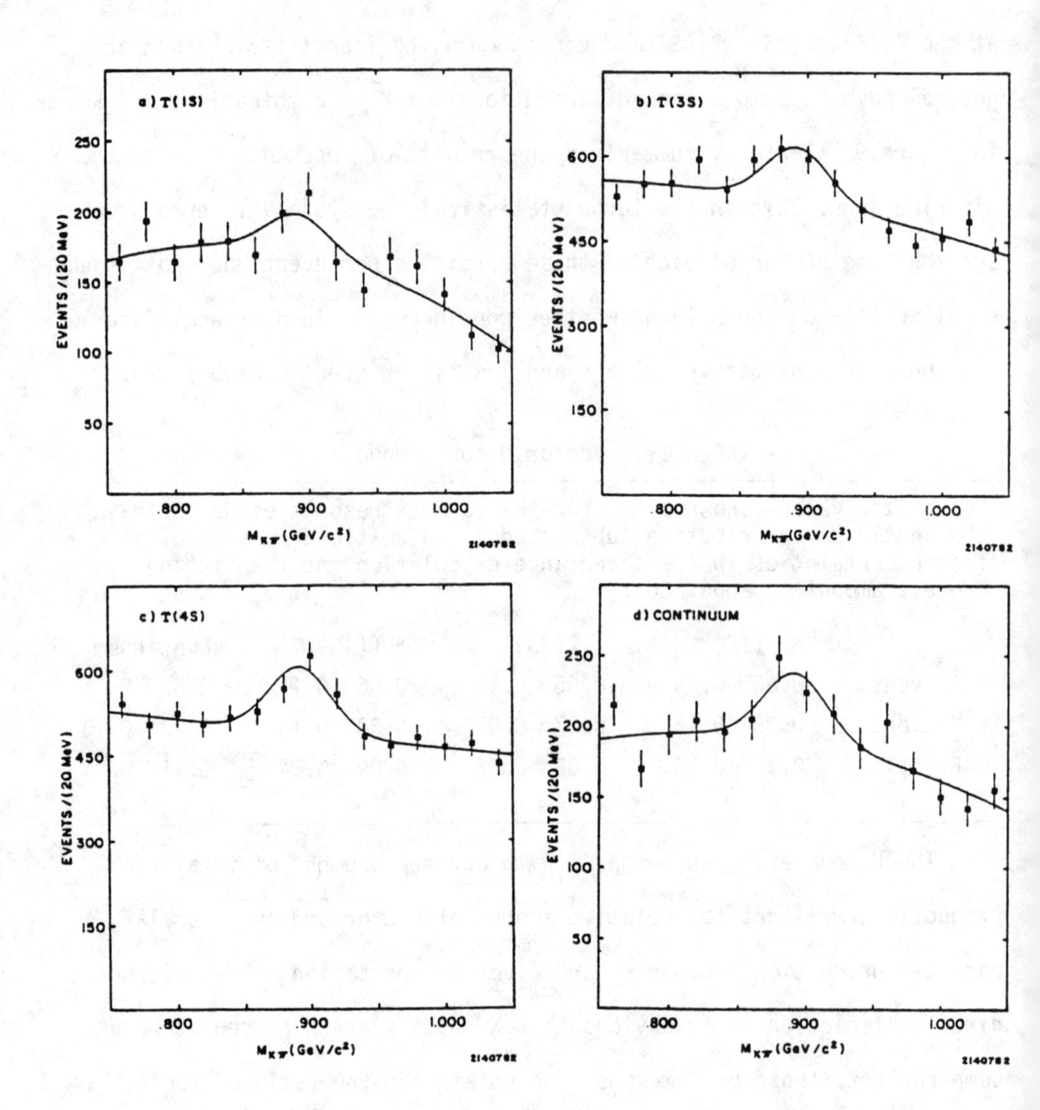

Fig. 9. Sample $K^{\pm}\pi^{\pm}$ distributions.

which produces an unknown fraction of K^* mesons.

VI. $\Upsilon(3S) \to \pi^+\pi^-\Upsilon(1S)$

QCD describes the hadronic transitions between heavy quark bound states in terms of soft gluon emission, with the gluons subsequently decaying into hadrons. The decay rates can be calculated in terms of a multipole expansion of the gluon fields, while the properties of the $\pi\pi$ system are constrained by PCAC and current algebra.[8] This approach successfully[9,10] predicts the properties of the decays $\Upsilon(2S) \to \pi^+\pi^-\Upsilon(1S)$ and $\psi' \to \pi^+\pi^-\psi$. With the approximately 9400 nb^{-1} of data recently collected at the $\Upsilon(3S)$, we can test these theoretical ideas in the $\Upsilon(3S) \to \pi^+\pi^-\Upsilon(1S)$ decay.

We determine the branching fraction of $\Upsilon(3S) \to \pi^+\pi^-\Upsilon(1S)$ in two ways The first involves computing the missing mass, M_X, against pairs of oppositely charged pions observed at the $\Upsilon(3S)$ as shown in Fig. 10. A 4 standard deviation enhancement can be seen at the $\Upsilon(1S)$ mass. Fitting this peak to a second degree polynomial plus one of several different signal representations yields 1070 ± 105 events, which, when corrected for trigger efficiency and acceptance, becomes

$$B(\Upsilon(3S) \to \pi^+\pi^-\Upsilon(1S)) = 5.4 \pm 1.3 \pm 0.5\% .$$

We also determine the branching fraction by measuring the decay sequence

$$\Upsilon(3S) \to \pi^+\pi^-\Upsilon(1S), \quad \Upsilon(1S) \to e^+e^- \text{ or } \mu^+\mu^- .$$

We obtained 34 events of this type on the $\Upsilon(3S)$ peak and 0 events on the nearby continuum by scanning suitable events. This sample includes 5 events found in our low field (0.42 T) data. A subset of 22 (13 $\pi\pi ee$ and

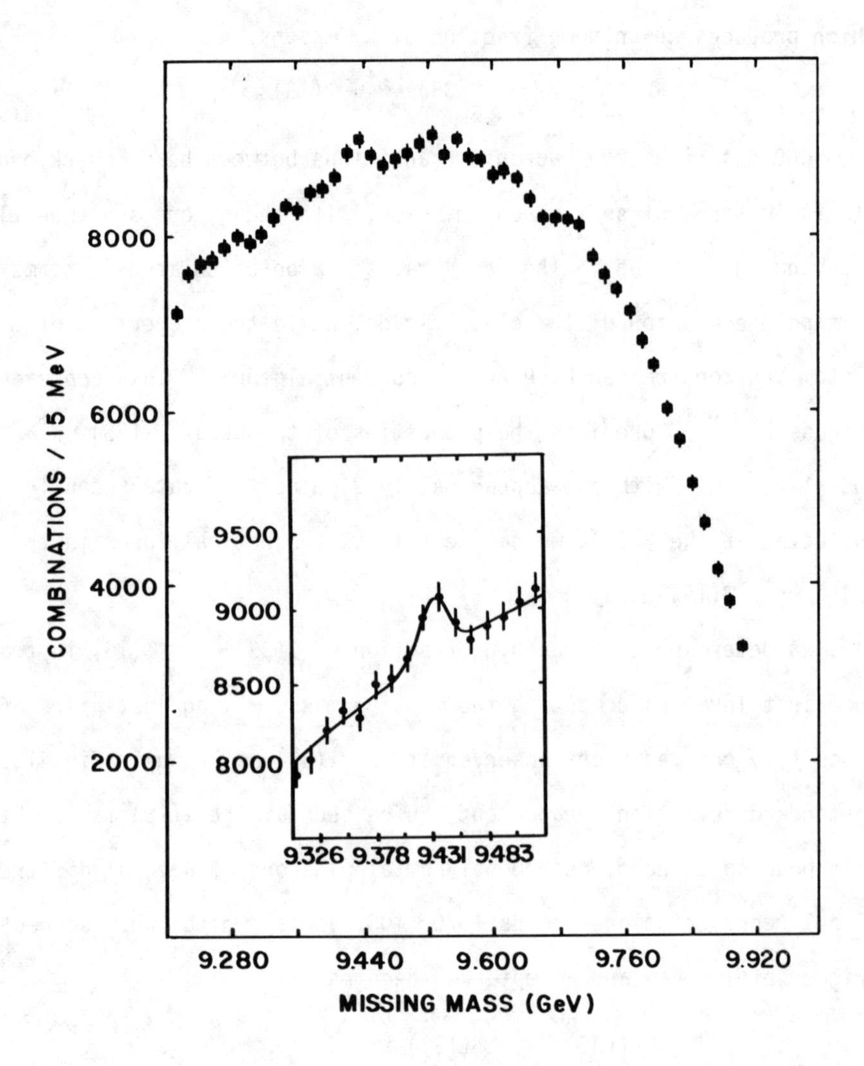

Fig. 10. $\pi^+\pi^-$ missing mass distribution for data taken on the $\Upsilon(3S)$.

9 $\pi\pi\mu\mu$) events, chosen on the basis of fiducial cuts, was used to compute the branching fraction. This yields

$$B(\Upsilon(3S) \to \pi^+\pi^-\Upsilon(1S)) = 4.4 \pm 1.2 \pm 0.5\%$$

where we have used the world average value $B(\Upsilon(1S) \to \mu^+\mu^-) = 0.033 \pm 0.005$ to compute the branching fraction. The two techniques have different statistical and systematic errors, so we quote the average of the measurements

$$B(\Upsilon(3S) \to \pi^+\pi^-\Upsilon(1S)) = 4.9 \pm 0.9 \pm 0.5\%$$

This value can be compared to the branching fraction predicted by Kuang and Yan in their multipole expansion. Their values lie somewhere between 1.3 - 3.4%, depending on the particular $b\bar{b}$ potential chosen.

The 34 events can be used to study the properties of the $\pi\pi$ system. The $\cos\theta_{\pi\pi}$ distribution, shown in Figure 11, is consistent with the

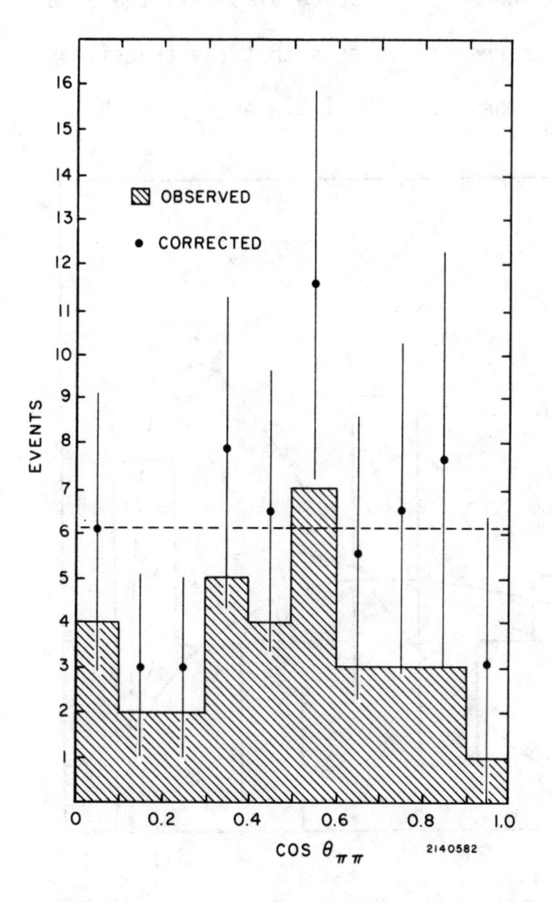

Fig. 11. $\cos\theta_{\pi\pi}$ distribution for the $\pi\pi$ system recoiling from the $\Upsilon(1S)$.

prediction of a flat angular distribution, but other angular distributions cannot be ruled out.

The matrix element for the dipion system is predicted by PCAC and the multipole expansion framework[8] to have the form

$$M \propto A q_1 \cdot q_2 + B q_{10} q_{20}$$

where q_1 and q_2 are the pion four momenta. The dipion mass spectra[9,10] for $\Upsilon(2S) \rightarrow \pi\pi\Upsilon(1S)$ and $\psi' \rightarrow \pi\pi\psi$ suggest $B = 0$ (curve (a) in Fig. 12). Our data for $\Upsilon(3S) \rightarrow \pi\pi\Upsilon(1S)$ can be fit by $A/B = -0.15 \pm 0.12$ (curve (b) in Fig. 12), but a distribution given by phase space also fits the data reasonably well (curve c). This result suggests that the transition amplitude might be sensitive to the total available energy or the heavy quark wavefunction.

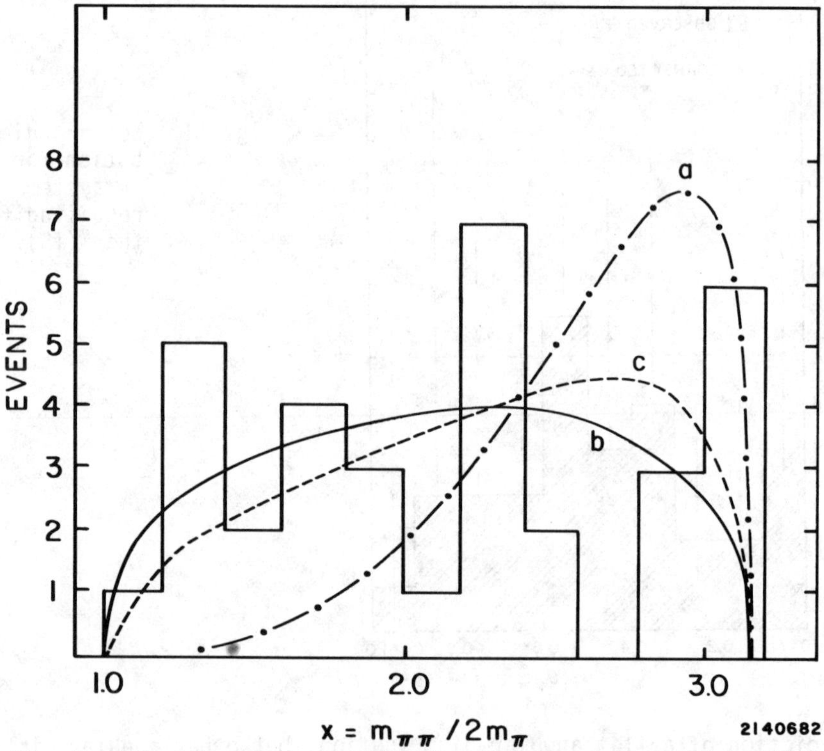

EVENTS

$x = m_{\pi\pi}/2m_\pi$

2140682

Fig. 12. $X_{\pi\pi}$ ($X_{\pi\pi} = M_{\pi\pi}/2M_\pi$) distribution for the $\pi\pi$ system recoiling from the $\Upsilon(1S)$. The curves are explained in the text.

VII. D^* PRODUCTION

The fragmentation functions of light quarks hadronizing into pions and kaons are known experimentally to fall rapidly with z, the fraction of quark energy carried by the meson. In contrast, theoretical models utilizing QCD hold that the fragmentation functions of heavy quarks should be relatively independent of z. Inclusive production of charmed mesons in e^+e^- annihilation provides a convenient laboratory for measuring these hadronization processes because the initial charmed quark has a unique energy. Thus, the inclusive charmed production cross section per unit energy is directly proportional to the charm fragmentation function.

However, detecting D^0 or D^+ mesons at center of mass energies significantly above $D\bar{D}$ threshold is made difficult by both large combinatoric backgrounds and low efficiencies for kaon identification. We choose instead to use the $D^{*\pm}$ (henceforth we will use the D^{*+} to refer to both charge states) meson as a probe of charm production by exploiting the cascade decay

$$D^{*+} \rightarrow D^0\pi^+ , \qquad D^0 \rightarrow K^-\pi^+ .$$

This is because we can measure the quantity $\Delta M = M_{K\pi\pi} - M_{K\pi}$ in the region of the $D^{*+} - D^0$ mass difference ($0.1453\ GeV/c^2$)[10] to a precision of approximately $1\ MeV/c^2$. By making a very tight constraint on ΔM, we can reduce the combinatoric background by as much as 2 to 3 orders of magnitude. Figure 13 shows the $M_{K\pi}$ mass distribution for $K\pi\pi$ combinations satisfying the requirement $0.1438 < \Delta M < 0.1468\ GeV/c^2$ and $z = E_{D^*}/E_{beam} > 0.7$. A clear D^0 signal of 32 ± 8 events (obtained by

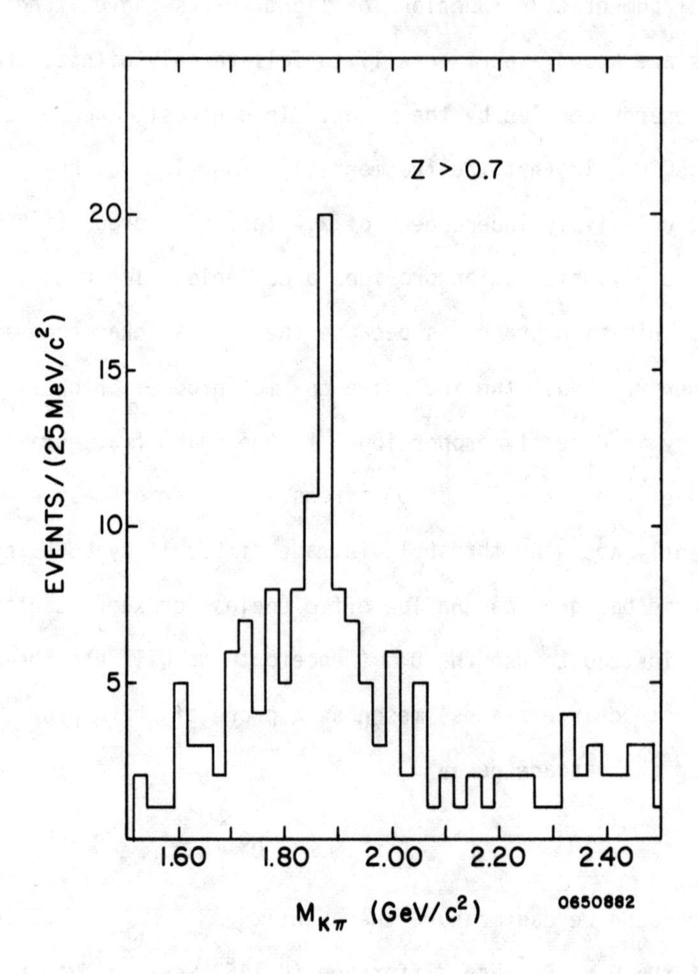

Fig. 13. $M_{K\pi}$ distribution for $K^{\mp}\pi^{\pm}\pi^{\pm}$ combinations satisfying $0.1438 \; M_{K\pi\pi} - M_{K\pi} < \; 0.1468$ GeV/c^2 and $z > 0.7$.

fitting) can be seen. Reversing the procedure, we show in Figure 14 the ΔM distribution under the requirement $1.830 < M_{K\pi} < 1.920$ GeV/c^2 and $z > 0.7$. The sharp enhancement seen in this plot peaks at the D^{*+}-D^0 mass difference of 0.1453 GeV/c^2.

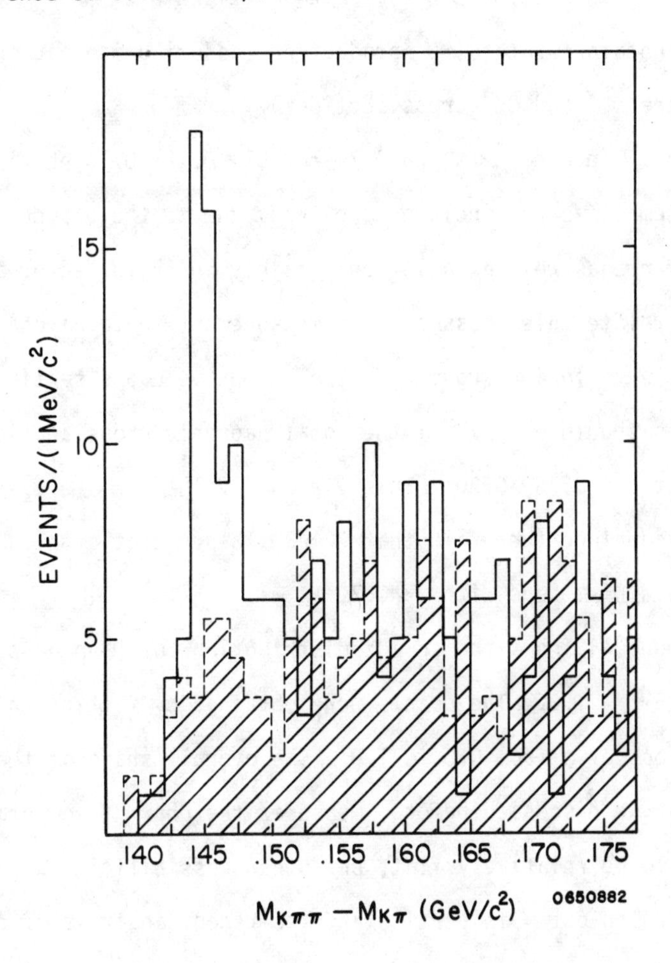

Fig. 14. $M_{K\pi\pi} - M_{K\pi}$ distribution for $K^{\mp}\pi^{\pm}\pi^{\pm}$ combinations satisfying $1.830 < M_{K\pi} < 1.920$ GeV/c^2 and $z > 0.7$. The shaded region was obtained by using $M_{K\pi}$ regions on either side of the D^0 position.

To convert this raw signal into a cross section, we need to correct for both the $K\pi\pi$ detection efficiency, which ranges between 18 and 30%, and the branching fractions in the cascade decay. For the latter we use the SPEAR values[6] $B(D^{*+} \to D^{o}\pi^{+}) = 0.44 \pm 0.10$ and $B(D^{o} \to K^{-}\pi^{+}) = 0.03 \pm 0.006$. We then obtain for total D^{*} production (assuming isospin invariance to equate the D^{*+} and D^{*o} cross sections)

$$\sigma(e^{+}e^{-} \to D^{*} + X,\ z > 0.7) = 0.57 \pm 0.12 \pm 0.19 \text{ nb}$$

The systematic error includes the error due to the branching fraction measurements as well as a 15% uncertainty in the detector acceptance.

To relate this cross section measurement to total charm production we use the following argument. The inclusive cross section into charmed particles should be 4/10 of the total hadronic cross section, which we measure to be 3.75 ± 0.30. This yields a value for $\sigma(e^{+}e^{-} \to c + X)$ of 1.5 nb. We conclude therefore that $38 \pm 8 \pm 13$ per cent of all charm productio results in D^{*} mesons with $z > 0.7$.

Figure 14 shows the $sd\sigma/dz$ distribution for D^{*} production in the range $0.5 < z < 1.0$. The points were obtained by fitting the $M_{K\pi}$ spectrum for regions of z from 0.5 to 1.0. The overall shape of the distribution, which, as we discussed before, measures the charm fragmentation function, appears to be relatively flat, but other possibilities cannot be completely ruled out. In particular, the recent analysis of data at W = 5.2 GeV by a SPEAR group[6] suggests that the charm fragmentation function appears to be falling at a rate not very dissimilar to that describing pions and kaons from light quarks.

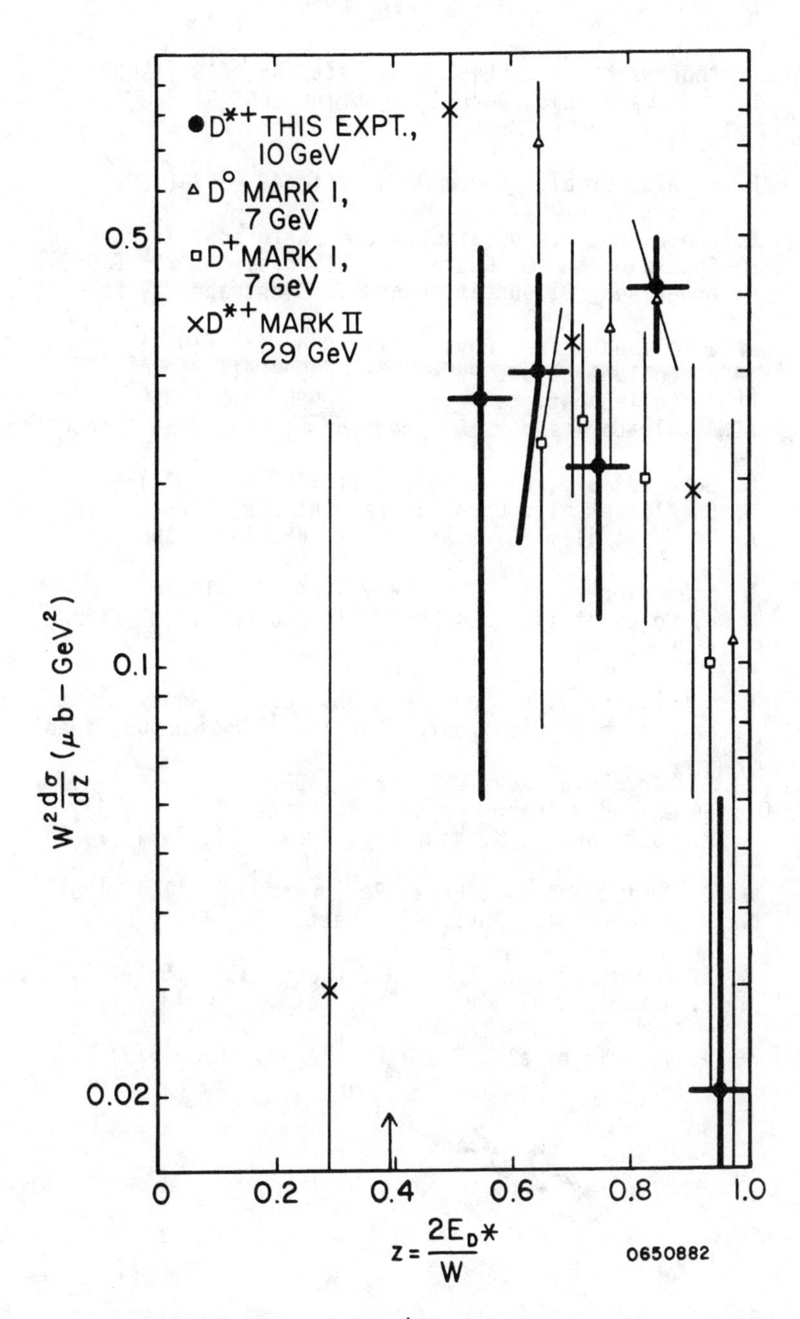

Fig. 15. sdσ/dz distribution for D^{*+} production compared to other experiments.[11]

REFERENCES

1. D. Andrews et al., Phys. Rev. Lett. 45, 219 (1980);
   D. Andrews et al., Cornell preprint CLNS 82/538, 1982 submitted to Nucl. Instr. Meth.

2. M. S. Alam et al., Phys. Rev. Lett. 49, 357 (1982).

3. B. Andersson, G. Gustafson and C. Peterson, Z. Phys. C1, 105 (1979);
   B. Andersson and G. Gustafson, Z. Phys. C3. 223 (1980);
   B. Andersson, G. Gustafson and T. Sjostrand, Z. Phys. C6, 235 (1980).

4. JADE Collaboration, Phys. Lett. 88B, 171 (1979);
   TASSO Collaboration, Phys. Lett. 89B, 418 (1980);
   PLUTO Collaboration, Phys. Lett. 95B, 313 (1980);
   LENA Collaboration, DESY preprint 81-008, 1981 (unpublished).

5. C. Bebek et al., Phys. Rev. Lett. 46, 84 (1981);
   M. S. Alam et al., Cornell preprint CLNS 81/513, 1981 (unpublished);
   A. Brody et al., Phys. Rev. Lett. 48, 1070 (1982).

6. R. Schindler et al., Phys. Rev. D24, 78 (1981);
   M. H. Coles et al., SLAC PUB-2916 and LBL 14402 (1982), submitted to Phys. Rev. D.

7. CLEO Collaboration, Contributions to the Twenty First International Conference on High Energy Physics, 1982 (unpublished).

8. T. M. Yan, Phys. Rev. D22, 1652 (1980);
   L. Brown and R. Cahn, Phys. Rev. Lett. 35, 1 (1975);
   Y. P. Kuang and T. M. Yan, Phys. Rev. D24, 2874 (1981).

9. J. J. Mueller et al., Phys. Rev. Lett. 46, 1118 (1981).
   G. Mageras et al., Phys. Rev. Lett. 46, 1115 (1981).

10. G. S. Abrams et al., Phys. Rev. Lett. 33, 1153 (1974);
    G. S. Abrams et al., Phys. Rev. Lett. 34, 118 (1975).

11. P. A. Rapidis et al., Phys. Lett. 84B, 507 (1979).

RECENT RESULTS FROM CUSB ON UPSILON SPECTROSCOPY

P. M. Tuts
S.U.N.Y. at Stony Brook, Stony Brook, NY 11794

ABSTRACT

We report on results obtained by CUSB at CESR on (1) the first observations of the hadronic transitions $\Upsilon''\to\Upsilon'\pi\pi$, and $\Upsilon''\to\Upsilon\pi\pi$ where the final Υ or Υ' subsequently decays to $\mu^+\mu^-$ or e^+e^-, (2) the first observation of the P-wave bound states (2^3P_J for J=0,1,2 or χ_b') in the $b\bar{b}$ system, in both the inclusive photon spectrum $\Upsilon''\to\gamma\chi_b'$, and the exclusive channels $\Upsilon''\to\gamma\chi_b'\to\gamma\gamma\Upsilon$, $\Upsilon''\to\gamma\chi_b'\to\gamma\gamma\Upsilon'$ where the final Υ or Υ' subsequently decays to $\mu^+\mu^-$ or e^+e^-, and (3) the results of an axion search in the decay $\Upsilon\to\gamma a$. Branching ratios for these processes are presented.

INTRODUCTION

There has been a great deal of progress in experimental heavy quark spectroscopy in the past few months. In this article we will discuss some of the novel results obtained in Υ spectroscopy by the CUSB Collaboration[1] (Columbia University - SUNY at Stony Brook - Louisiana State University - Max Planck Institut at Munich - Cornell University) at the Cornell Electron Storage Ring (CESR). The field began with the observation[2] of the J/ψ, and was enriched by the observation[3] of the Υ. The Υ system ($b\bar{b}$ bound states) provides us with a remarkable testing ground for our understanding of the b quark, its decays, potential models, and QCD.

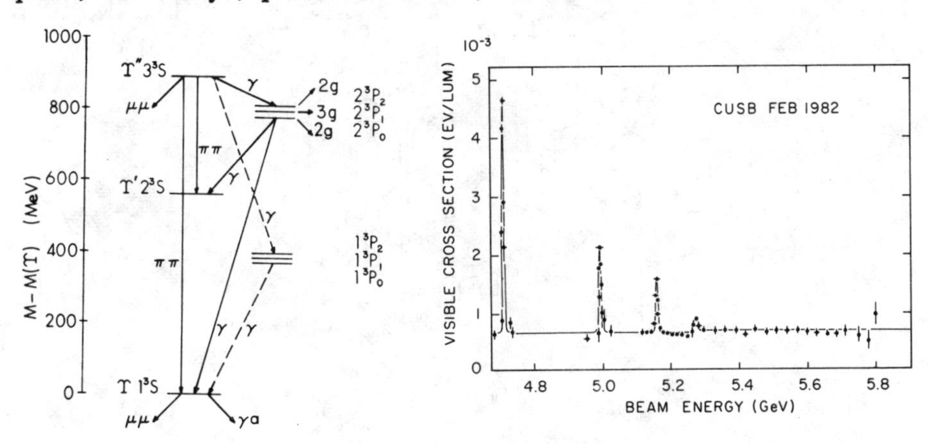

Fig. 1. The Υ'' transitions.

Fig. 2. $\sigma(e^+e^-\to hadrons)$ from CUSB, showing the four ^3S states.

Until now, the rich spectrum expected in the Υ system (see Fig. 1) was limited to the observation of the $\Upsilon'\to\Upsilon\pi\pi$ transition[4] and the first four triplet S-wave states[5,6] (n^3S) shown in Fig.2 ; these are

094-243X/82/930155-14$3.00

the bound $b\bar{b}$ states that can be directly produced in e^+e^- annihilations (i.e. those with $J^{PC}=1^{--}$). The other states can be reached by electromagnetic or hadronic transitions from the primary n^3S states.

Table I Summary of CUSB data reviewed here.

Region	Hadronic Events	Continuum Subtracted	Integrated Luminosity (pb^{-1})
ϒ	16,200	13,800	1
ϒ''	64,7000	37,300	15
Continuum	12,000	---	4

The results were mostly obtained from running taken from November 1981 to February 1982 (summarised in Table I) using the CUSB NaI and lead glass electromagnetic calorimeter. The CUSB detector has been described elsewhere[5,7], so we will limit ourselves to a brief description of the main elements, which are shown schematically in Fig. 3.

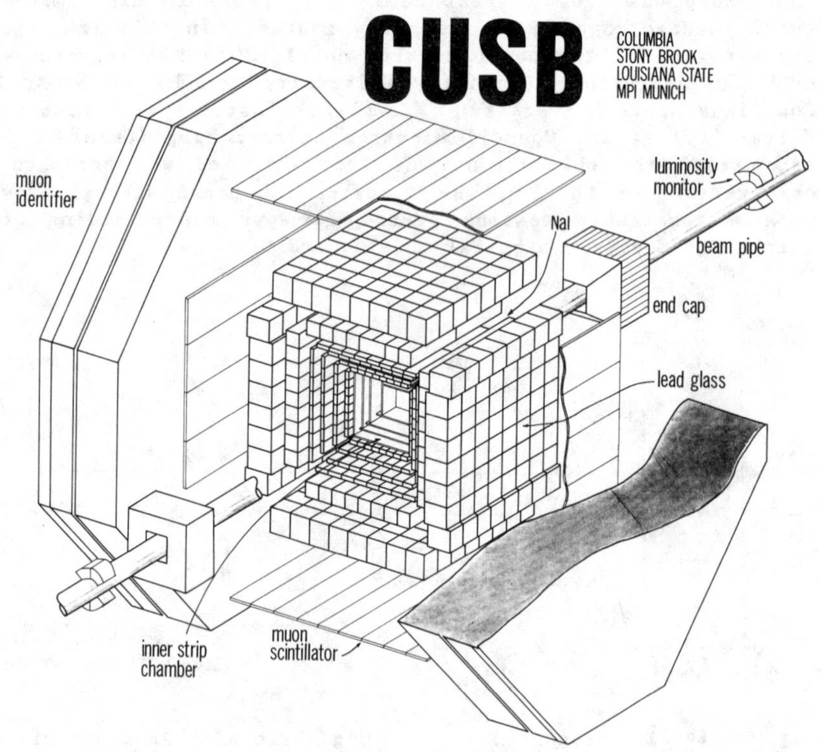

Fig. 3. Schematic view of the CUSB detector.

Starting from the innermost element the CUSB detector has (1) four planes of Inner Strip Chambers (ISC) which provide charged particle tracking over $\Delta\Omega$~62%, (2) 324 elements of NaI (8 r.l.) arranged in five planes as shown ($\Delta\Omega$~ 60%), (3) four planes of strip chambers located between NaI planes (not shown in Fig. 3), (4) 256 lead glass blocks (7 r.l.) surrounding the NaI array, (5) 168 NaI crystals in two endcaps which increase the coverage to $\Delta\Omega$~ 85%, (6) 27 scintillation counters which provide a muon trigger with $\Delta\Omega$~ 32%, (7) a muon identifier consisting of two magnetised iron toroids with 12 drift chamber planes covering $\Delta\Omega$~ 25%.

In addition to making precision electromagnetic energy measurements, the CUSB detector is capable of making precise total energy measurements (i.e. including the hadronic energy). A typical observed total energy distribution is shown in Fig. 4, with a FWHM of 28% (note that the fraction of the observed total energy to the total cm energy is plotted). From these precise total energy measurements (shown in Fig. 5 after continuum subtraction) it is possible to rule out exotic decays of the b quark[8].

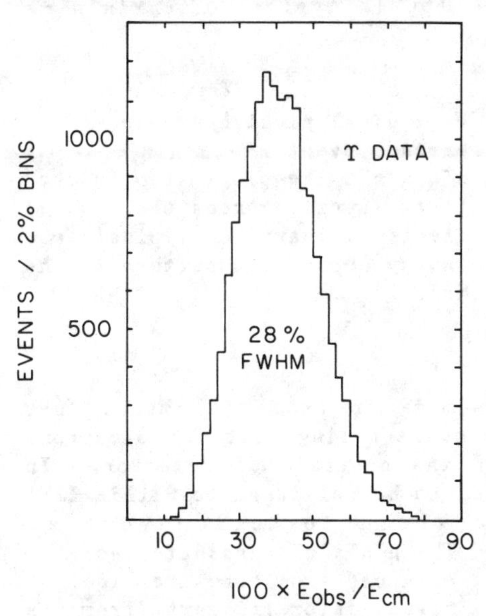

Fig. 4. Typical distribution of the observed total energy fraction for Υ events.

The lower observed total energy for the Υ''' is in agreement with Monte Carlo calculations of increased missing energy due to an increase in the the number of neutrinos from the weak decays of the B meson.

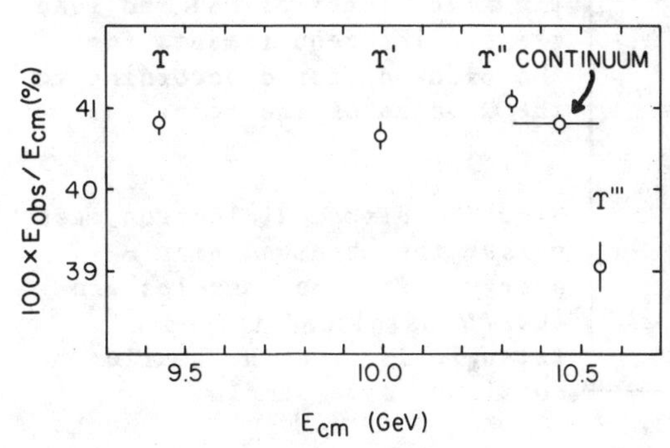

Fig. 5. The observed total energy fraction for the four Υ resonances, and the continuum. Note the zero suppressed vertical scale and the low Υ''' value. The average of the continuum and resonances excluding the Υ''' is (40.94±.06)%, whereas the Υ''' is at (38.44±.32)%.

It is inconsistent with Monte Carlo predictions for exotic decays (e.g. b->qll) of the b quark which lead to an observed energy fraction of less than 34%, in total disagreement with the observed value of 38%.

HADRONIC TRANSITIONS

The mass splittings[5,6,7] $\Upsilon''-\Upsilon'$ and $\Upsilon''-\Upsilon$ are 330 MeV and 889 MeV respectively, thus the hadronic transitions $\Upsilon''\to\Upsilon'\pi\pi$ and $\Upsilon''\to\Upsilon\pi\pi$ are possible, with Q values of 610 MeV and 51 MeV respectively. We have observed transitions of this type[9], where the final Υ' or Υ decays to e^+e^- or $\mu^+\mu^-$. The $e^+e^-\pi^+\pi^-$ candidates were obtained by scanning for events resembling Bhabha scatters with additional tracks present in the ISC (a typical $\Upsilon''\to\Upsilon'\pi^+\pi^-\to e^+e^-\pi^+\pi^-$ event is shown in Fig. 6).

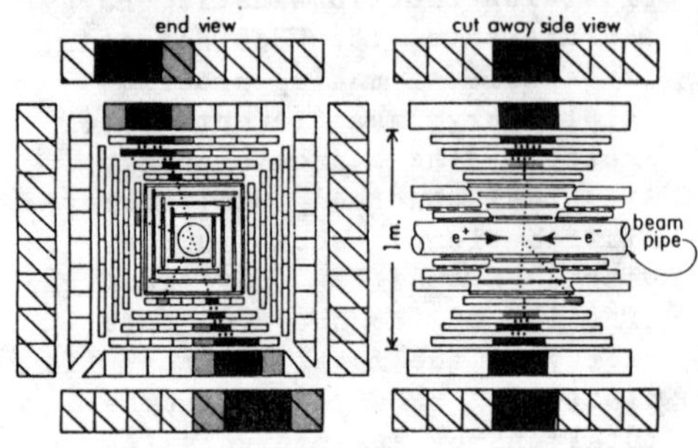

Fig. 6. Typical hadronic cascade event $\Upsilon''\to\Upsilon'\pi^+\pi^-\to e^+e^-\pi^+\pi^-$. The pions stop in the first layer, whereas the electrons leave the typical em energy deposition pattern in the NaI.

Further fiducial volume cuts were made in order to insure that shower energy leakage was kept low by requiring that the electrons be more than 5° from the edges of the central NaI detector. In addition, the electrons were required to be collinear to within 22°, with ISC tracks pointing to the interaction vertex. The $\mu^+\mu^-\pi^+\pi^-$ candidates were obtained from opposite side scintillator triggers (from the nearly collinear muons) which leave the characteristic minimum ionizing particle energy deposition pattern[7] in the five planes of NaI and lead glass. The requirements for the pions differed according to the Q value of the reaction.

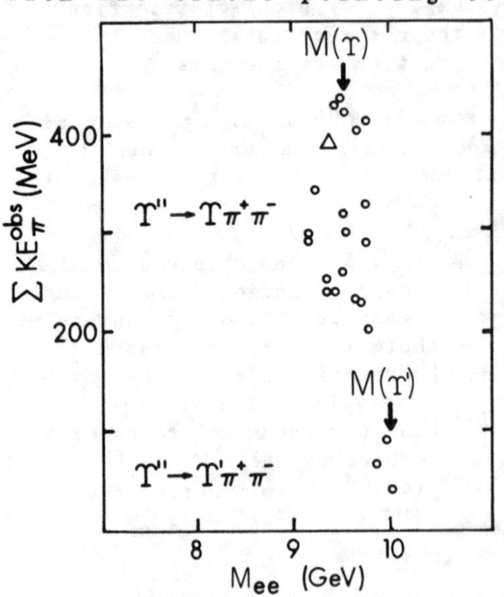

Fig. 7. Fitted dielectron mass versus the observed pion energy. The open circles are the Υ'' data, and the open triangle is from the smaller continuum data sample.

For $\Upsilon''\to\Upsilon\pi^+\pi^-$ decays at least one minimum ionizing pion must be seen in the ISC, and it must penetrate two or more NaI layers with the characteristic energy deposition pattern. For $\Upsilon''\to\Upsilon'\pi^+\pi^-$ decays both heavily ionizing pions must be seen in the ISC with little penetration (less than 2 layers) into the central detector. All candidates were then fit to two constraints (2C) and required to have a confidence level greater than 1%. A clear separation of the two categories of decays ($\Upsilon''\to\Upsilon(e^+e^-)\pi^+\pi^-$ and $\Upsilon'(e^+e^-)\pi^+\pi^-$) can be seen in Fig. 7 which shows the observed pion track energy versus the dielectron mass obtained from the 2C fit. In addition to clustering near the appropriate masses (M_{ee}=9480 MeV for $\Upsilon''\to\Upsilon\pi^+\pi^-$ and M_{ee}=9960 MeV for $\Upsilon''\to\Upsilon'\pi^+\pi^-$), the three lower events from $\Upsilon''\to\Upsilon'\pi^+\pi^-$ show heavier ionization in the ISC as expected. We have studied the QED backgrounds by scanning continuum data, from which we estimate the background to $\Upsilon''\to\Upsilon(e^+e^-)\pi^+\pi^-$ to be 1 event and to $\Upsilon''\to\Upsilon'(e^+e^-)\pi^+\pi^-$ to be .2 events.

Table II Summary of observed hadronic transitions (80% for muons).

Transition	Events	Acceptance (%) Ref.10	Acceptance (%) Phase Space	Product BR
$\Upsilon''\to\Upsilon(ee)\pi^+\pi^-$	19-1	24.2	22.4	$(1.3\pm.3\pm.2)\times10^{-3}$
$\Upsilon''\to\Upsilon(\mu\mu)\pi^+\pi^-$	7	14.4	13.1	$(1.1\pm.4\pm.3)\times10^{-3}$
$\Upsilon''\to\Upsilon'(ee)\pi^+\pi^-$	3	10.1	9.6	$(.5\pm.3\pm.2)\times10^{-3}$
$\Upsilon''\to\Upsilon'(\mu\mu)\pi^+\pi^-$	1	6.0	5.6	$(.4\pm.4\pm.2)\times10^{-3}$
$\Upsilon''\to\Upsilon(ee)\pi^o\pi^o$	5	13.6	10.9	
$\Upsilon''\to\Upsilon(\mu\mu)\pi^o\pi^o$	2	7.7	6.0	
$\Upsilon''\to\Upsilon'(ee)\pi^o\pi^o$	4	12.2	11.9	
$\Upsilon''\to\Upsilon'(\mu\mu)\pi^o\pi^o$	0	6.6	6.4	

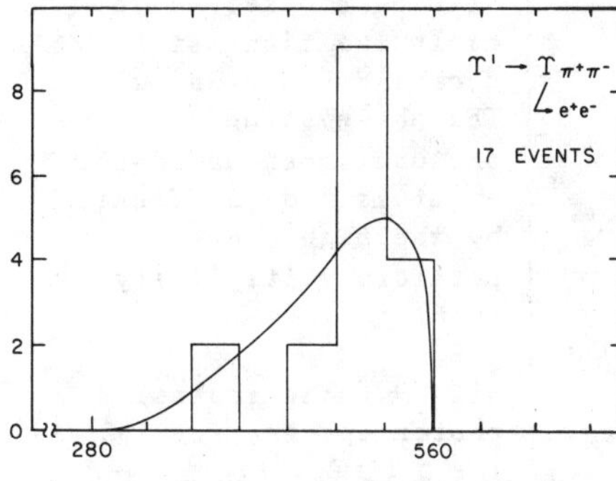

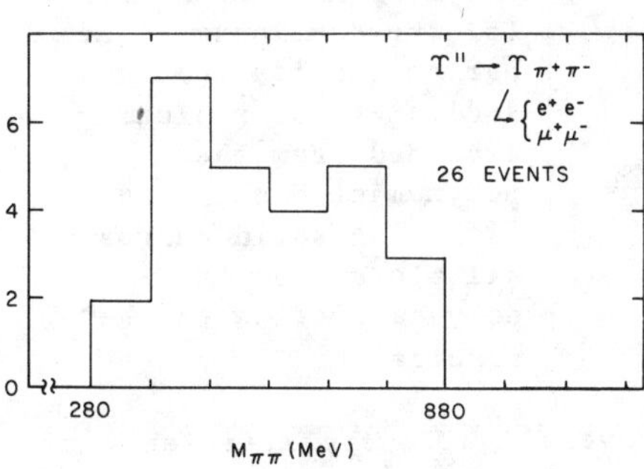

Fig. 8. Dipion mass spectra for hadronic decays of (a) Υ' (Ref. 4) and (b) Υ''.

Other backgrounds were estimated to be negligable after imposing the kinematic fitting procedure. The efficiencies were calculated by Monte Carlo, and vary according to the dipion mass distribution (Table II lists the efficiency for phase space and high dipion mass distributions). Using the number of Υ'' resonance events (efficiency 65%) from Table I, the product branching ratios listed in Table II, and using[10] $B_{\mu\mu}(\Upsilon)=.033\pm.006$ together with the CUSB derived value[4] of $B_{\mu\mu}(\Upsilon')=.020\pm.005$ we find $BR(\Upsilon''\to\Upsilon\pi^+\pi^-)=(3.7\pm1.2)\%$ $BR(\Upsilon''\to\Upsilon'\pi^+\pi^-)=(2.3\pm1.5)\%$. The $\Upsilon''\to\Upsilon(e^+e^-$ or $\mu^+\mu^-)$ $\pi^+\pi^-$ candidates were fit with the additional constraint that the dilepton mass be M(Υ).

The dipion mass was obtained from this procedure, and is shown in Fig. 8b. It is noteworthy that this dipion mass spectrum is markedly different from that obtained in the decays[11] $\psi'\to\psi\pi\pi$ and[4] $\Upsilon'\to\Upsilon\pi^+\pi^-$ (as shown in Fig. 8a) which peak at high dipion masses[12]. In addition Table II lists $\Upsilon''\to\Upsilon\pi^0\pi^0$ and $\Upsilon''\to\Upsilon'\pi^0\pi^0$ candidates, which are consistent with the number expected from the charged π analysis.

INCLUSIVE PHOTONS

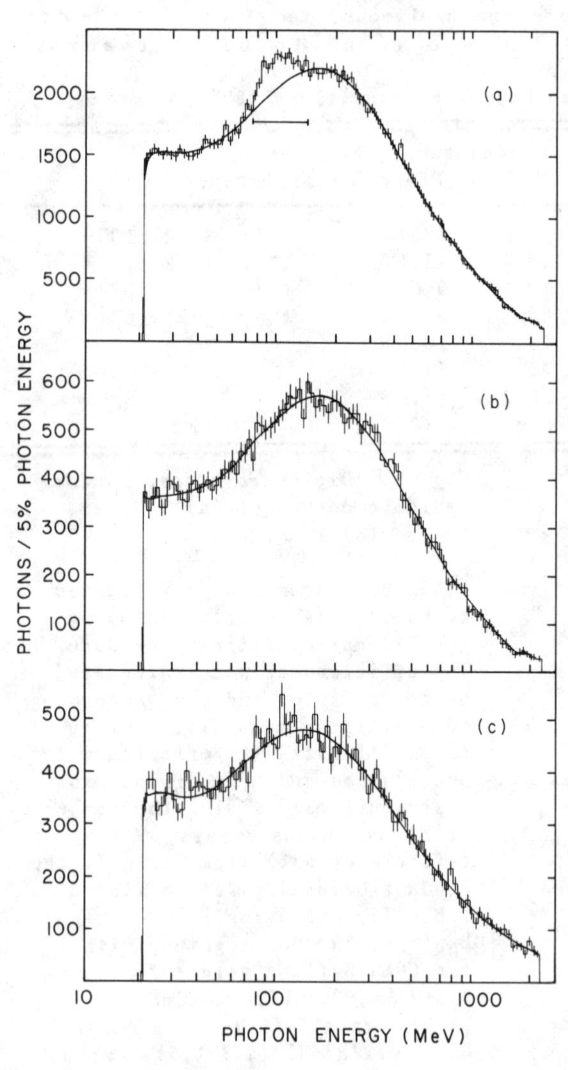

All potential models[13] predict the electric dipole (E1) transitions from the 3^3S to the 2^3P states of the Υ system. Indirect evidence for these P-wave states was inferred from a study of thrust distributions[14] for the Υ, Υ', and Υ''. These transitions have now been directly observed for the first time in the inclusive photon spectrum[15] from the decay $\Upsilon''\to\gamma\chi_b'$ (they have also been observed in exclusive final state decays[16] - see below). The observation of photons in an hadronic event is made difficult by the high average particle multiplicity (~16).

Fig. 9. The inclusive photon spectra for (a) the Υ'', (b) the Υ, and (c) the continuum. The bar on the top spectrum indicates the region excluded from the polynomial background fit. The solid curves on all plots are the polynomial fits to the spectra.

The photon finding algorithm has an overall recovery efficiency of

17% for photons in the 100 MeV region. The observed photon spectra for the Υ'', Υ and continuum regions are shown in Fig. 9, from which it is apparent that there is an excess of photons in the 100 MeV region for Υ'' events (Fig. 9a). We have used the photon spectra of Figs. 9b and 9c to model the spectrum for Υ'' events (excluding the electric dipole and hadronic cascade contributions), since the expected Υ'' spectrum is due to π^0's from two and three jet events. From the thrust analysis[14] we have determined that a combination of 58.4% Υ spectrum and 41.6% continuum spectrum models the Υ'' background spectrum. The contribution from π^0 cascades of the Υ'' is small (15% of the observed effect) and can be calculated from the known branching ratios. The background curve (solid line) in Fig. 9a has been computed in that way, where the Υ and continuum spectra used were polynomial fits to the Υ and continuum spectra (the solid curves in Figs. 9b and 9c). This proceedure leads to an excess of ~2,000 photon events in the 100 MeV region over a background of ~37,500 events, a ten sigma effect.

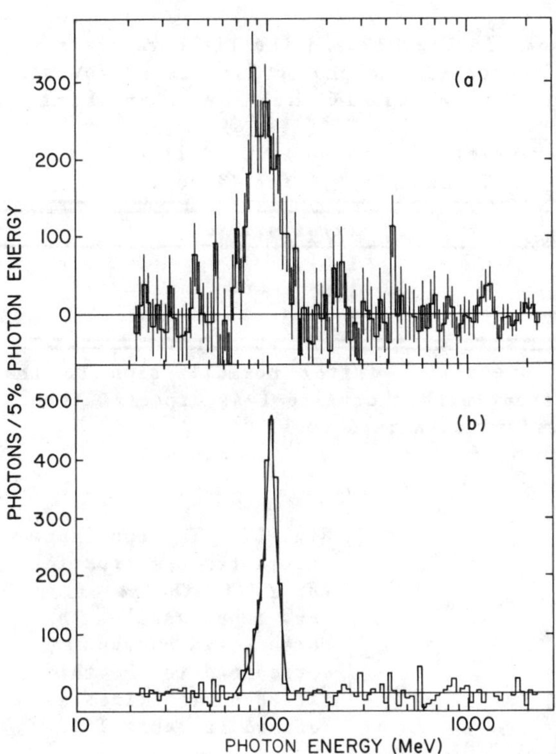

In order to reduce the dependence on the poor statistics of the Υ and continuum data samples, we have also fit the entire Υ'' photon spectrum to a polynomial, excluding the region of the excess. In this case the excess is 2289 photon candidates, and is stable against changes of the excluded region. While most of the photons from the π^0 cascade decays have been subtracted in this way, more careful analysis shows that 130 candidates are still present in the subtracted spectrum of Fig. 10a (they are subtracted in the final analysis).

Fig. 10. (a) The subtracted photon spectrum for the Υ'', (b) the MC computed resolution function for ~100 MeV photons.

Comparison of the excess with the expected resolution function obtained by superimposing ~100 MeV Monte Carlo generated photons on real hadrons is shown in Fig. 10b, and leads to the conclusion that the observed excess of events is inconsistent with a single photon line. In addition, the Monte Carlo method was checked against real π^0 events (see Fig. 11) and Monte Carlo π^0 events, from which an energy scale for ~100 MeV photons was obtained.

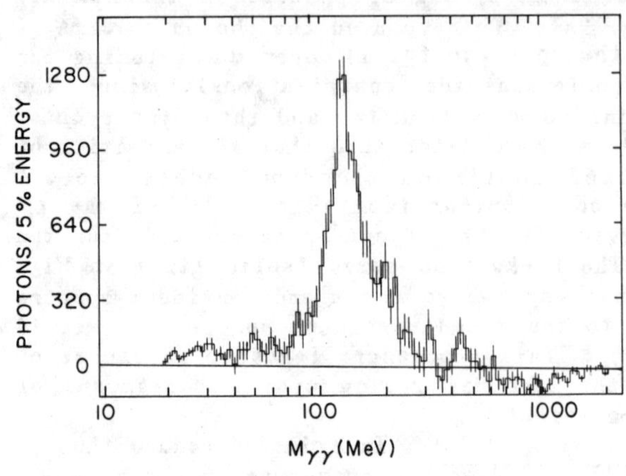

Fig. 11. The diphoton mass spectrum, for real events, after the combinatorial background subtraction, showing a clear π^0 mass peak.

A fit to three lines is shown in Fig. 12 and the fit parameters are available in Table III. The cog of the photon line is 98 MeV with the fine structure splitting of 16 MeV between lines.

Table III The fit parameters for three assumed lines. The c.o.g. is 97.7±1 MeV, and χ^2=16.4 for 14 dof.

Final State	Photon Energy (MeV)	$\Gamma/k^3(2J+1)$
$^3P_2$	84.3±1.9	1.0±.3
$^3P_1$	99.0±3.6	1.0
$^3P_0$	115.0±4.2	1.4±.5

In addition the measured intensities (after normalization to the $k^3(2J+1)$ factor) are consistent with the value 1 as expected (where the second line was arbitrarily normalised to 1).

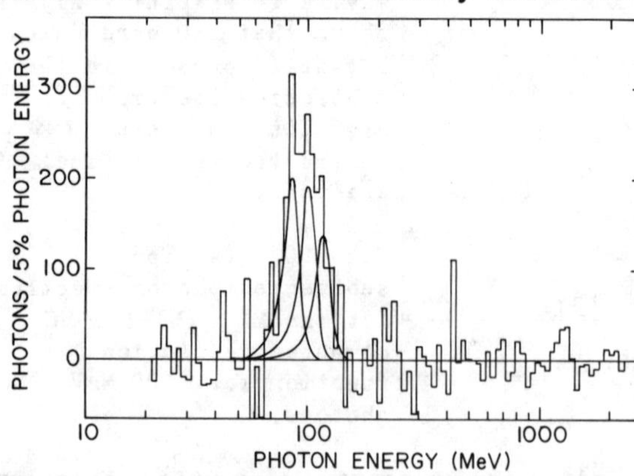

Fig. 12. The continuum subtracted spectrum for the Υ'' (with the error bars suppressed). The three solid curves correspond to the three fitted photon lines as defined in Table III.

The Monte Carlo also yields the recovery efficiency, and thus the branching ratio of:

$$BR(\Upsilon'' \to \gamma\chi_b') = (34 \pm 3 \pm 3)\%$$

which when combined[10] with $\Gamma_{ee}(\Upsilon'')$, $B_{\mu\mu}(\Upsilon'')$ and the hadronic decays discussed above gives $\Gamma(\Upsilon''\to\gamma\chi_b') = 6.5\pm0.8\pm0.6$ KeV.

EXCLUSIVE PHOTON DECAYS OF THE Υ''

We have searched for double photon cascade decays[16] of the Υ'' by searching for events of the type $\gamma\gamma(e^+e^-$ or $\mu^+\mu^-)$. Assuming an isotropic distribution for the photons, and $1+\cos^2\theta$ for the dileptons, we find the acceptance to be 17% and 12% for the dielectron and dimuon events respectively. The criteria used to obtain these events are described in ref. 16, in brief, events were initially scanned for nearly collinear dielectrons ($<17^o$) or dimuons ($<15^o$) and were additionally required to contain between 2 and 5 energy clusters (>7 MeV) in the calorimeter. These events were then handscanned for photon showers which were required to have more than 50 MeV for each photon with a sum between 150 MeV and 1,200 MeV. For the dielectron events one can carry out a 4C fit, the resultant dielectron invariant mass spectrum is shown in Fig. 13.

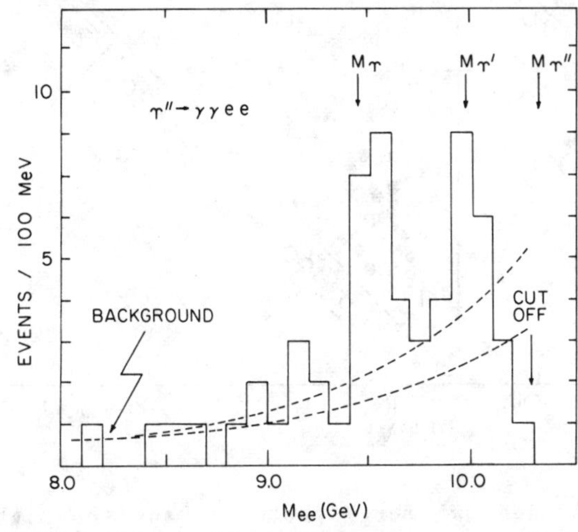

Fig. 13. The dielectron invariant mass as obtained from a 4C fit, showing clear enhacement at the Υ and Υ' masses. The dashed lines represent the range of the estimated background.

It is quite clear that there are double radiative Υ'' transitions present from $\Upsilon''\to\gamma\chi_b'\to\gamma\gamma\Upsilon$(to e^+e^- or $\mu^+\mu^-$) and $\Upsilon''\to\gamma\chi_b'\to\gamma\gamma\Upsilon'$(to e^+e^- or $\mu^+\mu^-$) decays. Figure 14 shows the scatter plot of the higher photon energy versus the lower photon energy for the sample of dielectron (14a) and dimuon (14b) events. The solid lines on the plot indicate the regions for transitions to the Υ and Υ' (including Doppler broadening). It is quite clear that the data cluster within the 2σ bands (dotted lines on the scatter plots) at 100 MeV for the lower energy photon and 230 or 760 MeV for the higher energy photon as expected from double radiative Υ'' decays via the χ_b'. We have studied the backgrounds from QED processes (e.g. $e^+e^-\to\tau^+\tau^-\to(\mu\nu\nu)(\rho\nu)\to\mu\pi\gamma\gamma$ where the π is misidentified) using MC and continuum data, and estimate the background to be less than two events. The π^o cascade decays (where two of the resultant photons are undetected) are calculated to contribute only two events to the entire scatter plot. Thus we consider the $\gamma\gamma\mu^+\mu^-$ sample to be essentially background free. The background problem is more severe in the $e^+e^-\gamma\gamma$ case. The dominant background is from the double Bremsstrahlung Bhabha process. Various methods for estimating this background lead to consistent results of 6±2 events within the $\Upsilon''\to\gamma\gamma\Upsilon$ band and 4±2 events within the $\Upsilon''\to\gamma\gamma\Upsilon'$ band.

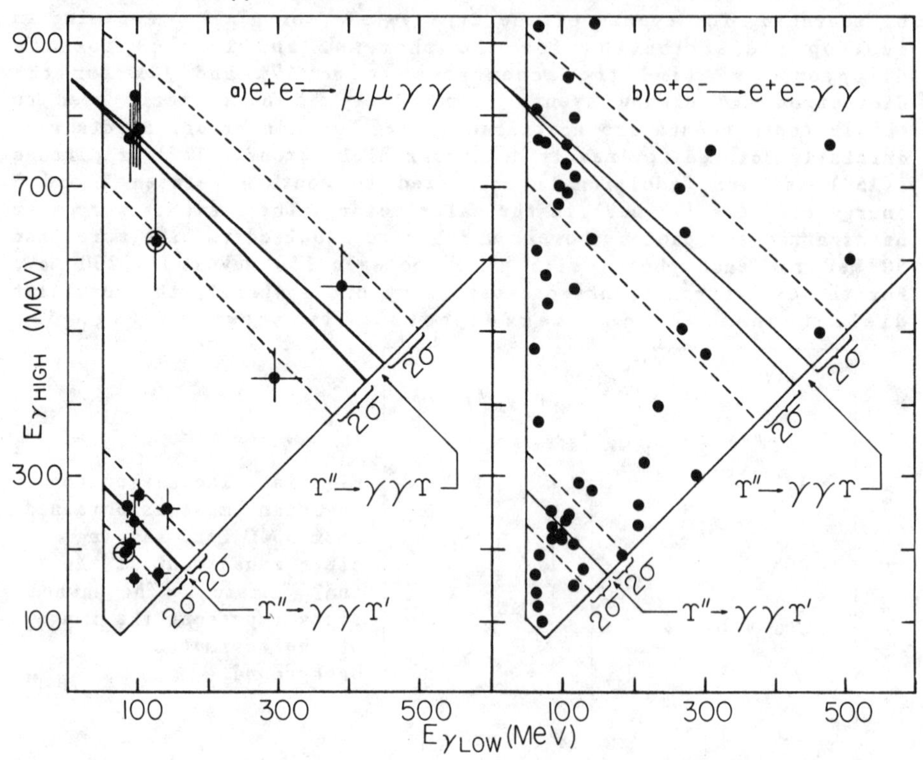

Fig. 14. Scatter plots of the low energy photon versus the high energy photon for (a) $\Upsilon''\to\gamma\gamma\mu^+\mu^-$ events, and (b) $\Upsilon''\to\gamma\gamma e^+e^-$. The outer solid line boundaries represent the kinematic limits, the diagonal solid lines represent the regions for $\Upsilon''\to\gamma\gamma\Upsilon'$ and $\Upsilon''\to\gamma\gamma\Upsilon$ (showing the range due to Doppler broadening). The dashed lines indicate the 2σ bands obtained from the resolution function. The encircled points in (a) were measured with the high energy γ in a detector region of slightly degraded energy resolution.

From the observed events we derive the product branching ratios:

$$BR(\Upsilon''\to\gamma\chi_b')\times BR(\chi_b'\to\gamma\Upsilon')=6.2\pm2.8\%$$

$$BR(\Upsilon''\to\gamma\chi_b')\times BR(\chi_b'\to\gamma\Upsilon)=4.5\pm1.6\%$$

which are consistent with theoretical predictions[17] of 3.3% and 2.8% respectively. We can also derive an upper limit from events observed in the 400-500 MeV region for double radiative decays via the χ_b:

$$BR(\Upsilon''\to\gamma\chi_b)\times BR(\chi_b\to\gamma\Upsilon)<2.7\%\ (90\%\ C.L)$$

Assuming that three lines are present (consistent with the resolution of 3.9%/ $E^{1/4}$ obtained for isolated photon showers), and constraining the $\gamma\gamma$ energy to the appropriate mass difference (i.e. 324 MeV or 856 MeV) we find the three photon energies to be 84±3±2, 99±2±2 and 119±5±2 MeV. These values are in good agreement with those obtained in the inclusive photon analysis (Table II) leading to the masses for the χ_b' states shown in Table IV.

Table IV Summary of $\Upsilon''\to\gamma\chi_b'$ transition from the inclusive and exclusive photon analyses.

State	Photon Energy (MeV) Inclusive	Exclusive	Masses (GeV)
2^3P_2	84.3±1.9	84±3	10.238±.002
2^3P_1	99.0±3.6	99±2	10.223±.003
2^3P_0	115.0±4.2	119±5	10.208±.005

We have also plotted the events in the 100 MeV region for both the inclusive and exclusive photons in Fig. 15. Note that the normalizations are different, but that the agreement is quite remarkable, adding strength to the claim that the χ_b' states have been observed and measured.

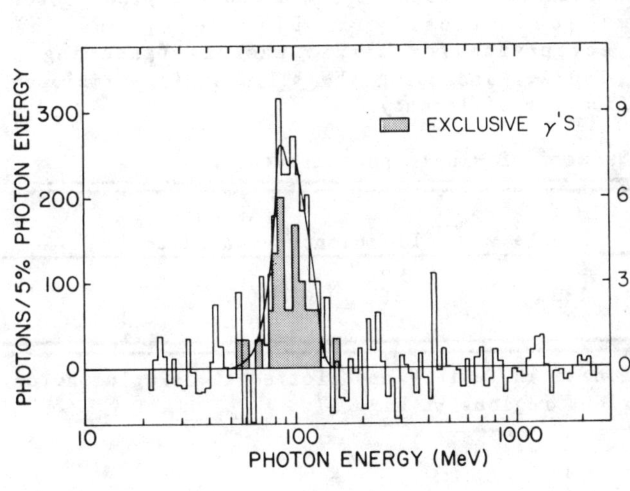

Fig. 15. Superposition of the photon spectra for $\Upsilon''\to\gamma\chi_b'$ for the inclusive and exclusive photon analysis. The shaded histogram is from the exclusive analysis, the other from the inclusive analysis (see Fig. 12). The solid line is the fit to the inclusive photon spectrum for three photon lines. The left vertical scale is for inclusive photons, the right is for exclusive photons.

AXION SEARCH

Original attempts to solve the 'strong CP' problem in a natural way lead to the prediction of a light neutral pseudoscalar particle named the axion[18], whose signature $\Upsilon\to\gamma a$ could be readily distinguished in the CUSB detector[19]. We have since learned that axions may indeed be 'invisible'[20]. A relatively model independent prediction for the product of branching ratios of $\Upsilon\to\gamma a$ and $J/\psi\to\gamma a$ can be made for the 'visible' axions, which with reasonable assumptions is $(1.6\pm0.3)\times10^{-8}$. The signature for the decay $\Upsilon\to\gamma a$ would be a single photon of energy $E_\gamma=(M^2-M_a^2)/2M$. Since the axion lifetime is sensitive to the axion mass, we are only sensitive to axions of mass less than 10 MeV (for which the axions decay outside our detector).

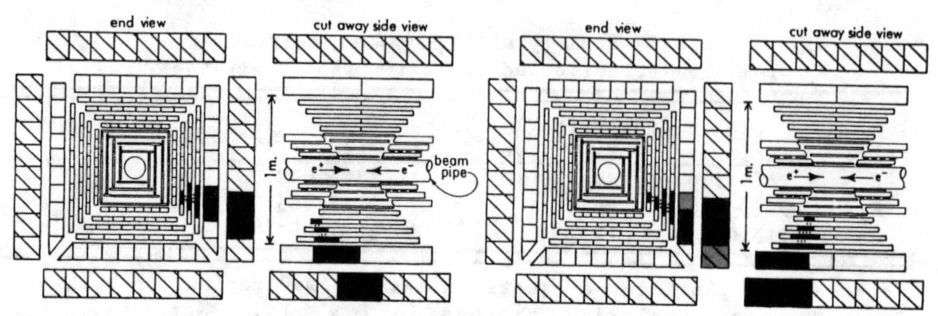

Fig. 16.(a) Rejected photon candidate due to vertex cut, (b) accepted photon candidate with E_γ=1.6 GeV.

The radial segmentation of the CUSB detector makes it well suited to searching for photon showers that originate at the interaction vertex (an example of both rejected and accepted photon candidates is shown schematically in Fig. 16(a) and (b) respectively). The acceptance for single photons (assuming a $1+\cos^2\theta$ distribution) is 32%, including the solid angle efficiency and the photon identification efficiency.

Table V Summary of single photon events.

Region	Hadronic Events	Single γ	Efficiency	Axion Candidates
Υ	16,511	7	.68	0
Υ'''	67,000	133	.66	0
Continuum	8,556	29	.57	0

We have summarised the data in Table V and plotted the single photon candidates for the various regions in Fig. 17.

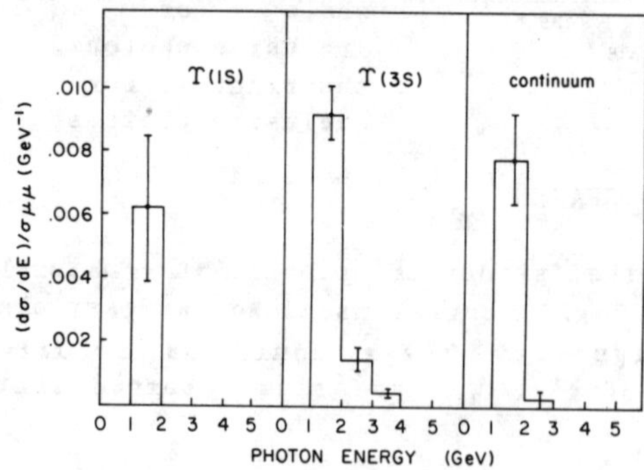

Fig. 17. Single photon spectra for Υ, Υ''' and continuum data relative to $\sigma_{\mu\mu}$.

Note that there are no candidates within four standard deviations (FWHM of the energy resolution is 6% for 5 GeV photons) of the beam energy. Using the upper limit for $J/\psi \to \gamma a$ of 1.4×10^{-5} and our combined values for the Υ and Υ'', we obtain a value for the product branching ratio:

$$BR(\Upsilon \to \gamma a) \times BR(J/\psi \to \gamma a) < 0.6\times10^{-9} \text{ (90\% CL)}$$

which is at least an order of magnitude below the theoretical prediction, finally putting to rest the 'visible' axion.

ACKNOWLEDGEMENTS

I would like to thank J. Lee-Franzini for helpful discussions during the preparation of this talk, and the CESR staff for the impressive improvements in running conditions. This work is supported in part by the National Science Foundation.

REFERENCES

1. The members of the CUSB collaboration are: T.Böhringer, P.Franzini, K.Han, G.Mageras, D.Peterson, E.Rice, J.K.Yoh (Columbia), J.E.Horstkotte, C.Klopfenstein, J.Lee-Franzini, R.D.Schamberger Jr., M.Sivertz, L.J.Spencer, P.M.Tuts, (Stony Brook), R.Imlay, G.Levman, W.Metcalf, V.Sreedhar (LSU), G.Blanar, H.Dietl, G.Eigen, E.Lorenz, F.Pauss, H.Vogel (MPI), S.W.Herb (Cornell).
2. J.J.Aubert et al., Phys. Rev. Lett. 33, 1404 (1974), J.E.Augustin et al., ibid. 33, 1406 (1974).
3. S.W.Herb et al., Phys. Rev. Lett. 39, 252 (1977).
4. G.Mageras et al., Phys. Rev. Lett. 46, 1115 (1981), J.Mueller et al., ibid. 46, 1181 (1981), B.Niczyporuk et al., Phys. Lett. 100B, 95 (1981).
5. T.Böhringer et al., Phys. Rev. Lett. 44, 1111 (1980), G.Finocchiaro et al., ibid. 45, 222 (1980).
6. D.Andrews et al., Phys. Rev. Lett. 44, 1108 (1980), 45, 219 (1980).
7. E.Rice et al., Phys. Rev. Lett. 48, 906 (1982), P.Franzini and J.Lee-Franzini, Phys. Rep. 81, 239 (1982).
8. P.M.Tuts, contributed paper to the XXIth International Conference on High Energy Physics, Paris (1982).
9. G.Mageras, contributed paper to the XXIth International Conference on High Energy Physics, Paris (1982).
10. R.D.Schamberger, Proceedings of the 1981 International Symposium on Lepton and Photon Interactions at High Energies, Bonn, edited by W.Pfeil (Physikalisches Institut Universitat, Bonn, 1981) p.217.
11. M.Oreglia et al., Phys. Rev. Lett. 45, 959 (1980).
12. L.S.Brown and R.N.Cahn, Phys. Rev. Lett. 35, 1 (1975), T-M.Yan, Phys. Rev. D22, 1652 (1980).
13. There are many potetial models described in the literature, we only list one: E.Eichten et al., Phys. Rev. Lett. D21, 303 (1980).
14. D.Peterson et al., to be published in Phys. Lett. B.

15.K.Han et al., submitted to Phys. Rev. Lett..
16.G.Eigen et al., submitted to Phys. Rev. Lett..
17.Y-P.Kuang and T-M.Yan, Phys. Rev. D24, 2874 (1981).
18.R.D.Peccei and H.R.Quinn, Phys. Rev. Lett. 38, 1440 (1977),
S.Weinberg, ibid. 40, 223 (1978),
T.Goldman and C.M.Hoffman, ibid. 40, 220 (1978).
19.M.Sivertz et al., to be published in Rapid Communications-
Phys. Rev. D (August,1982).
20.M.Dine, these proceedings.

e^+e^- COLLISIONS: RESULTS FROM PETRA

E. Hilger
Physikalisches Insitut der Universität Bonn

ABSTRACT

A selection of most recent results obtained by the five experiments CELLO, JADE, MARK J, PLUTO, and TASSO at PETRA is presented. The many interesting new results include, for example, the observation of effects of 2nd order QCD, details of inclusive hadron production, and the measurement of the forward-backward asymmetry in lepton pair production due to weak and electromagnetic interference.

INTRODUCTION

Since the time of the BONN conference last summer the total amount of integrated luminosity collected at the e^+e^- storage ring PETRA has been tripled to now more than 50 pb^{-1} per interaction region. The remarkable performance of the machine with the mini-beta focussing system has got the experimenters used to 500 nb^{-1}/day and more and has created a wealth of new data. At present PETRA energies such luminosities correspond to yields of about 30 muon pair (and τ pair) events and of roughly 120 hadronic annihilation events per day and per experiment.

There are five detectors at PETRA. Four of them - JADE, MARK J, the improved PLUTO, and TASSO - are taking data at present.

This is a report on results that have become available very recently. Complying with the motto of this conference, nothing that has been shown already at the BONN conference[1] shall be repeated. However, there is much new material and I have to be brief on most of the issues presented.

The topics to be considered are

* e^+e^- - hadrons
  - total cross section
  - charge of quark jets
  - tests of 2nd order QCD
  - inclusive hadrons
  - π^o production
  - $\pi^\pm$ production
  - identified charged hadrons
* search for scalars
* beauty life-time
* electroweak interference.

Since this is my personal selection, my apologies go to my friends whose work I have omitted from my presentation. Reports given at this years meetings at MORIOND[2] may serve as supplements.

TOTAL CROSS SECTION FOR $e^+e^- \to$ hadrons

The total cross section for e^+e^- annihilation into hadrons, σ_{tot}, is a quantity of fundamental importance. In the quark-parton-model (QPM) σ_{tot} measures the sum of the squares of the quark charges e_q

$$R = \sigma_{tot} / \sigma_{\mu\mu} = 3 \sum_q e_q^2$$

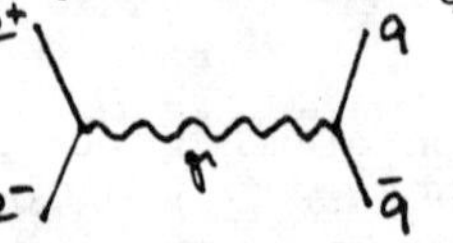

where using the quantity R instead of σ_{tot} conveniently removes the major 1/s dependence of the cross section. At high energies the simple QPM picture is modified by effects of the strong and the weak interactions:

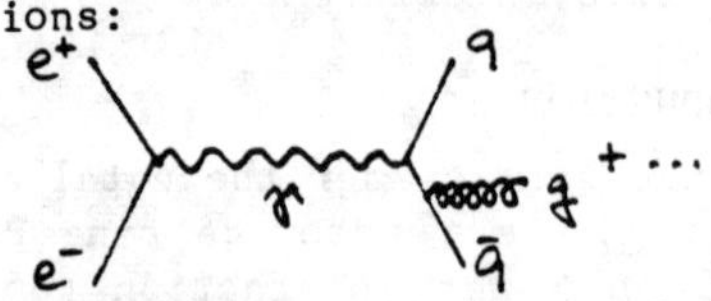

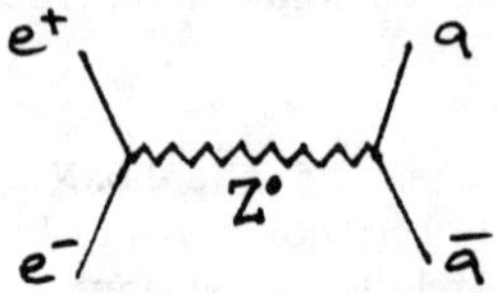

These modifications will change R from its QPM value. Also, a possible structure of the quarks would show up as an energy dependence of R. At energies around $\sqrt{s}$ = 34 GeV, where the bulk of the data has been collected at PETRA, such interesting phenomena will appear in R at the level of a few percent:

QCD (1st order)	(6-8) %,
α_s running	1.5 %,
weak interaction ($\sin^2\theta_W = 0.23$)	3 %.

Thus the expected modifications are small and their detection requires measurements of R with high precision.

Each PETRA experiment now typically has analysed 16,000 hadronic events, 3000 at 14 GeV, 2000 at 22 GeV, 10,000 at 34 GeV. The statistical errors are therefore small, 1 to 2 %. The experimental uncertainty of R measurements is dominated by systematic errors, which used to be on the order of 7 to 10 % typically.

All PETRA groups are trying to understand and reduce the systematic uncertainties. Recently the TASSO group has presented precision data on R with overall systematic errors of 5% [3]. A careful study of all possible sources of error in the quantities that make up the total cross section

$$\sigma_{tot} = 1/L \cdot (N_{observed} - N_{BG})/A,$$

has led to the following error assignments at $\sqrt{s}$ = 34 GeV:

		system. error
N_{BG}	background events	1.5%
L	luminosity	3.5%
A	acceptance	
	trigger	1.0%
	rad. corr's	2.5%
	selection criteria	2.5%
total systematic error (added up in quadrature)		5.2%.

Of this, 4.5% are overall normalization and 2.7% point-to-point uncertainty. Separation of overall normalization error and the point-to-point uncertainty is very important when studying the s-dependence of R.

The new TASSO data - shown in Fig. 1 together with older data from the other PETRA experiments and the low energy data[1] - are consistent with R being constant from 14 GeV upwards.
The average value for this energy range is $R = 4.01 \pm 0.03 \pm 0.20$.

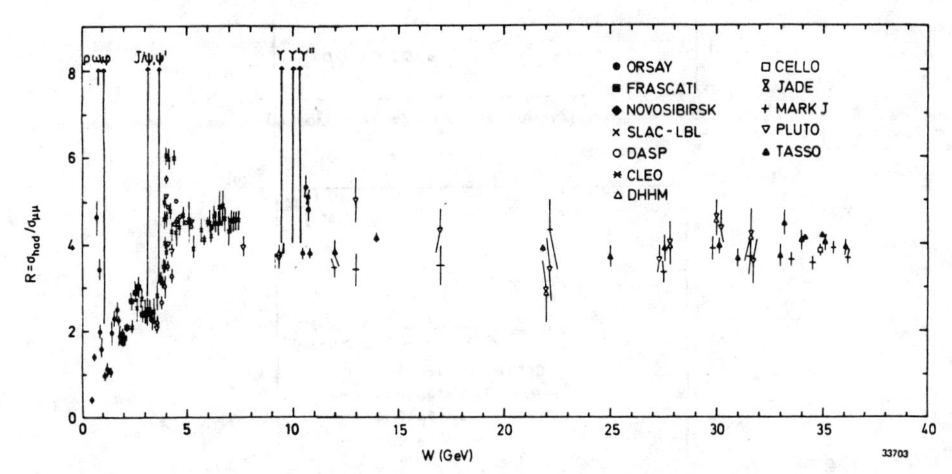

Fig. 1 : Compilation of R values from the different PETRA experiments together with data from lower energies. Only statistical errors are shown.

By fitting to the form factor parametrization

$$R(s) = R_o \, (1 \mp s/(s-\Lambda_\pm^2))^2$$

limits on the size of the quarks are obtained of $\Lambda_+ > 186$ GeV and $\Lambda_- > 363$ GeV. These limits are very similar to those obtained for lepton structure and correspond to a length of less than 10^{-16}cm.

Fits to the R-data from 14 to 36.7 GeV have been made by TASSO using the complete theoretical expression[5] for R accounting for

- QCD (up to 2nd order) $\langle C_2 = 1.39(\overline{MS})\rangle$,
- α_s running,
- weak interactions (Glashow, Weinberg, Salam) [4]
- finite quark masses.

The χ^2 has been defined such as to account fo the separation of systematic uncertainty in point-to-point and normalization errors:

$$\chi^2 = \sum_i \frac{(R_i - R(\alpha_s, \sin^2\Theta_W))^2}{\sigma_i^2}$$

where R_i are the measured values with errors σ_i (statistical and point-to-point systematic).

A two-parameter fit yielded α_s (Q^2 = 1000 GeV^2) = $0.18 \pm 0.03 \pm 0.14$ and $\sin^2\Theta_W = 0.40 \pm 0.15 \pm 0.02$. Setting $\sin^2\Theta_W = 0.228$ (PDG[6]) the fit yielded $\alpha_s = 0.24 \pm 0.05 \pm 0.13$. These results are consistent with previous measurements and new preliminary ones from JADE and MARK J.

The latter results as well as the TASSO data are shown in Fig. 2.

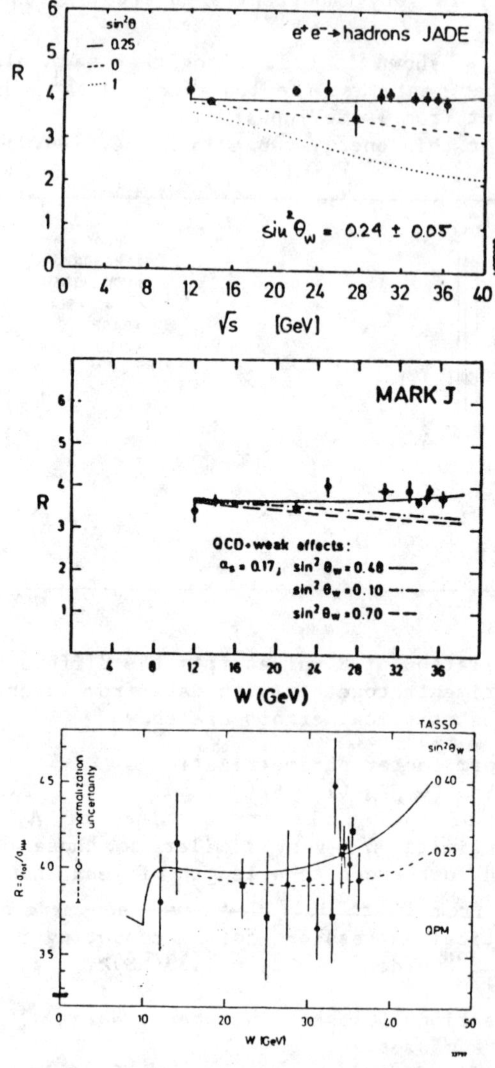

Fig. 2 : The ratio R as measured by JADE, MARK J, and TASSO together with the results of best fits (full lines).

Summing up this paragraph: Measuring R with overall 5% systematics and point-to-point errors less than 3% means progress. Covering a wide range in c.m. energy between flavor thresholds with good quality data is very important when studying modifications of R via its s dependence. However, going to higher energies will enhance the effects and will make measurements of R more decisive.

CHARGE PROPERTIES OF QUARK JETS

It is a well established feature of hadron production in e^+e^--annihilation that the bulk of the events shows the signature predicted by the QPM: two back-to-back jets reflecting the primary production of a pair of quarks which subsequently fragment into hadrons.

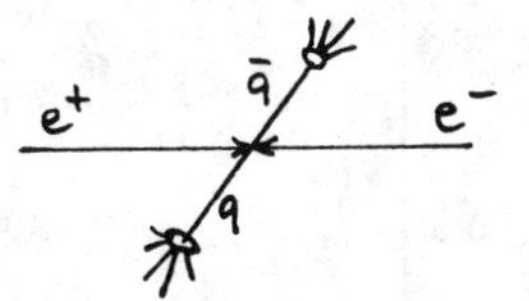

For some properties of the primary quarks it has been shown that they are preserved in the hadronization step and can be read off the final state of hadrons: We know (i) that these quarks have spin 1/2 from the angular distribution of the jet axis[7] and (ii) that these quarks are oppositely charged from the observed longe range charge correlations in rapidity space[8].
The ultimate goal is to measure all quantum numbers of the primary quarks and thus identify them. A new study of the charge distribution of quark jets carried out by PLUTO is an interesting attempt in that direction.

A simplified picture of the jet development from $q\bar{q}$ in Fig. 3 may help understand the quantity "jet charge", Q_{jet}, which is defined as the sum of all primary particles in a jet with positive rapidity.

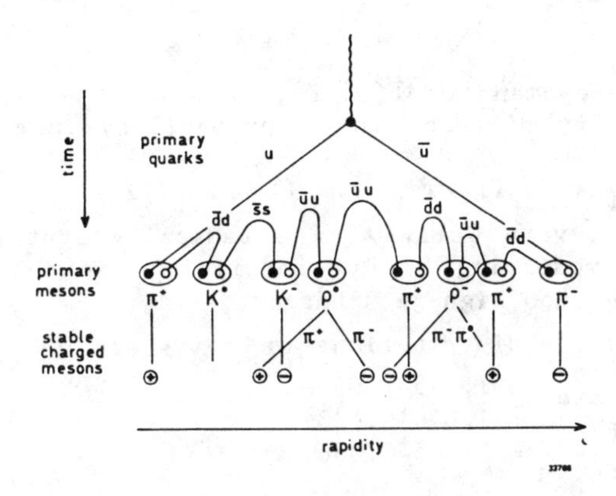

Fig. 3 : Schematic graph of jet development in $e^+e^- \rightarrow q\bar{q}$.

A measurement of this quantity is impeded by secondaries that cross the jet boundary at y = 0 and by measurement distortions. Thus in general $Q_{jet}^{observed} \neq Q_{jet}^{true}$.
The PLUTO study shows that Q_{jet}^{true} may be determined with the help of a good fragmentation model.
The PLUTO analysis is based on 750 hadronic events of the 2-jet category, obtained at c.m. energies around 30 GeV, with the event axis well inside the detector acceptance to minimize particle losses. For these data the jet charge distribution is shown in Fig. 4 as the histogram.

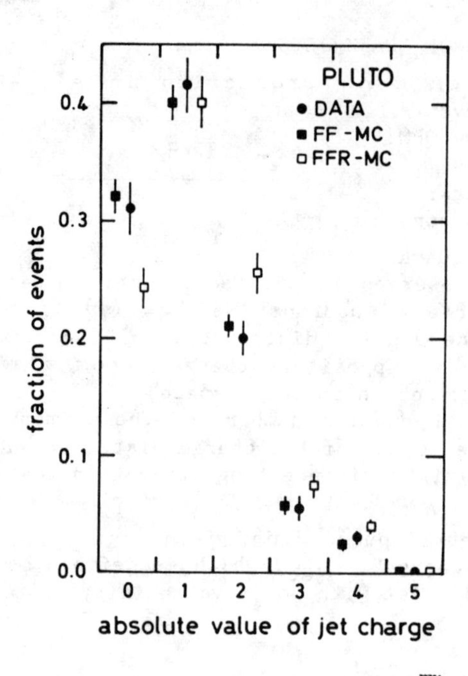

Fig. 4 : Distribution of the absolute value of the jet charge. The data are compared to the FF model and to its randomized version FFR.

The popular Field-Feynman model[9] (FF), in which the strict order in rapidity is partially disturbed by a probability function called fragmentation function,

$$f(z) = 1 - a_F + 3a_F(1-z)^2 , \quad z = (E + P)_h/(E + P)_q, \quad a_F \approx 0.57,$$

reproduces the data very well. A more radically statistical model (FFR) obtained by modifying FF by randomizing the charge order of primary mesons gives too high jet charges.

The mean values for the distributions in Fig. 4 are

$$\langle|Q_{jet}|\rangle_{data} = 1.045 \pm 0.034$$
$$\langle|Q_{jet}|\rangle_{FF} = 1.044 \pm 0.023$$
$$\langle|Q_{jet}|\rangle_{FFR} = 1.243 \pm 0.036.$$

PLUTO now transforms the observed jet charge into the true jet charge using the FF model. In such a model the true jet charge is, of course, known in each simulated event. Since the FF model only allows one quark charge to be exchanged across the jet boundary the true jet charge can either be 0 or ± 1. From the model calculations it is obtained that jets with charge ± 1 are measured with $\langle|Q_{jet}^{obs}|\rangle_{\pm 1}$ = 1.138 $\pm$ 0.31, and those with charge 0 as $\langle|Q_{jet}^{obs}|\rangle_0 = 0.932 \pm 0.033$.

These values are used to find for the actual data by means of a linear interpolation

$$\langle|Q_{jet}^{true}|\rangle_{data} = 0.55 \pm 0.17 \pm 0.19.$$

This small value of the jet charge implies a strong correlation to the charge of the primary quark. It strongly supports the picture adopted by Field & Feynman[9] that only one quark line links one jet to the opposite one.

PLUTO has looked also for charge correlations. Two-particle charge correlations were studied, within one jet and between the leading particles in opposite jets. Both, short range and long range correlations are seen confirming previous findings of TASSO[8].

Further demonstration of the evidence is given in Fig. 5 by the rapidity distribution of the quantity $\tilde{A}$, which is defined on the basis of the probability of charge compensation between particles at different rapidities, $P(y',y)$, and integrated over all test particles in the high negative rapidity interval indicated in the figure. The definition has been chosen such that for zero charge compensation $(P = 0) \rightarrow (\tilde{A} = 0)$.

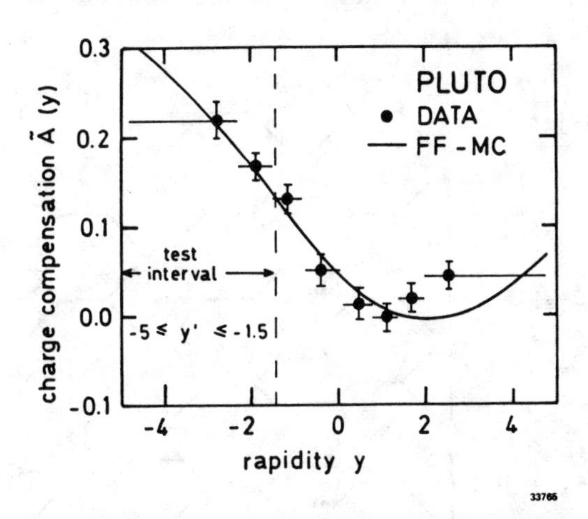

Fig. 5 : Compensation of the charge of particles in the test interval $-5 < y' < -1.5$ as a function of rapidity.

The y distribution of $\tilde{A}$ clearly shows short range and long range correlations.

Furthermore, using its small measured jet charge and the measured leading particle correlation as well as from elaborate Monte Carlo studies PLUTO arrives at the conclusion that the sign of the charge of the primary quark can be determined with a reliability greater than 50%.

EVIDENCE FOR EVENTS WITH 4-JET STRUCTURE

I will now discuss the evidence reported recently by the JADE group[10] for the existence of events with a 4-jet structure.

In addition to the predominant 2-jet events a significant percentage of planar 3-jet events has been found in high energy e^+e^--annihilation into hadrons[11]. These events are generally interpreted as due to hard gluon bremsstrahlung by quarks, as predicted by QCD. However, within this framework (see Fig. 6) higher order diagrams give rise to events with 4 partons in the final state leading to 4-jet events at a rate proportional to α_s^2.

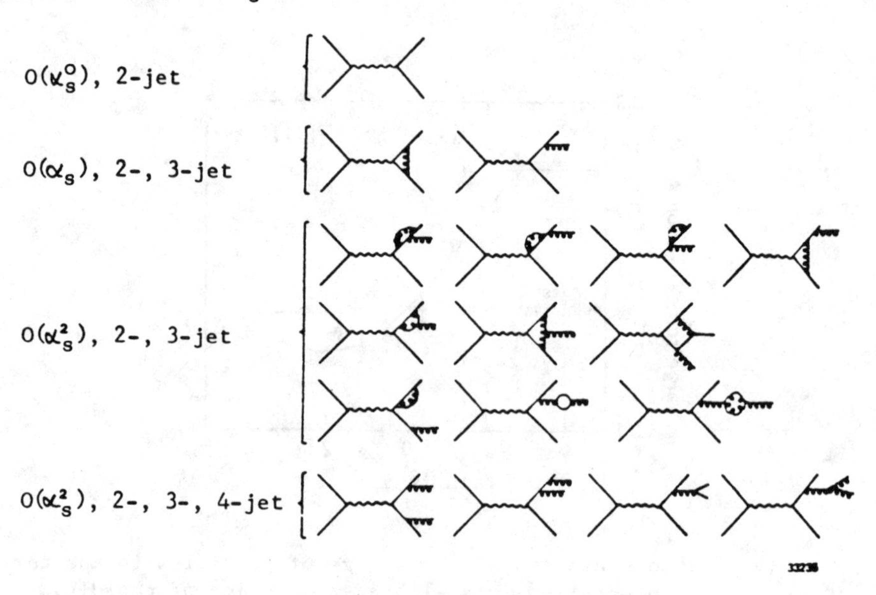

Fig. 6 : Leading and next to leading order QCD diagrams for hadron production.

The two event shape parameters used in the JADE 4-jet analysis are acoplanarity and tripodity. They are defined as

Acoplanarity: $$A = 4 \min \left\{ \frac{\sum_i |\vec{p}_i^{\perp}|}{\sum_i |\vec{p}_i|} \right\}^2 ,$$

$\vec{p}_i$ particle or parton momenta,

$\vec{p}_i^{\perp}$ components perpendicular to plane minimizing the bracket;

Tripodity : $$D_3 = 2 \max \left\{ \frac{\sum_i |\vec{p}_i^T| \cos^3(\vec{n}, \vec{p}_i^T)}{\sum_i |\vec{p}_i|} \right\} ,$$

$\vec{p}_i^T$ projections on plane perpendicular to thrust

$\vec{n}$ unit vector in this plane, maximizing the bracket.

Acoplanarity is standard. Tripodity[12] measures the symmetry of the momentum distribution in the plane normal to the thrust axis.
Before fragmentation, 2-jet and 3-jet events have $A = 0$ and $D_3 = 0$. 4-jet events in general have $A > 0$. As far as tripodity is concerned the events fall in two classes: one has $D_3 = 0$, the other, characterized by one parton (or jet-) axis coinciding with the thrust axis, generally has $0 \leq D_3 \leq 0.324$.

The event sample is selected with the following criteria:
$30 \leq \sqrt{s} \leq 35$ GeV, $E_{vis} \geq 12$ GeV, $|\cos \Theta_{thrust}| < 0.8$,
missing momentum $|\sum_i p_i| < 10$ GeV/c.

It is important to note that the full sample is used for the analysis. No attempt is made to isolate 4-jet candidate events.
For each event 4 axes are determined by means of a cluster algorithm[13]. 2560 events remain for this study.

A and D_3 are evaluated from the reconstructed axes and the events are plotted in the A-D_3 plane as shown in Fig. 7. The two classes of topology become very obvious.

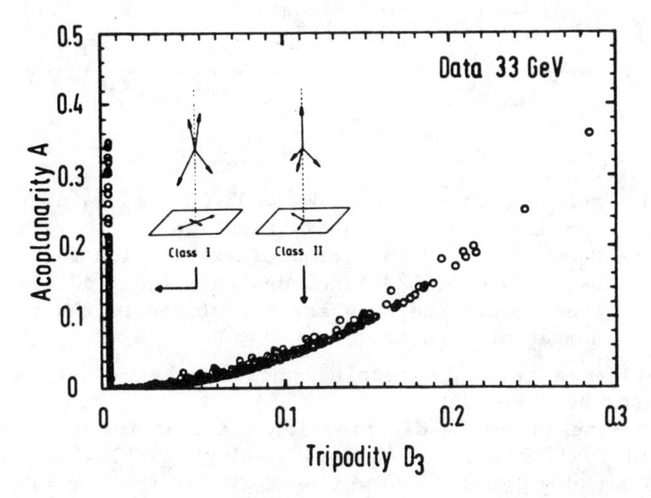

Fig. 7 : Scatter plot of acoplanarity A versus tripodity D_3 for multihadron events (JADE). The data form two bands as expected from the kinematics of four momentum balanced axes. The corresponding configurations in space are indicated.

The projections of the two classes on the D_3-axis and on the A-axis were compared to MC generated 2-jet, 3-jet, 4-jet events. In the MC the 2-jet and 3-jet events are obtained with the LUND generator[14], 4-jet events are produced with the Ali generator[15]. The MC called L23 contains just 2-jet and 3-jet events. In L234 a fraction of 5% 4-jet events have been included.

The tripodity distribution is shown in Fig. 8a. The 2- and 3-jet MC - the dashed line - does not reproduce the data, whereas the 2-,3- and 4-jet MC L234 gives a good description. These findings are more pronounced in the acoplanarity distribution of class 1 events shown in Fig. 8b. The MC L23 gives too small acoplanarity values, while L234 agrees much better.

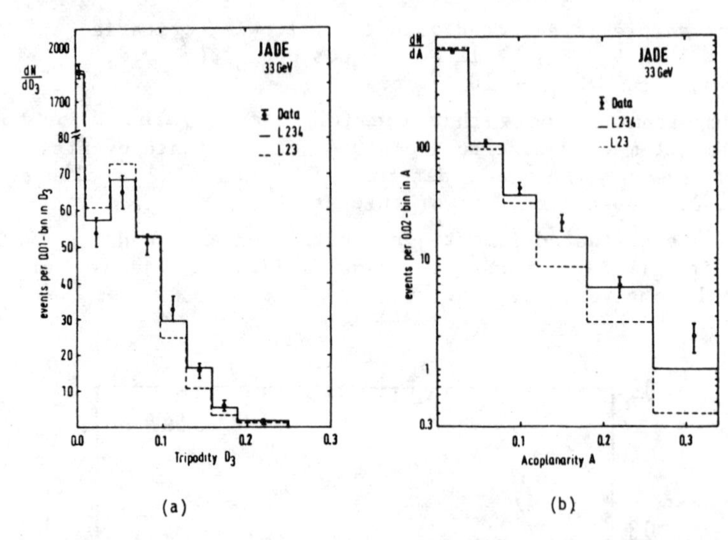

Fig. 8 : Multihadron events at $\sqrt{s}$ = 33 GeV: a) D_3 distribution for all events, b) A-distribution for the events with $D_3 < 0.01$. L23 is the model expectation for 2- and 3-jets alone, L234 includes additional 5% 4-jet events. L23 and L234 are normalized to the total number of events.

From this evidence the JADE people conclude that 4-jet events are needed to describe their data.
From a fit to the A- and D_3-distribution the fraction of 4-jet events is determined to $(7.2\pm1.2)\%$ with very good χ^2. Trying to do without 4-jets gives a bad χ^2 at 33 GeV, while at 22 GeV the fit is good.

Several checks have been made to find possible other sources of the observed effects, without success. Thus, the JADE people find significant deviations from expectations for 2- and 3-jet events alone at $\sqrt{s}$ = 33 GeV. The effects can be explained by inclusion of 4-jet events, for which 2nd order QCD is the most probable origin.

THRUST DISTRIBUTION IN 3-JET EVENTS

With the reduced statistical and systematic uncertainties of the data refined tests on the nature of the field theory for the strong interactions are possible. A new analysis of the thrust distribution in 3-jet events has been carried out by the JADE group[16].
In order to obtain a meaningful comparison with predictions from 2nd order QCD for jet cross sections an appropriate jet definition has to be used.

JADE employed the so-called mass method. The recipe is as follows: form all two particle masses $m_{jk} = (p_j+p_k)^2$ from the momenta p_i in an event, find the smallest $m_{jk} =: m_{lm}$, replace particles l and m by a single pseudoparticle with 4-vector $(|\vec{p}_l| + |\vec{p}_m|, \vec{p}_l+\vec{p}_m)$, repeat this procedure until for all combinations $m^2_{jk}/E^2_{vis} > y_{cut}$, which is set here to 0.04. The final number of pseudoparticles defines the jet multiplicity and their directions are taken as the jet directions.

For events with three reconstructed jets the relative energies $x_i = E_i^{jet}/E_{beam}$ are computed from the angles between the projected jet directions. The largest x_i is called x_{max}. Figs. 9a and b show the scaled cross sections $1/\sigma_{tot} \cdot d\sigma/dx_{max}$, corrected for detector resolution, initial state bremsstrahlung, and for the effects of fragmentation.

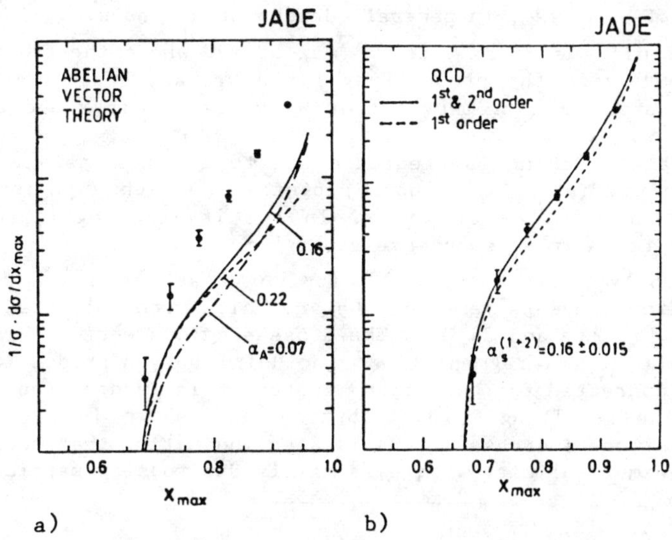

Fig. 9 : x_{max} distribution for jet definition with the mass method, compared with a) 2nd order predictions of an Abelian theory, and b) the 2nd order QCD best fit (and 1st order for the same value of α_s).

In Fig. 9a the lines are predictions from a candidate field theory of strong interactions of Abelian nature[17] which has been invented for the sake of contrast to the commonly adopted non-Abelian QCD. The different lines correspond to different values of the coupling constant in this Abelian theory, the solid line giving the largest possible cross section. Clearly, the yield of 3-jet events in this theory is by far too small to explain the data.

In Fig. 9b we return to QCD. The solid line is the cross section up to 2nd order for $\alpha_s^{(1+2)}$=0.16, which is in good agreement with the data. The 1st order contribution for this value of α_s (dashed line) is, however, not very different and can be made to agree completely by adjusting α_s slightly.

This study is preliminary and a quantitative comparison of the candidate theories is expected to be available soon.

SCALE BREAKING IN INCLUSIVE HADRON PRODUCTION

I now come to discuss inclusive hadron production and I will begin with those features that are obtained without particle identification.

The differential cross section for producing a hadron h with mass m, momentum p and energy E at the polar angle θ can be expressed in terms of two structure functions $\bar{W}_1$ and $\bar{W}_2$, which are closely related to those measured in lepton nucleon scattering[18]

$$s \cdot d\sigma/dx = 4\pi\alpha^2 \cdot \beta x \cdot (m\bar{W}_1 + x/6 \cdot \nu \bar{W}_2),$$

where x is the fractional energy of the hadron h and ν is the energy of the virtual photon viewed in the rest system of h. At high energies ($\sqrt{s} > 12$ GeV), $E \gg m$ and x may be replaced by $x_p = P/P_{beam}$. The functions $m\bar{W}_1$ and $\nu\bar{W}_2$, in general, depend on x_p and s.

The QPM predicts $s \cdot d\sigma/dx = 8\pi\alpha^2 \cdot \sum_q e_q^2 \; D_q^h(x)$, where the fragmentation functions D_q^h give the probability for a primary quark of a certain flavor to produce a primary hadron with fractional x and are independent of s.

In QCD scale breaking is predicted due to gluon emission. As s increases the radiation of gluons depletes the high x region and enhances particle yields at low x. In the leading log approximation this leads to an ln s dependence at high x.

TASSO recently[19] has produced inclusive cross sections from the 16,000 hadronic events mentioned before which were collected at energies $\sqrt{s}$ = 14, 22, and 34 GeV. The cross sections were obtained using the Hoyer-MC[20], FF fragmentation[9] including baryon production[21]. The data were corrected for background, detector influence, and radiative effects. The resulting x_p-distributions are shown in Fig.10. For $x_p < 0.2$ a strong increase with s is seen indicating that the observed increase in multiplicity with s is mainly due to slow particles.

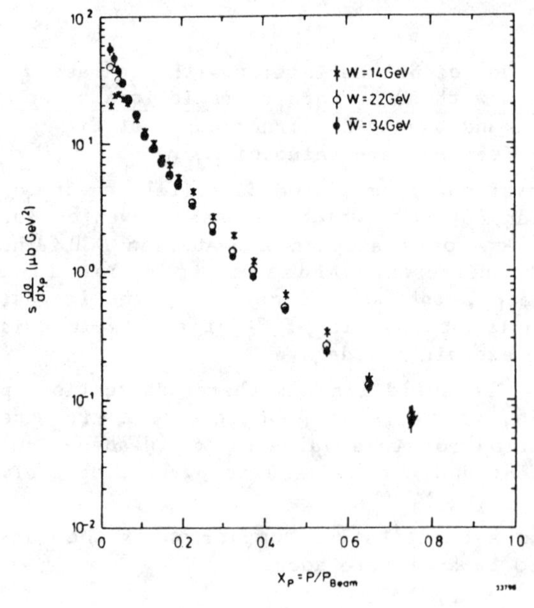

Fig.10 :
x_p distributions of scaled cross sections for inclusive charged particle production from TASSO.
The errors shown include statistical and point-to-point systematic uncertainties.

The s-dependence, in particular at higher x_p, is more clearly seen in Fig.11 where s-distributions are plotted for different x_p-bins.

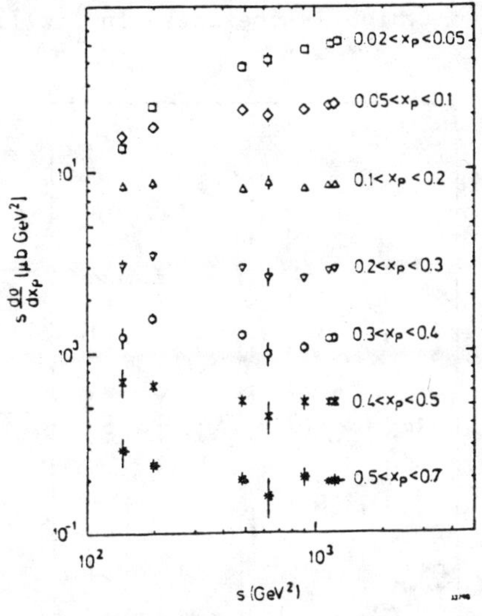

Fig. 11 :
s-distributions of scaled cross sections for inclusive charged particle production (from TASSO).

Fits to the cross sections in the QCD motivated parametrization $s \cdot d\sigma/dx_p = b(1+ c \cdot \ln(s/s_o))$, $s_o = 1$ GeV, yield with increasing x_p monotonically decreasing values for c. Going from 14 GeV to 34 GeV a 20% decrease is observed for $x_p > 0.2$. The data show a violation of scaling at a level of 4 to 10 standard deviations.

Adding the data from MARK I[22] and MARK II[23] obtained at lower energies the effect becomes more pronounced, compatible with an ln s behavior. This is seen from Fig.12. However, there is, of course, an uncertainty in the relative normalization of the experiments.

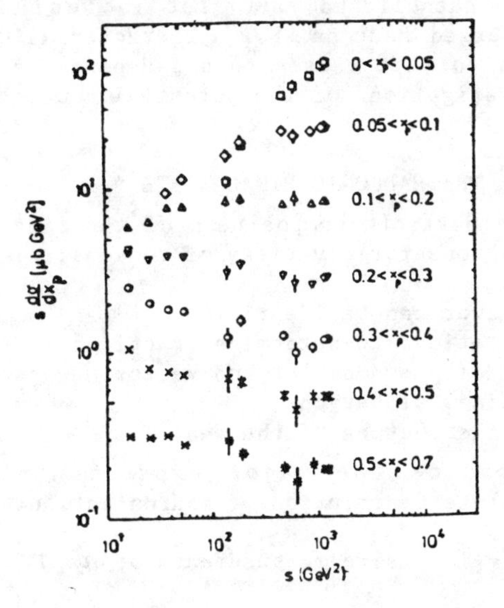

Fig. 12 :
Scaled cross sections for inclusive charged particle production (data from TASSO and from MARK I and MARK II at SPEAR).

Such a normalization problem does not exist for the PLUTO group who has taken data at DORIS as well as at PETRA with the same detector. Within the limitations of their present statistics they also observe an s-dependence in their combined data which is presented in Fig.13.

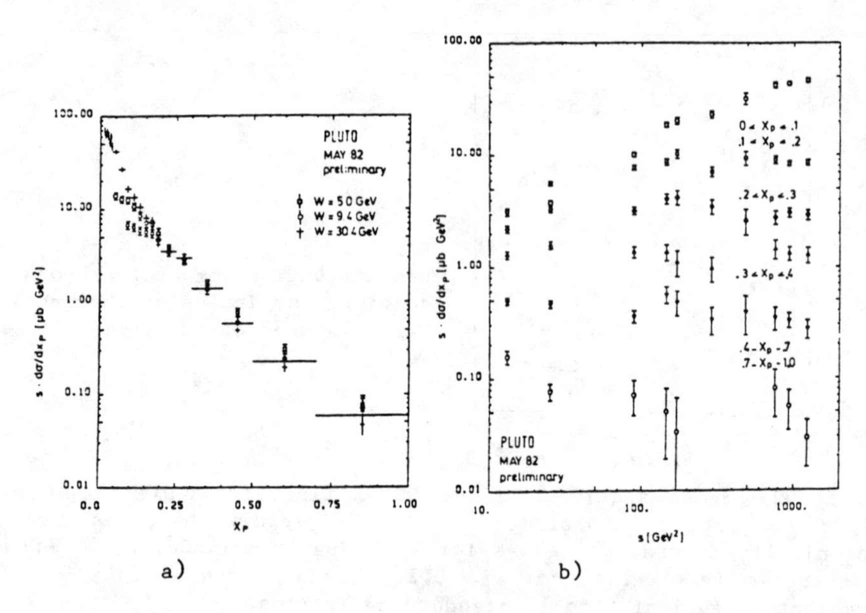

a) b)

Fig. 13: Scaled cross sections for inclusive charged hadron production from PLUTO at DORIS and at PETRA, a) x_p-distribution, b) s-distribution.

It is now an experimentally well established fact that inclusive cross sections of unidentified charged hadrons show a characteristic dependence on the c.m. energy. As to what causes this s-dependence, especially at higher x, more investigations of the potential sources are needed.

PARTICLE COMPOSITION OF THE HADRONIC FINAL STATE

It is very important to study the particle composition of the final state in e^+e^--annihilation into hadrons for a variety of reasons:

(i) to test QCD and scaling,

(ii) to see if the quark flavor can be identified,

(iii) to learn about fragmentation: to determine fractions of mesons and baryons, of pseudoscalar and vector mesons, and of the different kinds of baryons, and to study the flavor structure of the sea.

Particle identification should produce the major experimental evidence necessary for a more complete description of hadronization.

The detectors at PETRA now provide inclusive measurements of π^0, $\pi^\pm$, $K^0/\bar{K}^0$, $p/\bar{p}$, $\Lambda/\bar{\Lambda}$ in a wide x-range.

INCLUSIVE π^0 PRODUCTION

With the high granularity liquid argon detectors of CELLO and TASSO more detailed information has become available recently on neutral pion production. The first successful reconstruction of π^0's in multihadronic events at PETRA energies was achieved by TASSO[24], as reported at BONN[1]. Now CELLO has presented results as well[25].

An essential ingredience for π^0 reconstruction is a very good cluster separation, which in the liquid argon calorimeters is achieved by a fine strip structure. The two detectors have a similar two-cluster separation of around 50 mrad.

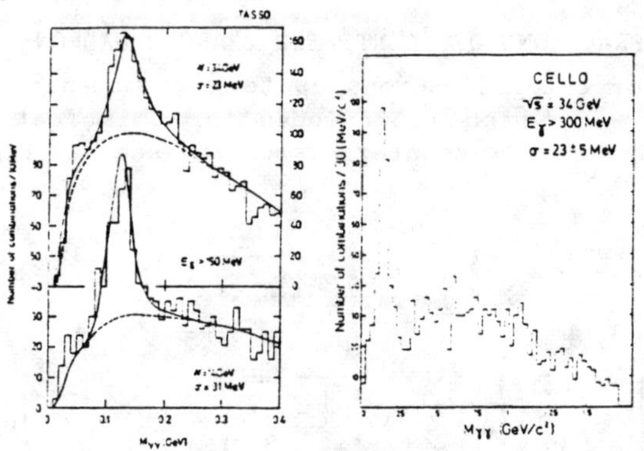

Fig.14 : Invariant mass of two photons (TASSO and CELLO).

The π^0 signals from TASSO and CELLO are shown in Fig.14 to have a similar mass resolution, $\sigma \approx$ 25 MeV. For the details of the π^0 detection I refer to the original papers[24,25].

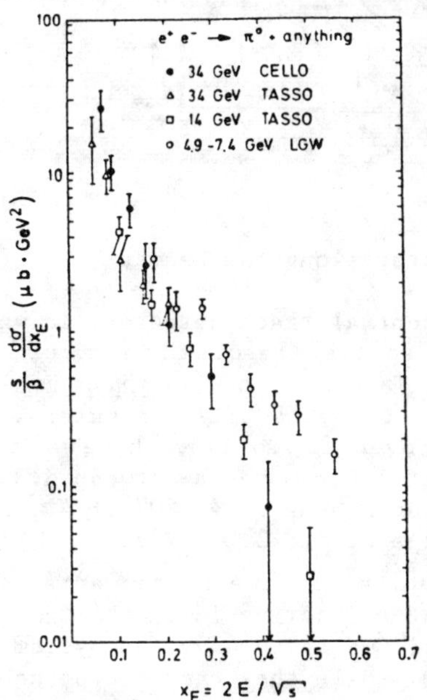

Fig. 15 :
Inclusive cross sections for π^0 production
(TASSO, CELLO and LEAD GLASS WALL)

The scaling cross sections $s/\beta \cdot d\sigma/dx$ for π^o production are shown in Fig.15. Here the data from TASSO and CELLO are compared with one another and with those from the LEAD GLASS WALL experiment[26] at SPEAR. For $x < 0.15$ CELLO agrees with TASSO, at both 34 and 14 GeV. For $x > 0.15$ CELLO obtains somewhat higher cross sections than TASSO.

At high x the data from PETRA (14 - 34 GeV) are roughly a factor of 2 below the low energy (4.9 - 7.4 GeV) data. Thus, in inclusive π^o production there is a strong energy dependence of a kind similar to that observed in inclusive charged hadron production.

FRACTIONS OF IDENTIFIED CHARGED HADRONS

Identification of charged hadrons up to high momenta is a domain of the TASSO detector, which is equipped for this feat with time-of-flight (TOF) and Cerenkov counters as can be seen in Fig.16.

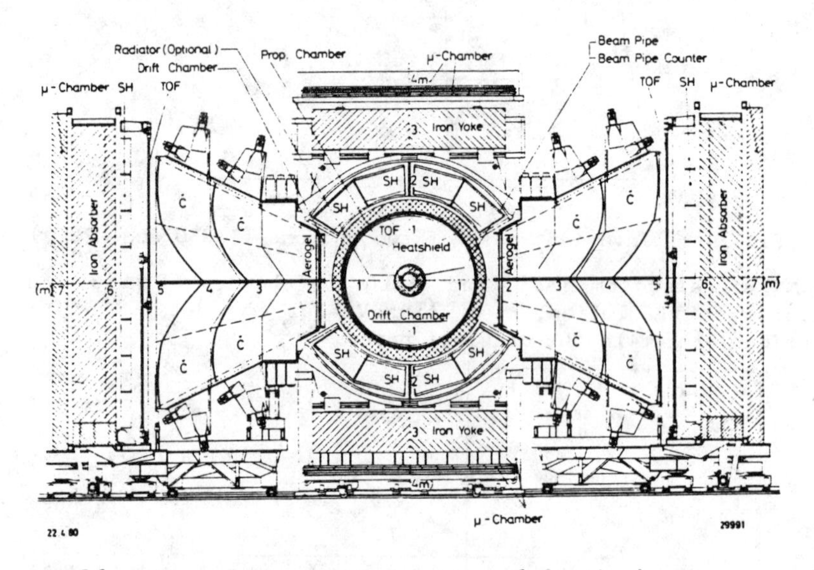

Fig.16 : View of the TASSO detector along the beams.

Particle momenta are measured in the central track detector, in which the cylindrical drift chamber has a spatial resolution averaged over the whole drift cell of 195μm. The momentum resolution is $\sigma_p/p = 0.017 \sqrt{1+p^2}$ (GeV/c). There are two sets of TOF counters. The inner TOF counters inside the coil cover 82% of 4π, have a time resolution of $\sigma = 380$psec, and are useful for momenta between 0.3 and 1.4 GeV/c. The TOF counters in the hadron arms cover 20% of 4π, have $\sigma = 450$psec, and are useful in the range 0.4 to 2.0 GeV/c.

There are three types of Cerenkov counters in the hadron arms using different radiator media with different refractive indices[27]: aerogel (1.024 ± 0.002), freon 114 (1.0014), and CO_2 (1.00043). The system has a geometrical acceptance of 19% of 4π, while the track acceptance is reduced by requirements to guarantee undisturbed measurements.

No accompanying track is allowed in the same counter cell nor an inelastic interaction in the coil nor an accompanying shower. These restrictions lead to track acceptances of (50±1.5)% at $\sqrt{s}$ = 14 GeV and of (40±1.0)% at $\sqrt{s}$ = 34 GeV within the solid angle of the Cerenkov counters.

The different radiators furnish different thresholds in momentum as can be seen in Fig.17.

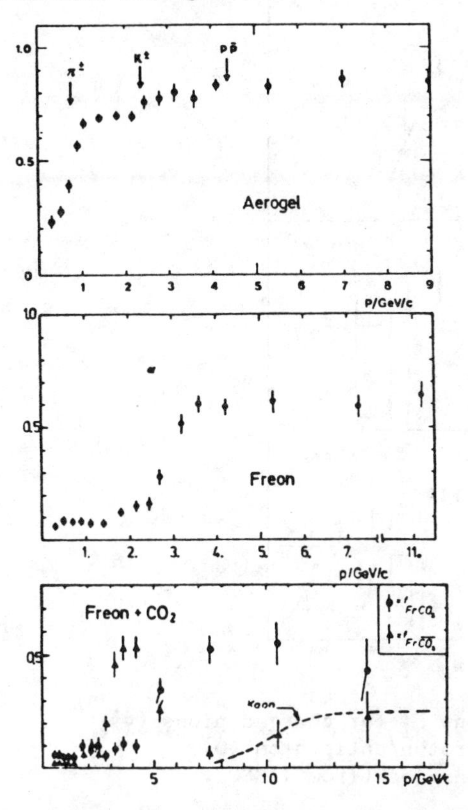

Fig. 17 :
Fractions of tracks that produce light in the three Cerenkov counter types of TASSO.

Table I contains the threshold momenta for particles of different mass. At high momenta the finite momentum resolution considerably broadens the threshold regions for the different particle species.

radiator	thresholds (GeV/c)		
	π	K	p
aerogel	0.6	2.2	4.2
freon 114	2.7	9.4	17.8
CO_2	4.8	16.9	32.

Table I : threshold momenta of Cerenkov counters

For a given momentum and particle a particular combination of flags set by the Cerenkov counter signals can be inferred from Table 1.

At present TASSO is the only detector at e^+e^- storage rings that identifies hadrons up to 17 GeV/c.

Particle fractions $f_i = N_i/N_{h^\pm}$, where $i = \pi^\pm$, $K^\pm$, $p/\bar{p}$ and $N_{h^\pm}$ is the number of charged hadrons, are obtained from the Cerenkov and TOF analysis. The results are plotted in Fig.18 versus the particle momentum for the three energies $\sqrt{s}$ = 14, 22, and 34 GeV/c.

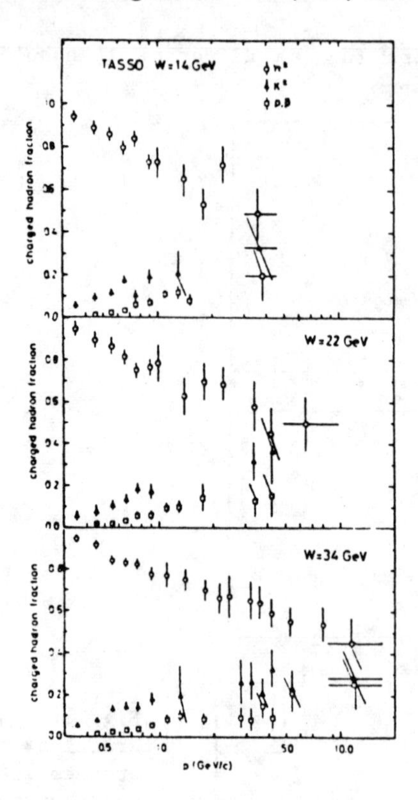

Fig. 18 : Particle fractions f_i for charged pions (ϕ), kaons (♦), and proton/antiproton (ɸ), at three c.m. energies (from TASSO).

The data are corrected for the decay of pions and kaons, for nuclear interactions and absorption in the material in front of the counters, and for the contamination due to prompt leptons and electrons from converted photons. Also the finite momentum resolution is taken into account. Within statistics equal numbers of positive and negative particles were observed for each particle species.

At very low momenta most of the charged hadrons produced are pions (>90% at 0.4 GeV/c). At higher momenta towards the kinematical limit the pion fraction decreases smoothly to about 50%, while kaon and proton/antiproton production becomes increasingly more important. At $\sqrt{s}$ = 34 GeV f_K and f_p both reach about 30%.

At this energy the average event contains 9.7±0.3 charged pions, 1.94±0.25 charged kaons, and 0.70±0.06 charged nucleons. The corresponding numbers at $\sqrt{s}$ = 14 GeV are 7.21±0.22, 1.25±0.16, and 0.40±0.05.

INCLUSIVE $\pi^{\pm}$ PRODUCTION

Exploiting its particle identification capability TASSO has presented new data on inclusive charged pion cross sections[28]. The statistical uncertainty is now very small, with several thousand tracks observed inside the Cerenkov counter acceptance. The data cover a wide momentum range, $0.3 < p_{\pi} < 10$ GeV/c, corresponding to $0.02 < x < 0.6$.

Combining the measured charged pion fractions shown in Fig.18 and the inclusive hadron cross sections from Fig.10 scaled cross sections for $\pi^{\pm}$ production are obtained. These are presented in Fig.19, which also contains low energy data measured with the DASP detector[29] at DORIS.

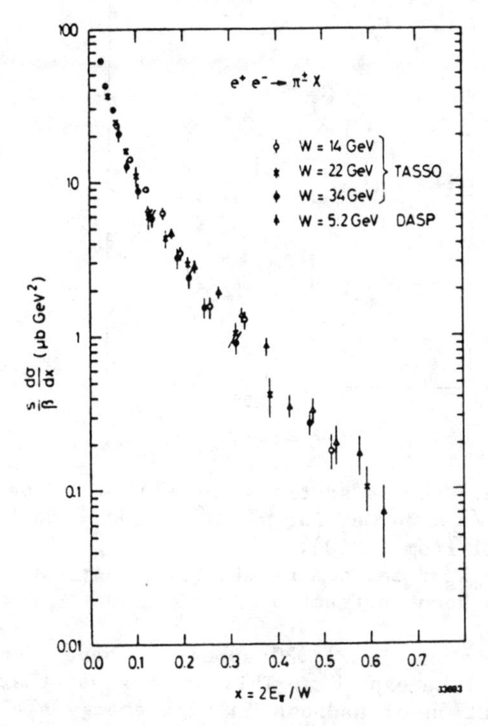

Fig.19 : Inclusive cross sections for $\pi^{\pm}$ (TASSO and DASP).

For $\pi^{\pm}$ the effect of scale breaking at large x is smaller than observed for unidentified charged hadrons and for π^0's, only two standard deviations in going from 5.2 to 34 GeV.

A clarification of the issue of energy dependence in x-distributions of different particle species should be easier, when other charged particles will have been studied.

In Fig.20 the $\pi^{\pm}$ data are compared with π^0 data[24] and with older data on $K^0/\bar{K}^0$ [30] and $\Lambda/\bar{\Lambda}$ [31] at $\sqrt{s}$ = 34 GeV. In the region of overlap, here limited to $1 < p < 4$ GeV/c, the ratio

$$2 \cdot \sigma_{\pi^0}/(\sigma_{\pi^+} + \sigma_{\pi^-}) = 1.01 \pm 0.20$$

is found. For $K^0/\bar{K}^0$ and $\Lambda/\bar{\Lambda}$ a comparison out to larger x is possible.

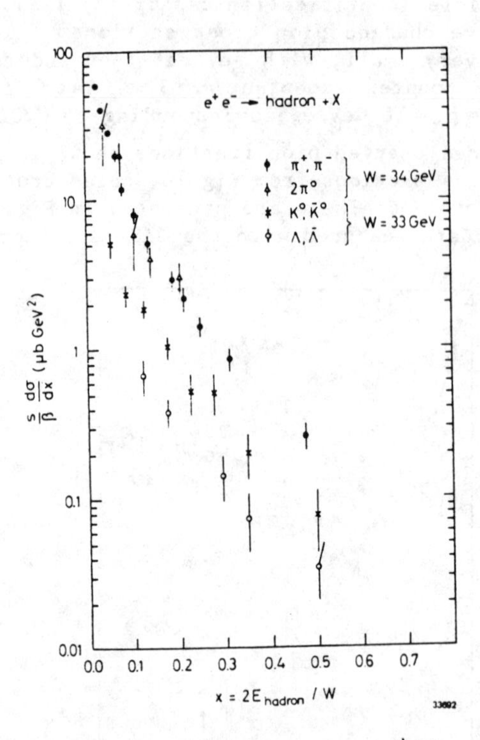

Fig. 20 : Scaled cross sections for $e^+e^- \rightarrow$ hadron + X at $\sqrt{s} \approx 34$ GeV for $\pi^\pm$, π^0, $K^0/\bar{K}^0$, and $\Lambda/\bar{\Lambda}$ (all from TASSO).
The K_s^0 decay contribution to the $\pi^\pm$ and π^0 yield has been subtracted.

The scaled cross sections in Fig.20 appear to have a universal slope falling off roughly like exp(-8x). This may suggest that the dominant mechanism for production of hadrons in high energy e^+e^- annihilation is the same for all types of particles.

SEARCH FOR CHARGED SCALARS

Electron-Positron annihilation is particularly well suited to search for new particles. Any charged and pointlike particle will be pair-produced with a sizeable cross section.

The PETRA experiments have performed sensitive searches for a great variety of hypothetical particles. So far the findings have been negative. However, the exclusion limits obtained are useful as constraints on various proposed symmetry schemes. The particular importance of scalar particles has been stressed recently by Okun[1]. Improved searches for unstable charged scalar particles ($S^{\pm}$) are presented by CELLO[32], JADE[33], MARK J[34], and TASSO[35].

The existence of such scalars is predicted in various models. There could be

- charged Higgs $H^{\pm}$ from schemes of spontaneous symmetry breaking with several Higgs doublets,
- charged technipions $P^{\pm}$ from models with dynamical symmetry breaking
- scalar leptons $s^{\pm}$, $t^{\pm}$ as supersymmetric partners of the leptons.

These particles would be pair-produced in e^+e^--collisions and are expected to decay predominantly to heavy fermions.

For $H^{\pm}$ (or $P^{\pm}$) the following reactions would be observed

(i) $e^+e^- \to H^+H^- \to (\tau\nu)(c\bar{s})$

$c\bar{s} \to$ hadrons

$\tau \to \begin{cases} l\nu\nu, \ l=e,\mu \\ \text{or} \\ \text{hadrons} + \nu \end{cases}$

or (ii) $e^+e^- \to H^+H^- \to (\tau\nu)(\tau\nu)$.

Scalar leptons $s^{\pm}$ (or $t^{\pm}$) would decay rapidly with 100% branching ratio into their spin 1/2 partners leading to reactions very similar to (ii):

(iii) $e^+e^- \to s^+s^- \to (\tau\lambda)\ (\tau\lambda)$

where λ stands for photinos or goldstinos.

These events have a distinctive signature as shown for reaction (i):

The two jets are acollinear and acoplanar with the beams due to the momentum carried off by the neutrinos or photinos or goldstinos. For scalar masses not too large reactions (ii) and (iii) would also lead to an excess of τ pairs over the QED expectation for $e^+e^- \to \tau^+\tau^-$.

I will briefly summarize the new results without going into the rather complicated details of the selection criteria which can be found in recent papers[32,33,34,35].

No evidence was seen for the existence of such scalar particles.

In all searches the branching ratios B_τ (B_τ = BR($S \to \tau\nu$)) and the mass m_S of the $H^{\pm}$ or $P^{\pm}$ have been varied. Fig.21 displays the regions in the (m_S, B_τ) plane, where the existence of a Higgs particle or technipion is excluded with 95% confidence level. The results are from JADE, CELLO and MARK J.

For B_τ = 100% the mass limits for $H^{\pm}$ and $P^{\pm}$ are identical to those for a scalar tau-lepton.

In summary: The existence of technipions or charged Higgs particles is excluded for masses between 4 and 14 GeV and branching ratios larger than 10%.
The existence of scalar supersymmetric partners of the leptons is excluded for masses less than 16 GeV.

Substantial improvement of the present exclusion limits - or a discovery - will come with higher beam energies.

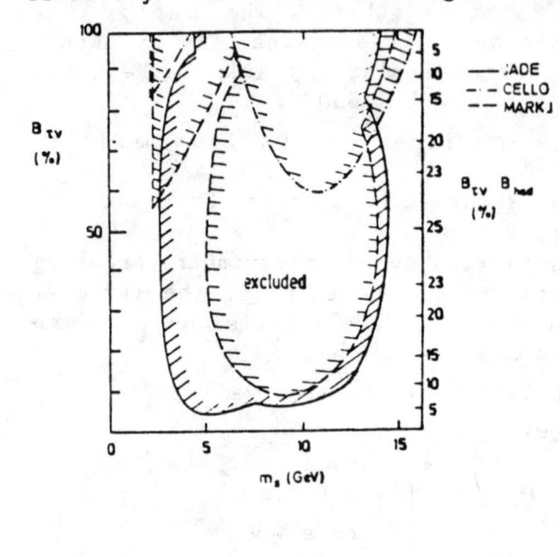

Fig. 21 :
90% c.l. limits of the branching ratio B_{τ} as a function of the mass m_s of the scalar particle (from CELLO, JADE, and MARK J).

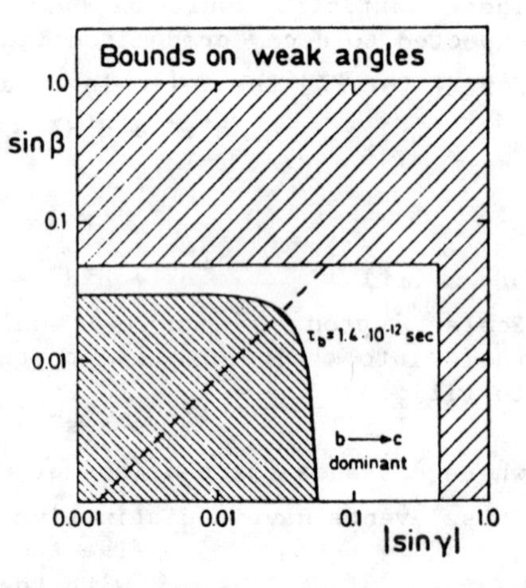

Fig. 22 :
Allowed region (unshaded) for the Maiani mixing angles.

NEW UPPER LIMIT FOR THE LIFETIME OF BOTTOM PARTICLES

An upper limit on the lifetime τ_B of hadrons with bottom quantum number was previously reported from JADE[1]. A refined analysis[36] now has enabled the group to substantially improve on this limit.

Hadronic events are used with a single muon of momentum larger than 1.4 GeV/c. Cuts are made to obtain a subsample with an enhanced fraction (50%) of $b\bar{b}$ events. From each of the remaining 27 events, the closest approach of the muon to the event vertex is determined in the plane normal to the beams. For this distance d_μ an upper limit is found of $\langle d_\mu \rangle < 0.26$mm. This converts to a limit for the lifetime of hadrons with beauty

$$\tau_B < 1.4 \cdot 10^{-12} \text{ sec} \quad (95\% \text{ c.l.}).$$

This result is used to put lower limits on the weak mixing angles of the Maiani scheme[37]. Fig.22 shows which region of the $\sin\beta$ - $|\sin\gamma|$-plane is still allowed.

TEST OF ELECTROWEAK THEORY IN e^+e^- ANNIHILATION

Probably the most exciting recent result from PETRA is the observation of interference effects of electromagnetic and weak interactions in e^+e^- collisions. The measurement of these effects provides a most important check on current gauge theoretical ideas which unify both forces.

Tiny electroweak interference effects of the order of 10^{-4} have been seen before, most convincingly in polarized electron deuteron scattering[38] as well as in atomic physics experiments[39]. In these studies the effects were small, because only small momentum transfers were involved, and thus rather difficult to measure. The strength of the weak force, however, is expected to increase as the square of the momentum transferred, q^2, in the interaction. In e^+e^- annihilation the squared momentum transferred is s, the squared sum of the energies of the colliding particles. At PETRA energies this represents a significant fraction of the predicted mass of the neutral weak boson Z^o.

First results on the observation of interference effects at PETRA were shown at the BONN conference[1]. Now much more data has been collected and analyzed by the experimental groups and improved results are available.

The standard theory developed by Glashow, Weinberg and Salam[4] predicts a neutral weak current contribution via Z^o exchange to the process $e^+e^- \rightarrow f\bar{f}$, where f can be any fundamental fermion. Lepton pair production, e.g. $e^+e^- \rightarrow \mu^+\mu^-$, in lowest order will proceed via the two diagrams

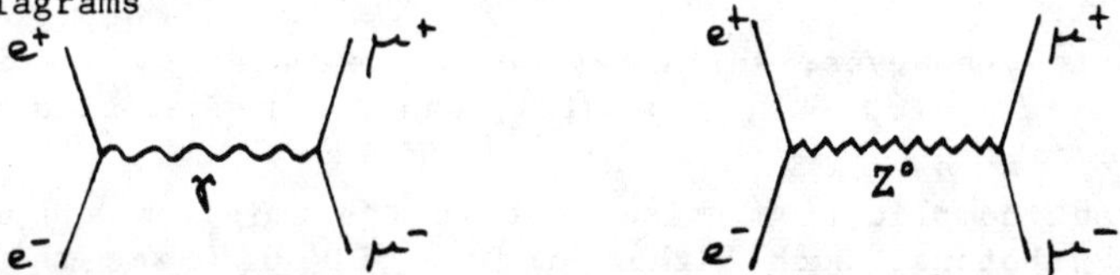

The Z^o exchange amplitude in addition to the usual exchange will alter the normalized QED cross section by R and will lead to a forward-background charge asymmetry A in the final state.

In the standard model the differential cross section for $\mu^+\mu^-$ production takes the form

$$d\sigma/d\cos\Theta = \pi\alpha^2/2s \cdot \left(B_1(1+\cos^2\Theta) + B_2\cos\Theta\right)$$

where Θ is the scattering angle between the incoming e^+ and the outgoing μ^+.

The coefficients B_1 and B_2 can be written as

$$\begin{array}{lllllll} B_1 = & 1 & + & 2v^2 D & + & (v^2 + a^2)^2 D^2 \\ B_2 = & & & 4a^2 D & + & 8v^2 a^2 D^2 \\ & \text{QED} & & \text{interf.} & & \text{weak} \end{array}$$

where the different origins of the terms are indicated.

D is conveniently defined as

$$D = G_F/8\pi\sqrt{2}\alpha \cdot s/(s/M_Z^2-1) = g \cdot s/(s/M_Z^2-1)$$

where $\Gamma_Z \ll M_Z$ has been assumed. g is $4.5 \cdot 10^{-5} GeV^2$.

The terms v^2 and a^2 are short-hand for products of the vector and axial-vector coupling constants of the leptons to the neutral current, here $v^2 = 4\ g_v^e g_v^\mu$ and $a^2 = g_a^e g_a^\mu$. If lepton universality holds: $g^e = g^\mu = g^\tau$. The standard model predicts $a^2 = (-1)^2$ and $v^2 = (-1 + 4\ \sin^2\Theta_W)^2$. Using $\sin^2\Theta_W = 0.23$ [6] one finds $v^2 = 0.006$ which makes $v^2 \ll a^2$.

The Z^o contribution leads to a deviation from the total cross section predicted by QED

$$\Delta R = (\sigma_{\gamma+Z^o} - \sigma_\gamma)/\sigma_\gamma = B_1 - 1 = 2v^2 + (v^2 + a^2)^2 D^2$$

and to a forward-backward charge asymmetry due to the presence of the $\cos\Theta$ term

$$A(\cos\Theta) := \frac{d\sigma/d\cos\Theta\ (\Theta) - d\sigma/d\cos\Theta\ (-\Theta)}{d\sigma/d\cos\Theta\ (\Theta) - d\sigma/d\cos\Theta\ (-\Theta)}$$

$$= B_2 \cos\Theta \ /\ B_1(1 + \cos^2\Theta).$$

Integration over the full range of the polar angle leads to

$$A = 3B_2/8B_1 = 3/2 \cdot (a^2 D + 2v^2 a^2 D^2)/(1 + 2v^2 D + (v^2 + a^2) D^2).$$

At present PETRA energies $s \ll M_Z^2$ and $D \approx -0.065$. Thus ΔR is expected to be only 0.5% in $e^+e^- \to \mu^+\mu^-$ and $e^+e^- \to \tau^+\tau^-$, which is too small to be mesured. For example, MARK J who have analyzed all their data up to Easter obtained 2435 $\mu\mu$ events, which corresponds to a statistical uncertainty of 2%, while they state a systematic error of 3%.

The asymmetry, however, which may be approximated by $A = 3/2 \cdot a^2 D$, is sizeable at $\sqrt{s} = 35$ GeV, $A \approx -10\%$, and can be detected with a few thousand events.

An important complication arises due to the emission and exchange of additional photons. Such higher order QED processes introduce asymmetries by themselves due to interference of diagrams with opposite C-parity of the μ pairs, which are on the few % level and of positive sign. These radiative effects are properly corrected for by including them in the Monte Carlo generators.

All PETRA experiments[40,41,42,43,44] have measured the reactions $e^+e^- \to \mu^+\mu^-$ and $e^+e^- \to \tau^+\tau^-$. Fig. 23 shows a fairly recent compilation of the data on the total cross sections. The agreement with the prediction from pure QED is excellent.

These data may be used to put limits on the lepton structure. With the standard parametrization of the lepton form factor $\Lambda_\pm$ values of 100 to 200 GeV are obtained.

The angular distributions measured by the four experiments CELLO, JADE, MARK J, and TASSO are shown in Fig. 24 (μ pair data in Fig. 24a and τ pair data in Fig. 24b)[45]. The predictions from pure QED are given as dashed lines. The negative asymmetry is clearly seen in each experiment. The full lines are fits to the data.

In the case of $\tau^+\tau^-$ there is considerably less statistics available since not all of the τ decay channels are observed. Thus the asymmetry is less pronounced in the angular distributions.

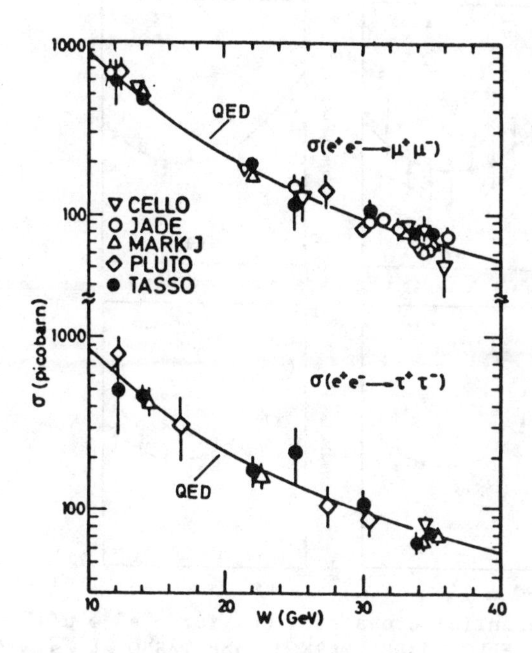

Fig.23 : Total cross sections for μ pair and τ pair production from PETRA experiments

The presently available asymmetry values from PETRA at $\sqrt{s}$ = 34 GeV, where the bulk of the data has been taken, are compiled in Table II, for both $e^+e^- \to \mu^+\mu^-$ and $e^+e^- \to \tau^+\tau^-$.

In μμ production a significant asymmetry is observed by JADE, MARK J, and TASSO. These groups each have collected roughly the same number of events. The differences in event numbers and in the statistical errors given in the table reflect the fractions of the data that have been analyzed to date. CELLO, due to the push-pull operation mode with PLUTO, has had less exposure.

Group	$N_{\mu\mu}$	$A_{\mu\mu}$ (%) (|cos θ|<1)	$N_{\tau\tau}$	$A_{\tau\tau}$ (%) (|cos θ|<1)
CELLO	387	-(6.4±6.4)	434	-(10.3±5.2)
JADE	1590	-(12.7±2.7)	569	-(4.7±4.2)
MARK J	2435	-(9.8±2.3)	550	-(9.±6.)
TASSO	1155	-(16.1±3.2)	262	-(0.4±6.6)

Table II : Charge asymmetries $A_{\mu\mu}$ for $e^+e^- \to \mu^+\mu^-$ and $A_{\tau\tau}$ for $e^+e^- \to \tau^+\tau^-$ at $\sqrt{s}$ = 34 GeV from the PETRA experiments (extrapolated to $|\cos\theta| < 1$).
The GWS model predicts -9.1 % for both reactions.

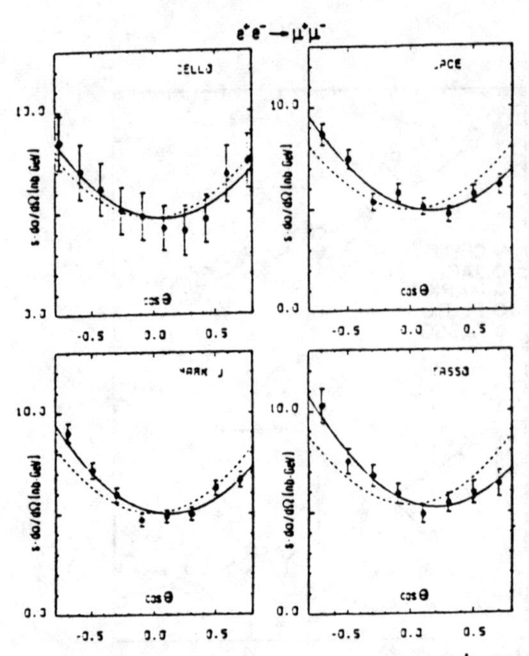

Fig. 24a : Differential cross sections for $e^+e^- \to \mu^+\mu^-$ from CELLO, JADE, MARK J, and TASSO at $\sqrt{s} \approx 34$ GeV.

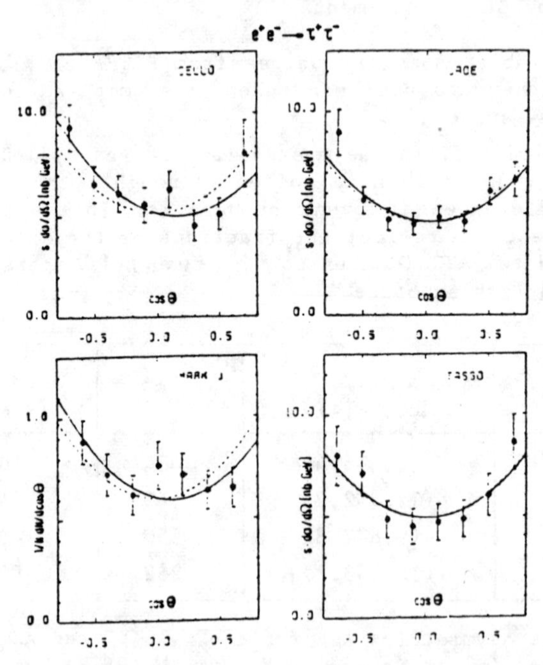

Fig. 24b : Differential cross sections for $e^+e^- \to \tau^+\tau^-$ from CELLO, JADE, MARK J, and TASSO at $\sqrt{s} \approx 34$ GeV.

There is also a fair amount of data taken at $\sqrt{s}$ = 14 and 22 GeV. The angular distributions in Fig. 25 obtained by MARK J[44] show very nicely how the asymmetry develops as the energy rises.

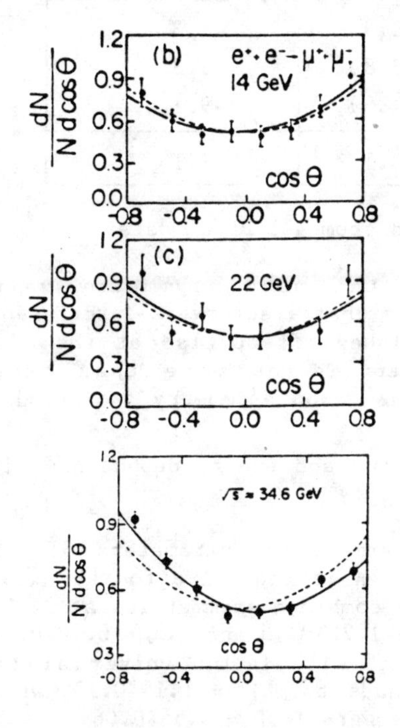

Fig. 25 :
Differential cross sections for muon pair production at $\sqrt{s}$ = 14, 22, and 34 GeV (MARK J).

All groups have made thorough studies of possible sources for a systematic bias in the asymmetry measurement.
The uncertainty in momentum measurement affects the determination of muon charge. A wrong charge sign for both muons could introduce a bias. TASSO[41], for example, estimates that in less than 0.1% of the events the wrong charge assignment was made for both muons. Similar limits are obtained by the other groups.
Mark J has made extensive studies of a possible acceptance asymmetry using cosmics. They conclude that the detector contribution to the charge asymmetry is less than 1%. Furthermore they have eliminated any residual detector asymmetry by alternating the polarity of the detector magnetic field.
All groups present convincing arguments that their muon identification and in particular their charge assignments are correct. From their studies they conclude that the total systematic errors are less than 1% in each experiment, and thus smaller than the present statistical uncertainties.

Since the statistical errors of the individual experiments are yet larger than the systematic ones it should be allowed to combine the data from all PETRA experiments and form average asymmetries. This is done in Table III.

Reaction	$\sqrt{s}$	Asymmetry (%)	GWS-Theory
	14	$+(4.\pm3.)$	-1.3
$e^+e^- \to \mu^+\mu^-$	22	$-(7.8\pm3.8)$	-3.3
	34	$-(11.9\pm1.5)$	-9.1
$e^+e^- \to \tau^+\tau^-$	34	$-(6.1\pm3.1)$	-9.1

Table III : Asymmetries combined from all PETRA data.

From this table, in particular from the muon asymmetry value obtained at $\sqrt{s}$ = 34 GeV, it is obvious that electroweak interference is now an experimentally well established effect also at the high $q^2 = s = 1200\ GeV^2$. The muon asymmetry is now more than seven standard deviations away from zero, the tauon asymmetry two standard deviations.

The measurements agree in absolute value and energy dependence with the prediction of the standard model[4].

The combined data may be used to determine the parameters of the electroweak interaction. The asymmetry A is sensitive to the axial-vector coupling: $A = 3/2 \cdot Da^2$. From the combined asymmetries at 34 GeV (using M_Z = 90 GeV) one obtains $a^2_{\mu\mu} = 1.22\pm0.15$ and $a^2_{\tau\tau} = 0.70\pm0.27$, thus $a_{\tau\tau}/a_{\mu\mu} = 0.76\pm0.31$, in agreement with lepton universality. Consequently combining both figures leads to $a^2_{\ell\ell} = 1.10\pm0.13$, while GWS predict $a^2_{\ell\ell} = 1$. With $g^2_a = a^2/4$ one gets $|g_a| = 0.55\pm0.06$.

In order to determine the vector coupling total cross section data have to be used as well. MARK J[44] (setting M_Z = 90 GeV) from a simultaneous fit to their data on $A_{\mu\mu}$ and $\sigma_{\mu\mu}$ obtains $g^2_v = 0.01\pm0.05$ and $g^2_a = 0.28\pm0.06$. TASSO[41] (using $M_Z = \infty$) gets $g^2_v = -0.11\pm0.13$ and $g^2_a = 0.53\pm0.10$. TASSO has also carried out a combined fit to their $\mu\mu$ and ee data[35] obtaining $g^2_v = -0.06\pm0.07$ and $g^2_a = 0.34\pm0.10$.

The determination of the weak coupling constants from high energy e^+e^- lepton pair production is very important since the results serve to single out one solution of the two possibilities left open by the neutrino scattering data[46]. This is shown in Fig. 26. The unique solution determined in this way is in good agreement with the predictions of the standard model[4], namely $g^2_v = 0.0016$ and $g^2_a = 0.25$.

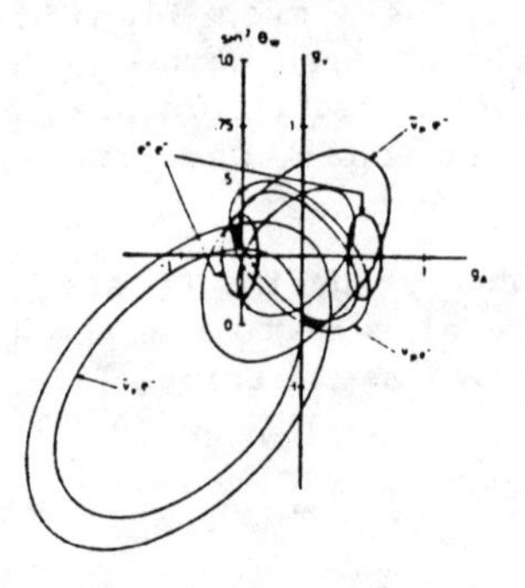

Fig. 26 :
Limits in the g_v-g_a-plane determined from elastic e scattering and e^+e^- annihilation.

Furthermore a fit to the standard model allows a determination of the only free parameter of the model, $\sin^2\theta_W$, from purely leptonic reactions at high q^2.
Recent results are

$\sin^2\theta_W = 0.25\pm0.12$ (CELLO[42,47], from ee, μμ, and ττ),
$\sin^2\theta_W = 0.25\pm0.15$ (JADE[40,48], from ee and μμ),
$\sin^2\theta_W = 0.27^{+0.06}_{-0.07}$ (TASSO[35], from ee and μμ).

These figures have to be compared to the combined result[6] from other leptonic reactions at much lower q^2, $\sin^2\Theta_W = 0.228\pm0.009$.

The effect of a Z^o propagator cannot yet be measured at PETRA. But assuming a single Z^o to be responsible for the electroweak effects a lower limit can be found of M_Z 51 GeV (95% c.l.).

The standard model may be extended to contain more than one neutral boson[49] by adding to the Hamiltonian another piece proportional to the electromagnetic current squared and to a constant C, which would appear also in the modified vector coupling as $g_V^2=(1-4\sin^2\theta_W)^2/4+4C$. Assuming $\sin^2\theta_W=0.228$ an upper limit on C can be derived. Currently the best value[35] for this limit is $C < 0.02$ (95%c.l.). Two specific models with two neutral bosons are based on $\{SU(2)xU(1)\}_{GWS}xU(1)$ and $\{SU(2)xU(1)\}_{GWS}xSU(2)$. Fig. 27 shows the restrictions on the neutral bosons in these models imposed by the new limit on C.

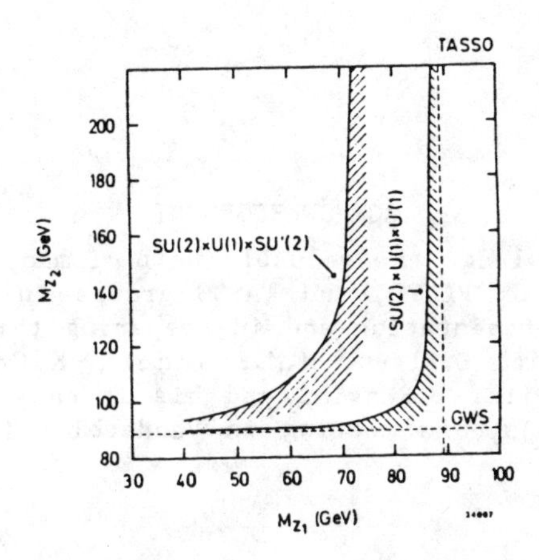

Fig. 27 : Limits on M_{Z1} and M_{Z2} in models with two neutral gauge bosons. The allowed regions are shaded. The mass of the GWS Z_o boson is indicated by dashed lines.

In conclusion, the parameters of the electroweak interactions as measured in lepton pair production by e^+e^- annihilation are all in accordance with the predictions of the standard electroweak theory. Thus, the model of Glashow, Weinberg and Salam appears to be in good shape, not only at low momentum transfers but also at the high q^2 values near 1200 GeV^2 reached with PETRA.

OUTLOOK

The potential of e^+e^- reactions at high energies as a testing ground for field theories of electromagnetic, strong, and weak interactions is excellent. Important results have been obtained and a lot more will come out in the future. Thus at PETRA the detectors as well as the machine are steadily being upgraded.

For the storage ring a firm improvement program exists. In the summer shutdown this year the radio frequency power will be doubled therewith increasing the maximum beam energy by about 2 GeV. Then, starting in December 1982 and extending into summer of next year the number of cavities will be increased considerably. Eventually this will allow to go up in beam energy by another 2 GeV.

There are encouraging prospects to go beyond even that since recently a superconducting test cavity installed in PETRA was operated successfully.

Certainly we experimenters are eagerly looking forward to much excitement as the energies rise.

ACKNOWLEDGEMENT

I gratefully acknowledge the valuable help of many colleagues of the CELLO, JADE, MARK J, PLUTO, and TASSO groups in collecting the material for the presentation and in preparing this report. Special thanks go to A.Böhm, E.Elsen, H.M.Fischer, K.Gather, J.Meyer, B.Naroska, and B.Wiik. B.Panvini and his friends organized a very interesting and enjoyable meeting in Vanderbilt from which I benefitted very much.

REFERENCES

1. Proceedings of the 1981 Intern. Symposium on Lepton and Photon Interactions at High Energies, ed. W.Pfeil, Bonn 1981
2. Proceedings of the XVIIIth Rencontre de Moriond, ed. J.Tran Tanh Van, Les Arcs 1982;
   A.Böhm, DESY 82-027 (1982);
   J.Olsson, Heidelberg preprint (1982);
   P.Steffen, DESY 82-039 (1982)
   J.Haissinski, Orsay preprint LAL 82/11 (1982);
   H.F.Kolanoski, BONN-HE-82-11 (1982)
3. TASSO Coll., R.Brandelik et al, Phys.Lett. 113B(1982)499
4. S.L.Glashow, Nucl. Phys. 22(1961)579;
   S.Weinberg, Phys.Rev.Lett. 19(1967)1264;
   A.Salam, Proceedings of the 8th Nobel Symposium, ed. N.Svartholm, Stockholm 1968
5. J.Ellis and M.K.Gaillard, Physics with High Energy e^+e^- Colliding Beams, CERN 76-18 (1976);
   J.Jersak, E.Laermann, P.W.Zerwas, Phys.Lett. 98B(1981)363;
   Th.Appelquist and H.D.Politzer, Phys.Rev. D12(1975)1404
6. Particle Data Group (PDG), Rev.Mod.Phys. 52(1980)
7. Mark I Coll., G.Hanson et al, Phys.Rev.Lett. 35(1975)1609
8. TASSO Coll., R.Brandelik et al, Phys.Lett. 100B(1981)357
9. R.Field and R.P.Feynman, Nucl.Phys. B136(1978)1
10. JADE Collaboration, W.Bartel et al, DESY 82-016 (1982)
11. TASSO Coll., R.Brandelik et al, Phys.Lett. 83B(1979)261
    MARK J Coll., D.P.Barber et al, Phys.Rev.Lett. 43(1979)830,
    PLUTO Coll., Ch.Berger et al, Phys.Lett. 86B(1979)418;
    JADE Coll., W.Bartel et al, Phys.Lett. 91B(1979)142
12. O.Nachtmann and A.Reiter. Heidelberg prepr., HD-THEP-82-1(1982)
13. M.C.Goddard, Rutherford Appleton Lab. prepr., RL-81-069 (1981)
14. B.Andersson, G.Gustafson, T.Sjöstrand, Phys.Lett.94B(1980)211;
    T.Sjöstrand, Lund prepr. LU TP 80-03(1980) and LU TP 82-3(1982)
15. A.Ali et al, Phys.Lett. 82B(1979)285; Nucl.Phys. B167(1980)454
16. JADE Collaboration, E.Elsen, private communication
17. M.Glück and E.Reya, Phys.Rev. D16(1977)3242;
    (For a recent comparison with data from deep inelastic lepton scattering see H.Abramowicz et al, Z.Physik C13(1982)199)
18. S.D.Drell, D.Levy, T.M.Yan, Phys.Rev. 187(1969)2159;
    Phys. Rev. D1(1970)1035,1617;
    T.M.Yan and S.D.Drell, Phys.Rev. D1(1970)2402
19. TASSO Collaboration, R.Brandelik et al, Phys.Lett. 114B(1982)65
20. P.Hoyer et al, Nucl.Phys. B161(1979)349
21. T.Meyer, Z.Physik C12(1982)77
22. MARK I Coll., J.Siegrist et al, SLAC-PUB-2831/LBL 13464 (1981)

23. MARK II Coll., R.Hollebeek in ref. 1; SLAC-PUB 2936 (1982)
24. TASSO Collaboration, R.Brandelik et al,Phys.Lett. 108B(1982)71
25. CELLO Collaboration, H.J.Behrendt et al, DESY 82-018 (1982)
26. D.Scharre et al, Phys.Rev.Lett. 41(1978)1005
27. H.Burkhardt et al, Nucl.Instr. & Meth. 184(1981)319; G.Poelz and R.Riethmüller, Nucl.Instr. & Meth. 195(1982)491
28. TASSO Collaboration, R.Brandelik et al, Phys.Lett. 113B(1982)98
29. DASP Collaboration, R.Brandelik et al, Nucl.Phys. B148(1979)189
30. TASSO Collaboration, R.Brandelik et al, Phys.Lett. 94B(1980)91
31. TASSO Collaboration, R.Brandelik et al, Phys.Lett. 105B(1981)75
32. CELLO Collaboration, H.J.Behrendt et al, DESY 82-021 (1982)
33. JADE Collaboration, W.Bartel et al, DESY 82-023 (1982)
34. MARK J Collaboration, B.Adeva et al, MIT Lab.f.Nucl.Science Techn.Report #125 (1982)
35. TASSO Collaboration, R.Brandelik et al, DESY 82-032(1982)
36. JADE Collaboration, W.Bartel et al, Phys.Lett. 114B(1982)71
37. L.Maiani, Proceedings of the 8th Intern. Symposium on Lepton and Photon Interactions at High Energies, Hamburg 1977; M.Gaillard and L.Maiani, Annecy preprint, LAPP-TH-09 (1979); V.Barger, W.Y.Keung and R.J.N.Phillips, Phys.Rev. D24(1981)1328
38. C.Y.Prescott et al, Phys.Lett. 77B(1978)347
39. Proceedings Intern. Workshop on Neutral Current Interactions in Atoms, Cargèse 1979, ed. W.L.Williams, Paris 1980
40. JADE Collaboration, W.Bartel et al, Phys.Lett.108B(1982)140
41. TASSO Collaboration, R.Brandelik et al, Phys.Lett 110B(1982)173
42. CELLO Collaboration, H.J.Behrendt et al, Phys.Lett.
43. CELLO Collaboration, H.J.Behrendt et al, DESY 82-020 (1982)
44. MARK J Collaboration, B.Adeva et al, MIT Lab. of Nucl. Science, Techn. Report #124 (1982)
45. This figure was prepared by B.Naroska.
46. G.Barbiellini in ref. 1
47. CELLO Coll., H.J.Behrendt et al, Phys.Lett. 103B(1981)148
48. JADE Collaboration, W.Bartel et al, Phys.Lett. 99B(1981)281
49. P.Q.Hung and J.J.Sakurai, Nucl.Phys. B143(1978)81; J.D.Bjorken, Phys.Rev. D19(1979)335; E.H.de Groot, G.J.Gounaris, and D.Schildknecht, Phys.Lett. 85B(1979)399; 90B(1980)427; Z.Physik 5(1980)127; D.Schildknecht, Proc.Int.School of Subnucl.Physics, Erice 1980 and Bielefeld preprint BI-TP 81/12 (1981); V.Barger, W.Y.Young, E.Ma, Phys.Rev.Lett. 44(1980)1169; Phys.Rev. D22(1980)727; Phys.Lett. 94B(1980)377

SOME RECENT RESULTS FROM THE MAC DETECTOR AT PEP*

MAC Collaboration:
Colorado-Frascati-Northeastern-SLAC-Utah-Wisconsin†
Presented by W. T. Ford, Colorado

ABSTRACT

Preliminary results are presented for non-radiative and radiative muon pair production and limits on the production of excited muon states. A new measurement of the tau lepton lifetime is presented. Calorimeter studies of multihadron production are described, with preliminary results for the total and energy-correlation cross section and inclusive muon production rates.

---

*Work supported in part by the Department of Energy, under contract numbers DE-AC02-76ER02114 (CU), DE-AC03-76SF00515 (SLAC), and DE-AC02-76ER00881 (UW), by the National Science Foundation under contract numbers NSF-PHY80-06504 (UU), NSF-PHY79-20020, and NSF-PHY79-20821 (NU), and by I. N. F. N.

†MAC Collaborators are: W. T. Ford, J. S. Marsh, A. L. Read, Jr., J. G. Smith, Department of Physics, University of Colorado, Boulder, CO 80309; A. Marini, I. Peruzzi, M. Piccolo, F. Ronga, Laboratori Nazionali Frascati dell' I.N.F.N. (Italy); L. Baksay, H. R. Band, W. L. Faissler, M. W. Gettner, G. P. Goderre, B. Gottschalk, R. B. Hurst, O. A. Meyer, J. H. Moromisato, W. D. Shambroom, E. von Goeler, Roy Weinstein, Department of Physics, Northeastern University, Boston, MA 02115; J. V. Allaby, W. W. Ash, G. B. Chadwick, S. H. Clearwater, R. W. Coombes, Y. Goldschmidt-Clermont, H. S. Kaye, K. H. Lau, R. E. Leedy, S. P. Leung, R. L. Messner, S. J. Michalowski, K. Rich, D. M. Ritson, L. J. Rosenberg, D. E. Wiser, R. W. Zdarko, Stanford Linear Accelerator Center, Stanford University, Stanford, CA 94305; D. E. Groom, H. Y. Lee, E. C. Loh, Department of Physics, University of Utah, Salt Lake City, UT 84112; M. C. Delfino, B. K. Heltsley, J. R. Johnson, T. Maruyama, R. M. Morse, R. Prepost, Department of Physics, University of Wisconsin, Madison, WI 53706.

INTRODUCTION

The MAC detector has been running at PEP for about a year and a half and has accumulated about 24 inverse picobarns of data, at center-of-mass energy 29 GeV. The detector (Fig. 1) which is described in some detail in the proceedings of the 1982 SLAC instrumentation conference[1], is designed to measure the total energy and its angular distribution in a hadron and electromagnetic calorimeter of nearly 4π acceptance. The calorimeter surrounds a large-acceptance drift chamber and solenoid magnet for charged-particle tracking, and is surrounded by drift chambers for muon tracking.

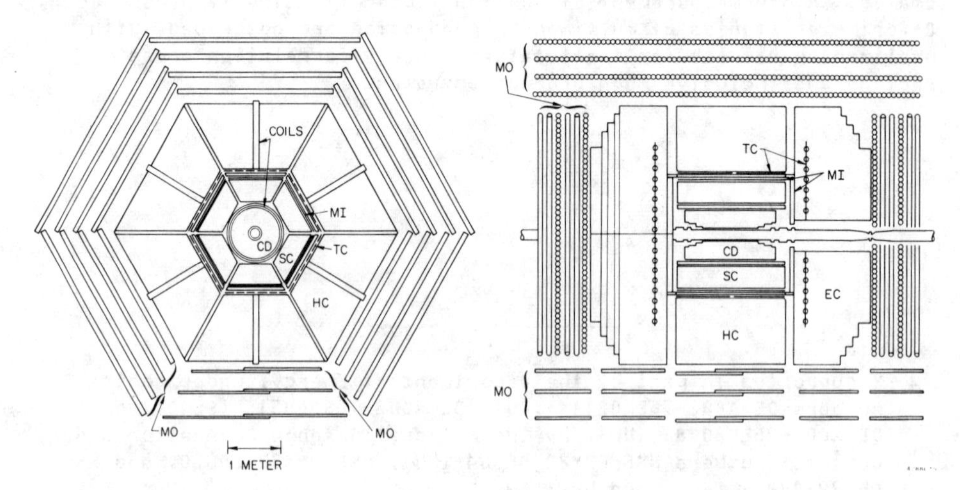

Fig. 1. MAC detector layout. The components labelled in the figure are: central drift chamber (CD), shower chamber (SC), trigger/timing scintillators (TC), central and endcap hadron calorimeters (HC, EC), and the inner and outer muon drift chambers (MI, MO). Also indicated are the solenoid and toroid coils.

COLLINEAR MUON PAIRS

The production of muon pairs is described according to the standard electroweak theory by

$$\frac{d\sigma}{d\cos\theta} = \frac{\pi\alpha^2}{2s}\left[(1+a_1)(1+\cos^2\theta) + 2a_2\cos\theta + \Delta QED(\alpha^3)\right] ,$$

where (1)

$$a_{1,2} = g_{V,A}{}^{e}\, g_{V,A}{}^{\mu}\, \frac{1}{\pi\alpha}\, \frac{G}{\sqrt{2}}\, \frac{-s}{1-s/M_Z{}^2} .$$

Equation (1) includes the contributions of the single photon annihilation diagram and of its interference with annihilation through the neutral weak intermediate vector boson, Z_0. The term $\Delta QED(\alpha^3)$ refers to the radiative corrections to the QED cross section, calculated to order α^3 in reference 2.

The sample consists of those events having two charged prongs forming a vertex in the beam intersection volume which are collinear within 10°, and calorimeter pulse heights along the trajectories that are consistent with either two muons or with one muon and one unresolved muon-plus-photon. Cosmic rays are rejected by requiring that each trajectory intersect a scintillator which produced a pulse with the correct arrival time. The sum of the two muons' momenta is required to be greater than 8 GeV to discriminate against $ee\mu\mu$ events. Events are rejected if the curvature measurement for one of the tracks gives the wrong sign[3], since we want to measure the charge asymmetry. The final sample contains 1239 events, for $\int L dt \simeq 20$ inverse picobarns.

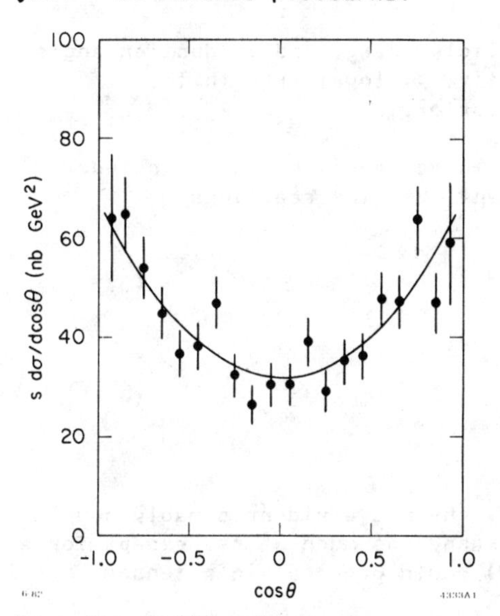

Fig. 2. Differential cross section for μ pair production.

The differential cross section for collinear μμ production is shown in Fig. 2. The curve accompanying the data points was obtained from a maximum likelihood fit of equation (1) to the data with a_2 adjusted. The asymmetry, $A_{\mu\mu}$, is related to a_2 by

$$A_{\mu\mu} = \frac{3}{4} a_2 , \qquad (2)$$

and we find

$$A_{\mu\mu} = -0.032 \pm 0.028 ,$$

which is about one standard deviation from either zero or the standard model prediction of -0.063. The fit χ^2 is 22 for 18 degrees of freedom. This result implies that the product of axial-vector coupling constants is $g_A^e g_A^\mu = 0.13 \pm 0.11$.

HIGHER-ORDER QED AND μ* SEARCH IN μμγ(γ)

By requiring two muons that are non-collinear by at least 10° and one or more observed photons of at least 1 GeV, we select radiative muon pairs which may be used to test the higher-order QED calculations required, for example, in the weak-electromagnetic asymmetry analysis described in the previous section. Table 1 gives a summary of the $\mu\mu\gamma$ and $\mu\mu\gamma\gamma$ event yields for samples meeting the above criteria and having satisfactory kinematic fits under the appropriate hypotheses. For comparison the prediction of QED calculations including terms of order α^3 (ref. 2) and α^4 (ref. 4) are given for the $\mu\mu\gamma$ reaction. The agreement with the α^4 term included is good. For the $\mu\mu\gamma\gamma$ process we list an estimate obtained by assuming $\mu\mu\gamma\gamma/\mu\mu\gamma \simeq \mu\mu\gamma/\mu\mu$.

	$\mu\mu\gamma$	$\mu\mu\gamma\gamma$
$\int L dt$ (pb^{-1})	24	12
Events found	133	5
Events expected:		
QED(α^3)	114	—
QED(α^4)	≃138	≃4

Table 1. Summary of event yields and predictions for $\mu\mu\gamma(\gamma)$ events.

The distributions in single-particle energy and production angle are given in Figs. 3 and 4, respectively, together with the predictions from the Kleiss Monte Carlo[2].

Possible deviations from QED in these reactions would include production of excited muon states, μ^*, via the reactions

$$e^+e^- \rightarrow \mu^+\mu^{*-}, \quad \mu^{*-} \rightarrow \mu^-\gamma \tag{3}$$

$$e^+e^- \rightarrow \mu^{*+}\mu^{*-}, \quad \mu^{*-} \rightarrow \mu^-\gamma, \quad \mu^{*+} \rightarrow \mu^+\gamma \tag{4}$$

In the pair-production process (4), the μ^*s would presumably couple to the virtual photon in the same manner as muon pairs, except for a possible form factor. Reaction (3) could proceed via a tensor coupling

$$\lambda e \sigma_{\mu\nu} F^{\mu\nu},$$

which for μ^* mass M leads to the cross section[5]

$$\frac{d\sigma}{d\Omega} = \lambda^2\alpha^2 \frac{(s-M^2)^2}{s^2}\left[(s+M^2)-(s-M^2)\cos^2\theta\right] \tag{5}$$

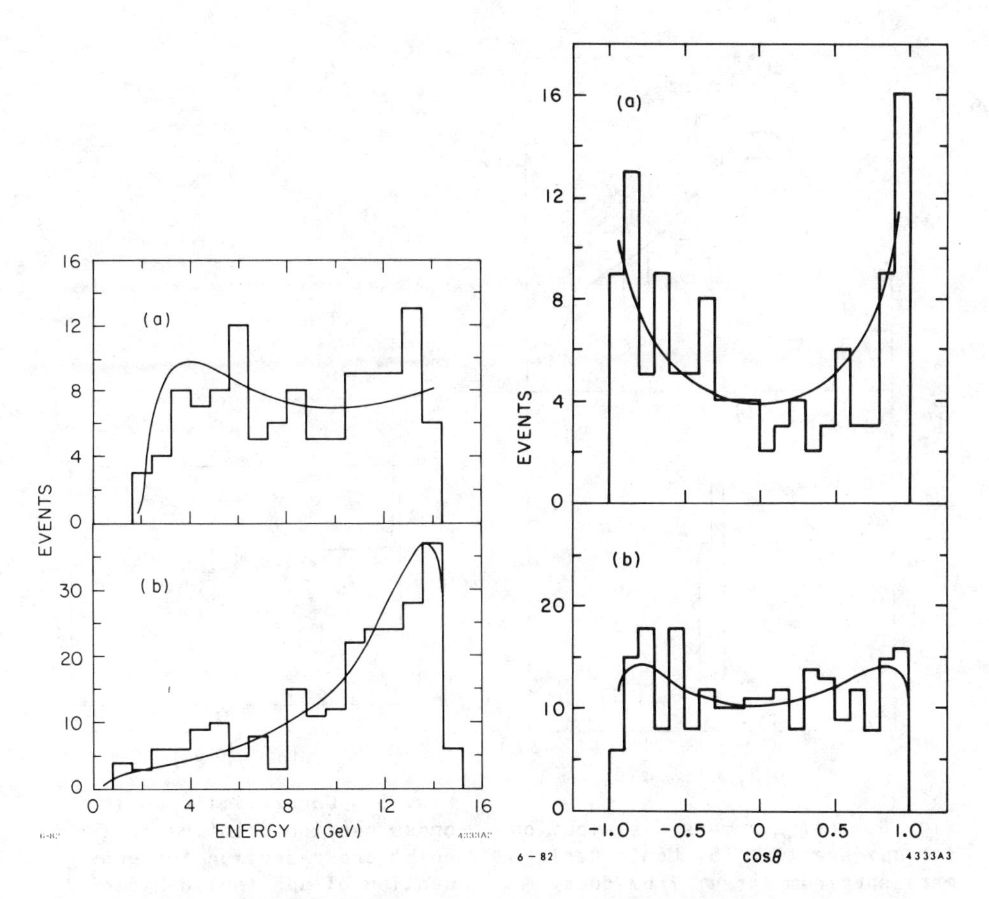

Fig. 3. Energy distributions for (a) photons and (b) muons for $\mu\mu\gamma$ events.

Fig. 4. Polar angle distributions for (a) photons and (b) muons for $\mu\mu\gamma$ events.

Figure 5(a) shows the distribution of the $\mu\mu\gamma$ events in the $\mu\gamma$ mass, together with the QED calculation. The mass rsolution is indicated in Fig. 5(b), the distribution of Monte Carlo events of reaction (3) with mass($\mu\gamma$) = 15 GeV/c^2. The lack of excess events in the data compared with the calculation can be translated into a limit on the production rate of μ^*s for each mass bin, as shown in Fig. 6. The bin width has been chosen to match the resolution.

For reaction (4) we would expect the correct pairing of muons with photons to yield two equal masses. None of our 5 $\mu\mu\gamma\gamma$ events meet this criterion within 2 standard deviations, so we have no candidates. The resulting limit as a function of μ^* mass is plotted also in Fig. 6.

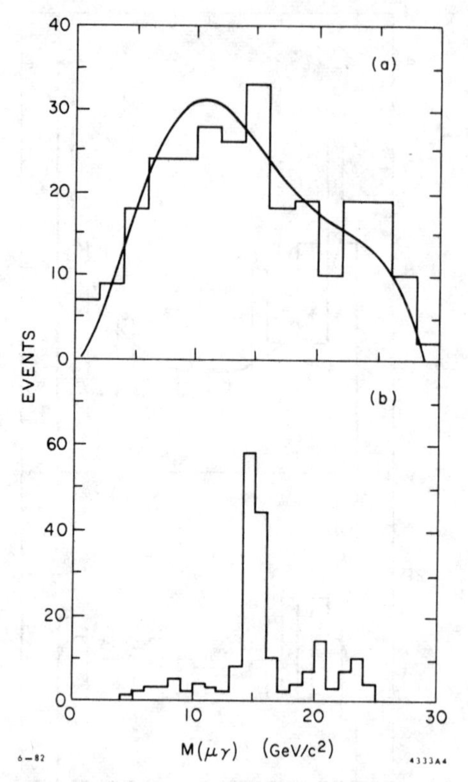

Fig. 5. (a) μ-γ mass distribution for μμγ events. (b) Monte Carlo mass spectrum for μγ from decay of a μ* of mass 15 GeV.

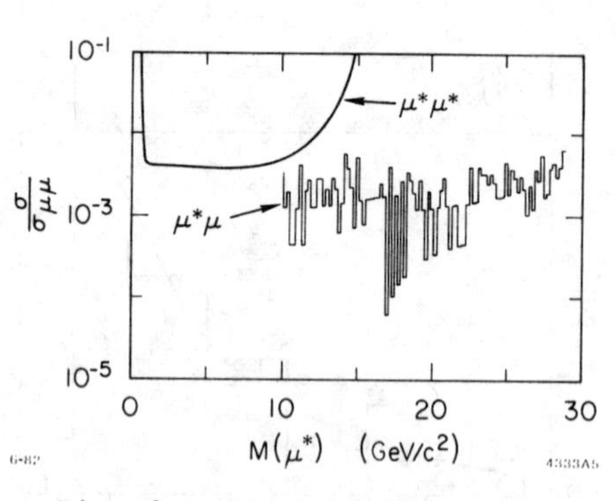

Fig. 6. Upper limits on the cross section relative to the point cross section for production of μμ* (solid histogram) and μ*μ* (dashed curve).

TAU LIFETIME

Tau leptons of 14.5 GeV are expected to travel about 0.7 mm on average before decaying. About 15-20% of the decay final states contain three charged prongs, whose vertex can be localized within about 3-4 mm in our detector. The error on the mean for a sample of N events is, of course, smaller by a factor $\sqrt{N}$, allowing a measurement of interesting significance if N is of order a hundred or more and systematic effects can be controlled.

We selected events having 4 or 6 prongs in narrow jets of 1 or 3 particles, and zero net charge. Each three prong decay candidate was subject to the following track quality conditions: an average of at least 7 hits per track, χ^2 for the vertex fit less than 15 for 3 degrees of freedom, and net charge equal to ±1. Events passing these criteria were examined visually; about half were rejected as obvious background, mostly single photon or two photon multihadrons, often containing extra unreconstructed tracks.

Remaining backgrounds were: (1) Bhabha electron pairs with an additional pair from conversion of a radiated photon that failed to be reconstructed as such; (2) $ee\tau\tau$ and ee(hadron) events with undetected electrons; (3) beam-gas interactions; and (4) 4- and 6-prong multihadron events.

To remove these backgrounds the following additional requirements were imposed: (1) the total energy computed from the calorimeter pulse heights less than the pure-electromagnetic equivalent of 24 GeV, to eliminate converting radiative electron pairs; (2) charged-track sphericity less than 0.03, which Monte Carlo calculations showed would retain almost all τ pairs while removing a large fraction of beam-gas, $ee\tau\tau$, and ee(hadron) events; (3) the net momentum of each triplet greater than 4 GeV/c to discriminate further against $ee\tau\tau$ events; and (4) the larger of the two jet invariant masses, determined from energy-flow in the calorimeters, less than 4.5 GeV/c² (the invariant mass distribution of the visible τ decay products is bounded by $m_\tau \simeq 1.8$ GeV, broadened by the hadron cascade in the calorimeter; from the corresponding distribution for detected multihadron events (see Fig. 7) we found that about 90% were excluded by this cut, independent of their charged multiplicity).

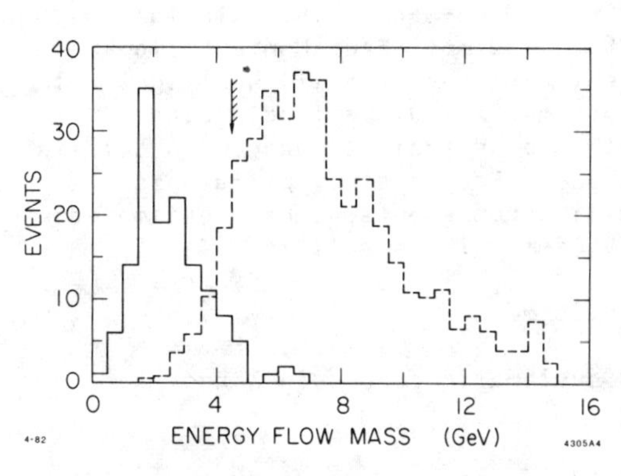

Fig. 7. Mass of the more massive jet for τ candidates (solid histogram and multihadron events (dashed histogram). The multihadron curve is normalized arbitrarily with respect to the τ curve.

From an integrated luminosity of 17 pb^{-1} we found 135 events. Table 2 shows the distribution of these events among the several decay modes of the second τ, together with theoretical estimates[6]. We find generally good agreement, except for an indication of fewer 3-prong and more 1-prong multihadron decays than predicted.

X	theoretical b.r.	theoretical yield	observed yield
e	0.18	23.5	23
μ	0.175	22.9	26
π	0.111	14.5	10
$\pi + n\pi^0$	0.334	43.7	55
3-prong	0.187	24.4	15
	0.987	129.0	129

Table 2. τ decay modes from $e^+e^- \to \tau(\to\text{3-prong})\tau(\to X)$. The first column lists theoretical branching ratios from ref. 6. The second column entries are branching ratio times 129, to be compared with the observed breakdown of the 129 events for which assignments could be made.

Fig. 8 shows the distribution in calculated error of the flight path. For the decay path distribution of Fig. 9 we include all events for which the error is less than 8 mm. From Fig. 9 we see that the flight path does appear to be non-zero. The calculated mean from Fig. 9(b) is $\langle x \rangle = 1.75 \pm 0.40$ mm. From Monte Carlo calculations we find good agreement with the data if the flight path is taken to be $\lambda = 1.20$ mm; for assumed $\lambda = 0$, we find a bias $\langle x \rangle_b = 0.55$. The stability of this bias against changes in details of the Monte Carlo has been checked. From a study of fake τs selected in various ways from multihadron events we have estimated the systematic error of λ to be 0.3 mm. The result is

$$\lambda = 1.20 \pm 0.39 \pm 0.30 \text{ mm}$$

Combining the statistical and systematic errors in λ and converting to a lifetime value gives

$$\tau_\tau = (4.9 \pm 2.0) \times 10^{-13} \text{ sec} ,$$

to be compared with the τ/μ unversality prediction of $(2.8 \pm 0.2) \times 10^{-13}$ sec.

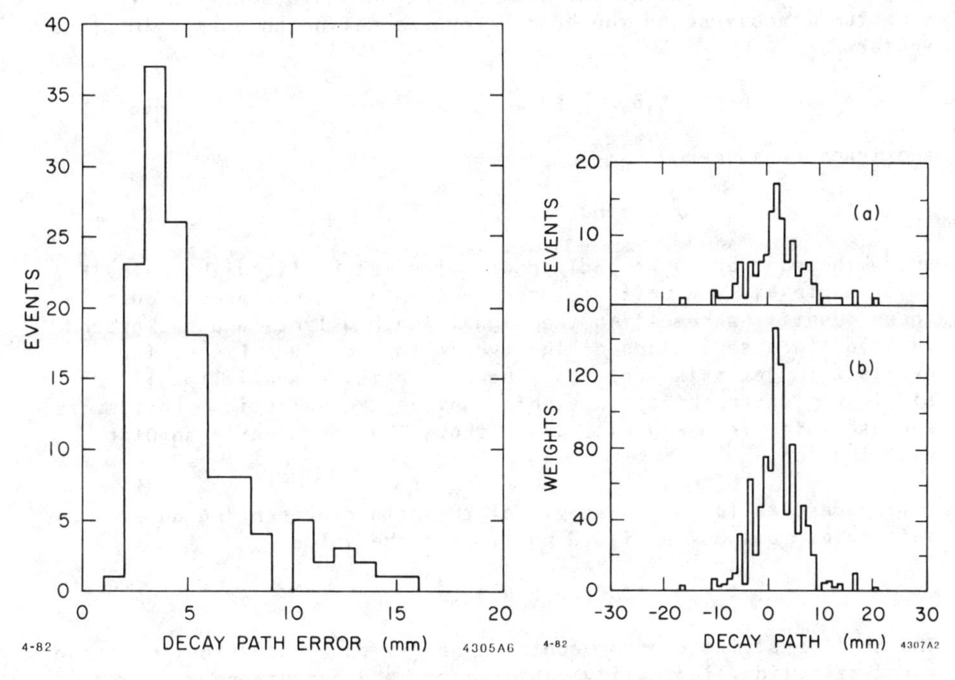

Fig. 8. Distribution of the decay path length error from vertex fits to three-prong τ decays.

Fig. 9. Decay length distribution for τ decays; (a) unweighted; (b) weighted by the reciprocal squared error.

HADRONIC CROSS SECTION (R)

The ratio R of the cross sections of ee→hadrons to ee→μμ is expected to deviate from the simple quark model prediction (3 2/3 at PEP energies) due to gluon radiative corrections, in a way calculable with quantum chromodynamics (QCD). The first order correction term is expected to be approximately 5%. A measurement of R with sufficient accuracy would be another important test of QCD. The nearly complete solid angle acceptance and calorimetric measurement of the total hadron energy make the MAC detector well suited to a precision measurement of R.

The primary difficulty involved in this measurement is to separate the one-photon annihilation signal from a large background dominated by two-photon annihilation. This is illustrated in Fig. 10, which shows the total energy distribution for a sample of events with at least five reconstructed tracks originating from a common vertex, consistent with coming from the interaction point. The low energy two-photon background is separated out by taking advantage of the fact that these processes tend to have a relatively large net momentum along the beam direction and a small total momentum transverse to the beam. Thus we define an energy imbalance vector

$$\vec{\beta} = \sum E_i \hat{n}_i / \sum E_i \ , \tag{6}$$

and transverse energy

$$E_\perp = \sum E_i \sin\theta_i \ , \tag{7}$$

where the sums are over individual calorimeter hits with energy E_i, polar angle θ_i, and unit vector direction $\hat{n}_i$. Appropriate cuts in these quantities result in the signal (dashed line) and background (dotted line) separation of the events in Fig. 10. Except for events with jet axis very near the beam axis, essentially all single-photon-annihilation events survive the selection process, as one can infer from Fig. 11, which shows the thrust-axis angular distribution.

The results for a sub-sample of the data representing an integrated luminosity of 5.5 pb^{-1}, give the value

$$R = 4.1 \pm 0.05 \pm 0.3 \ .$$

The total systematic error contains nearly equal contributions from event selection, luminosity measurement, and acceptance and radiative correction calculations. A value of 3.9 is expected from QCD with lowest order radiative corrections and α_s=0.17. This result is still very preliminary and we expect the systematic errors to decrease substantially in the near future.

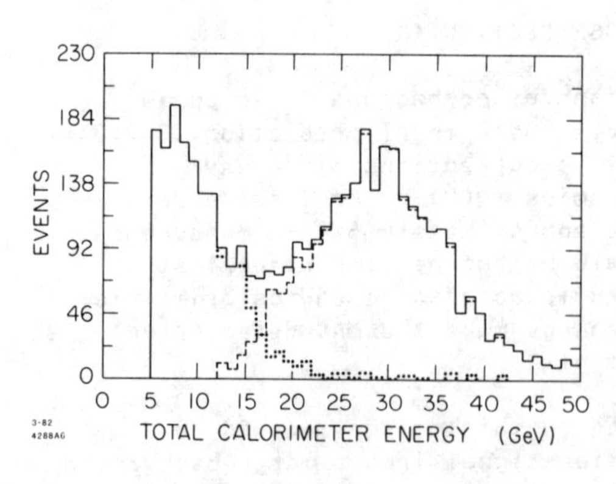

Fig. 10. Total energy for all ≥5-prong events. The dashed histogram represents events passing the selection criteria for multihadron events; events in the dotted histogram are the remainder.

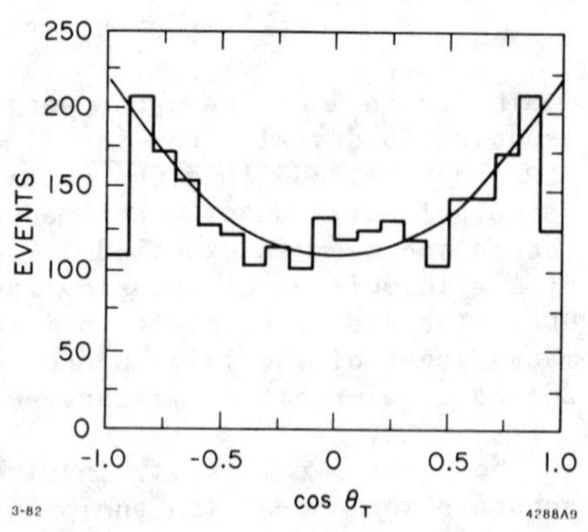

Fig. 11. Angular distribution of the thrust axis for multihadron events, determined from calorimeter energy.

ENERGY CORRELATIONS

Considering hadron production in more detail, a useful approach is to study the angular distribution of the final-state energy, or the "antenna pattern". The energy-weighted cross section can be computed in QCD and is free of divergences. The jettiness is measured by the energy correlation[7],

$$\frac{1}{\sigma}\frac{d\Sigma}{d\cos\chi} = \frac{1}{N\cdot\Delta\cos\chi}\sum_{n=1}^{N}\sum_{i,j}\frac{E_i E_j}{W_n^2}, \qquad (8)$$

where σ is the total cross section, χ is the angle between energy parcels i and j of energy E_i and E_j, N is the number of events and W_n the visible energy of event n ($W_n^2 = s$ for an ideal detector). The second sum is over all pairs having angular separation in the range $\Delta\cos\chi$ about χ.

In zeroth order (parton model), where events consist of two collinear jets, the correlation cross section reduces to delta functions at $\cos\chi = 1$ (self-correlation) and -1 (opposite-parton correlation). Single gluon emmision spreads the two peaks. Since the gluon usually makes a small angle with the quark that radiated it, we get a contribution near $\chi = 0$ from the correlation between this quark and the gluon. At comparable nearness to 180° we get contributions from the correlation of the second (anti-)quark and both the first quark and the gluon, so the cross section is larger in the backward than the forward hemishphere. (The self-correlations still feed the δ-function, so the calculation is not taken too close to $\cos\chi = \pm 1$.)

Fragmentation of the partons leads to contributions of comparable magnitude to single-gluon radiation. The authors of ref. 7 give an estimate of the dominant part of the fragmentation by considering two jet events, i.e., creation of two collinear partons followed by their fragmentation:

$$\left.\frac{1}{\sigma}\frac{d\Sigma}{d\cos\chi}\right|_{frag} = \frac{C\langle h_\perp\rangle}{\sin^3\chi} \qquad (9)$$

Regardless of the specific form of equation (9), the fragmentation to this order is symmetric in $\cos\chi$, which suggests that the asymmetry,

$$A(\cos\chi) = \frac{1}{\sigma}\left[\frac{d\Sigma}{d\cos\chi}(\pi-\chi) - \frac{d\Sigma}{d\cos\chi}(\chi)\right], \qquad (10)$$

is a particularly clean measure of the QCD effects.

The preliminary result of this experiment for the correlation cross section is shown in Fig. 12. The data have been corrected for resolution and acceptance effects determined with a Monte Carlo calculation. The sample was restricted to events having thrust axis at least 30° away from the beam axis to avoid the detector's blind spots near the poles. The solid curve in Fig. 12 is the sum of first-order gluon radiation, quark fragmentation and initial-state photon radiation contributions and is the result of fitting over the range $-.8 \leq \cos\chi \leq .8$, to the data adjusting α_s and the fragmentation coefficient, $C\langle h_\perp \rangle$. One can see from Fig. 12 that we get a poor fit. We may take the point of view that the discrepancy is to be blamed on the fragmentation model and try fit to the asymmetry, as shown in Fig. 13. In this case we get a good fit with $\alpha_s = 0.22 \pm 0.01$ (statistical error only). Systematic detector biases are still being studied.

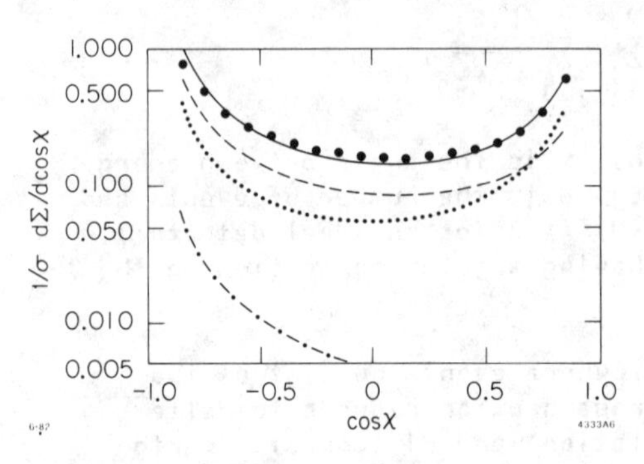

Fig. 12. Energy-correlation cross section. Dashed curve: QCD prediction from Ref. 7; dotted curve: fragmentation contribution; dot-dashed curve: radiative correction; solid curve: sum of all contributions.

More careful consideration shows that the fragmentation contribution will not be exactly symmetric, since fragmentation smearing of a three-jet event adds to the near-forward region from the correlations within each jet, but very little to the corresponding backward region. Instead, the interjet correlations merely smear the region already contributed by gluon emission. This effect may be considered to be higher order than α_s since it involves both gluon emission and fragmentation. In any case, the conclusive confrontation between experiment and QCD awaits a better understanding of fragmentation and probably higher-order QCD calculations as well.

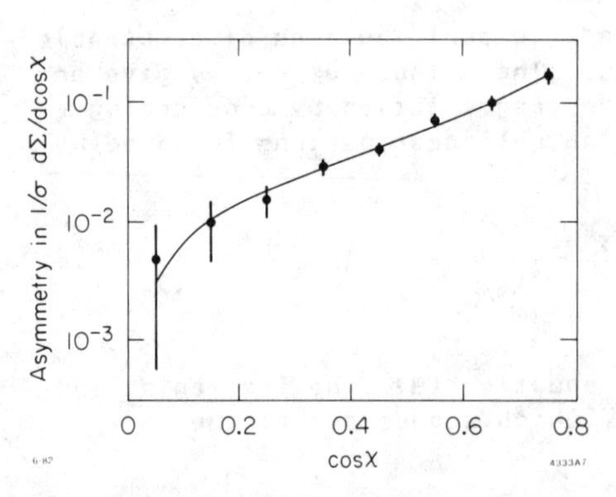

Fig. 13. Asymmetry of the energy-correlation cross section. The curve is the result of a fit of the QCD prediction to the data ($\alpha_s = 0.22$).

$e^+e^- \to \mu$ + HADRONS

Multihadron events which include identified leptons can be used for a variety of interesting physics objectives, which include: (1) measurement of the lifetime of the lightest meson containing b quarks (B), thereby obtaining information on the weak mixing angles; (2) measurement of a forward-backward asymmetry of large transverse momentum muons, which, for a pure b quark signal, would be expected to be a factor of three larger than the lepton pair asymmetry since the size of the asymmetry varies inversely as the parton charge; (3) measurement of the branching ratio for $B \to \mu + X$; (4) measurement of the fragmentation functions of heavy quarks; (5) search for production of new heavy quarks. Though the majority of this physics requires much larger data samples than are currently available, we give here a preliminary report on the status of our inclusive muon analysis and show our sensitivity to item (5) above.

Several features of the MAC detector are very favorable for obtaining a sample of multihadrons containing muons with low backgrounds from decay or punch-through of charged π or K mesons. The detector is compact, therefore minimizing the probability that a π or K decays before it interacts in the calorimeter. In addition, the matching of information in the central drift chamber, inner muon chambers, and outer muon chambers results in a poor χ^2 for most tracks resulting from K decays. Finally, punch-through background can be reduced to a very low level by examining the lateral spread and energy deposition of the track going through the hadron calorimeter.

A sample of multihadron events including an apparent muon was selected by requiring a visible track in the outer muon drift chambers (only those chambers covering 62% of the solid angle could be used for this analysis) correlating with a track in the hadron calorimeter which is μ-like in energy deposition and number of struck channels. Muon track information from the central and inner muon drift chambers is not yet used for this analysis. The momentum of each muon is determined by linking the outer muon chamber track with the interaction position measured by the central drift chamber, and the transverse momentum found relative to the thrust axis for the event. The $p_\perp$ distribution of 219 events from 2/3 of the total data sample is shown in Fig. 14 along with Monte Carlo

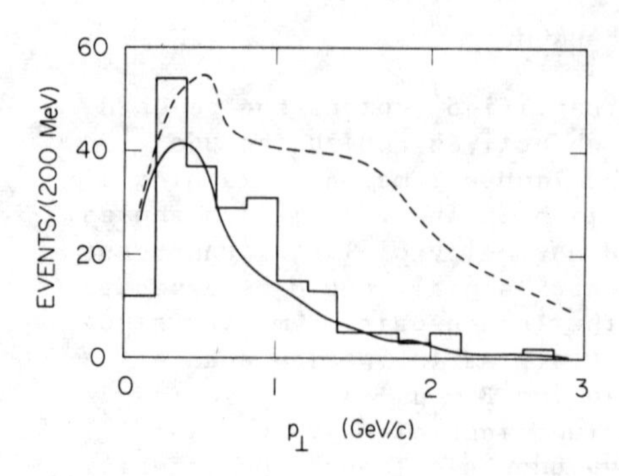

Fig. 14. Momentum perpendicular to the thrust axis for muons in multihadron events. Solid curve: calculated contribution from π, K, c, and b decay. Dashed curve: contribution of a 10-GeV t-quark with 10% muonic branching ratio.

calculated predictions including all known particles (solid curve) and adding a hypothetical sixth quark (dashed curve) of mass 10 GeV and charge 2/3 with a 10% branching ratio to muons. We can therefore exclude at the 95% confidence level the existence of such a new quark with a mass $m_q < 13.5$ GeV. Approximately 40% of the 50 events with $P_\perp > 1$ GeV/c are expected to result from the semileptonic decay of B mesons, while the remainder are roughly equally split between the semileptonic decays of charmed particles and background from π or K mesons.

REFERENCES

1. MAC Collaboration (W. T. Ford, et al.), SLAC-PUB-2894 (to be published in Proceedings of the 1982 International Conference on Instumentation for Colliding Beams, SLAC, edited by W. Ash).

2. The QED cross section is obtained from the Monte Carlo program described by F. A. Berends and R. Kleiss, Nucl. Phys. B177, 237 (1981), and includes all diagrams to order α^3.

3. Reconstruction in the outer drift chambers, which measure the bending of muons in the toroidal magnetic field of the calorimeter, was not yet available for the present analysis. For the central drift chamber alone, $\Delta p/p \simeq .85$ at 14.5 GeV/c, and the probability of mismeasuring the sign for one of the two muons is about 14%.

4. Y. S. Tsai, SLAC-PUB-2741 (1981).

5. A. Litke, Harvard University Ph.D. thesis, 1980 (unpublished).

6. F. J. Gilman and D. H. Miller, Phys. Rev. D17, 1846 (1978) N. Kawamoto and A. I. Sanda, Phys. Letters 76B, 446 (1978).

7. C. L. Basham, el al., Phys. Rev. D19, 2018 (1979); L. S. Brown and S. D. Ellis, Phys. Rev. D24, 2383 (1981).

RECENT RESULTS FROM MARK II[1] AT PEP*

Presented by

D. Schlatter, SLAC

Results on hadronic final states in e^+e^- annihilation are reported. The data were collected with the MARK II detector at the PEP storage ring at the Stanford Linear Accelerator Center, operating at a center-of-mass energy of $\sqrt{s}$=29 GeV. The MARK II detector, a 4.5 KG solenoid with cylindrical drift chambers, surrounded by a liquid argon calorimeter, has been described in detail in ref. 2. Hadronic events are selected by applying several cuts. There have to be at least 5 charged particles, each with momentum greater than 100 MeV, in an event. The total visible energy has to be larger than 8 GeV (or 15 GeV in the case of the energy correlation). The vertex position has to coincide with the beam crossing point. The data used for this report correspond to a total integrated luminosity of about 15 pb^{-1} collected in Spring 1981.

1. THE TOTAL CROSS SECTION

The total hadronic cross section at $\sqrt{s}$=29 GeV as expressed in terms of $R = \sigma_{had}/\sigma_{\mu\mu}$ is R = 3.90 ± 0.05 (statistical) ± 0.25 (systematic). Table I gives measurements of R from the MARK II detector at SPEAR[3] and PEP.

TABLE I. $R = \sigma_{had}/\sigma_{\mu\mu}$ with the MARK II Detector

The systematic error is 6% in all cases.

$\sqrt{s}$(GeV)	R	# Events
5.2	3.90 ± 0.02	44180
6.5	3.95 ± 0.05	11900
29.0	3.90 ± 0.05	4750

Within the systematic uncertainty of 6% there is no variation of R in the energy range from 5.2 to 29 GeV. The systematic error comes from the uncertainties in the background subtraction, event selection, radiative corrections and the luminosity measurements. The expected

*Work supported in part by the Department of Energy, contracts DE-AC03-76SF00515, W-7405-ENG-48, and DE-AC02-76ER03064.

0094-243X/82/930215-09$3.00

variations of R with energy are of the same order of magnitude (10% due to the onset of bottom production, 5% due to gluon bremsstrahlung and 3% due to electro weak interference) as the systematic uncertainty.

II. THE INCLUSIVE HADRON SPECTRUM

The inclusive cross section for hadrons, $sd\sigma/dx$, ($x = 2P/\sqrt{s}$) has been measured[4] both at PEP and SPEAR with the MARK II detector (Figure 1). The relative uncertainty among the three measurements in the

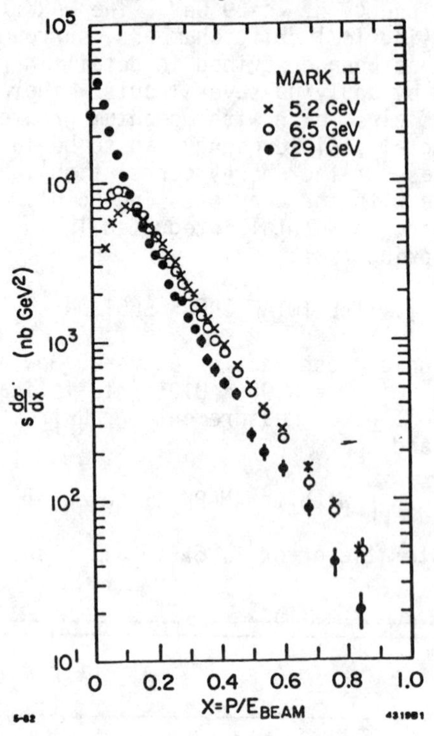

Fig. 1. $sd\sigma/dx$ at $\sqrt{s}$ = 5.2 GeV, 6.5 GeV and 29 GeV.

normalization is 10%. Strong scaling violations are observed.[4] At large x the cross section decreases with energy while at small x ($x < 0.15$) it increases. In Figure 2 the quantity $(1/\sigma d)\sigma/dx$ is plotted as a function of s for different bins of x together with data from the TASSO[5] group at PETRA. There is good agreement between the two experiments given the 10% uncertainty in the relative normalization.

Kinematic effects (in particular from the mass of the charm quark) as well as dynamic effects such as gluon radiation can cause scaling violations[6]. In figure 3 ratios of the inclusive cross sections at 29 GeV and 6.5 GeV from MARK II and at 34 GeV (35 GeV) and 14 GeV from TASSO are shown. A pure perturbative QCD calculation

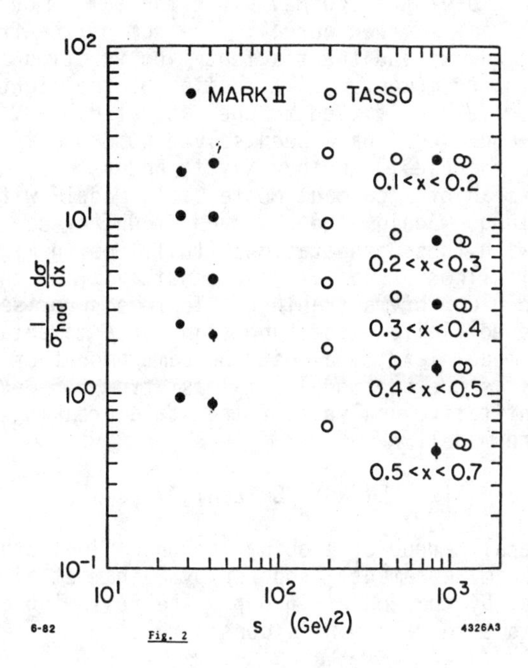

Fig. 2. $1/\sigma d\sigma/dx$ from MARK II and TASSO.

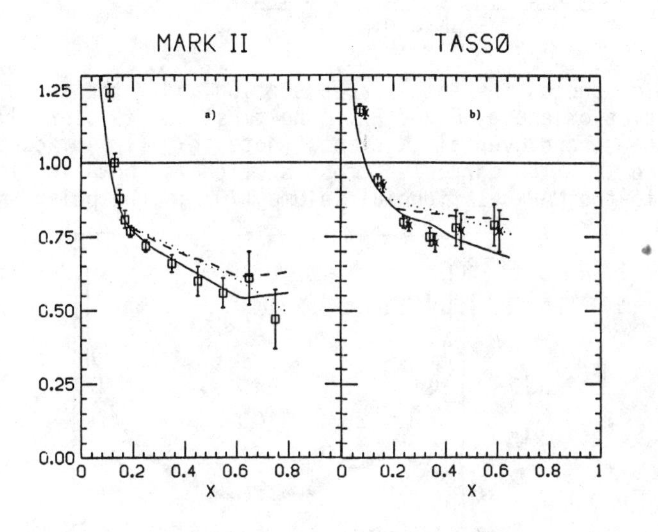

Fig. 3. Ratios of $1/\sigma d\sigma/dx$. A) MARK II for 29 GeV over 6.5 GeV. B) TASSO[5] 34 GeV (35 GeV) over 14 GeV. The full line is from a $q\bar{q}g$ Monte Carlo Model, the dotted line is is for $q\bar{q}$ two-jet Monte Carlo, the dashed line is an analytic calculation of perturbative QCD.

with a scaling parameter Λ = 200 MeV gives the same amount of scaling violations as the data (dashed curve). The sum of the fragmentation function of light quarks and the fragmentation function of the charm quark, folded with the momentum distribution of the light quarks from the charm decay, have been fitted to the data at 6.5 GeV. Then the Altarelli-Parisi equations[7] have been solved numerically to evolve the spectra to higher energies. Another way to understand the scaling violations is by mean of a cascade Monte Carlo Model[8] with single gluon bremsstrahlung. Again, the observed amount of scale breaking is in agreement with these expectations (full line in Fig. 3). The Monte Carlo model allows us to test the sensitivity of the inclusive hadron spectra to gluon bremsstrahlung. To some surprise the kinematic effects due to finite masses and transverse momenta in a pure $q\bar{q}$ fragmentation model lead to almost the same amount of scale breaking (dotted line) as the $q\bar{q}g$ model (at least from 6.5 GeV to 29 GeV). This makes a quantitative analysis of the scale breaking dependent on the details of the model.

III. ENERGY CORRELATIONS

Another general method of probing hadronic final states is the energy correlation measurement[9] proposed by Basham et al.[10] and previously studied by the PLUTO[11] group. The following cross section for the two particle correlation is considered:

$$\frac{1}{\sigma}\frac{d\,\Sigma}{d\cos\chi} = \frac{1}{N}\frac{1}{\Delta\cos\chi}\,\Sigma\,\Sigma\,\frac{E\,E'}{s} \qquad (1)$$

where σ denotes the total cross section and χ the angle between two particles of energy E and E'. The first sum is over all combinations and the second over all N events (note that $(1/\sigma)d\sigma/d\cos\chi$ is normalised to 1). The corrected cross section is shown in figure 4 normalised to the MARK II fiducial volume (70% in the polar angle and 86%

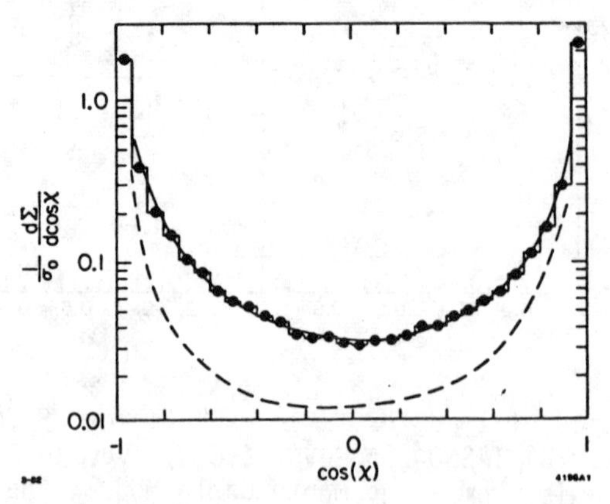

Fig. 4. Energy correlation cross section with the MARK II solid angle.

in the azimuth). Strong correlations inside a jet ($\chi < 40°$) and between opposite jets ($\chi > 140°$) are observed as expected from a two jet configuration. However this distribution is not symetric around 90°. Figure 5 shows the opposite to same-side asymmetry. Within the

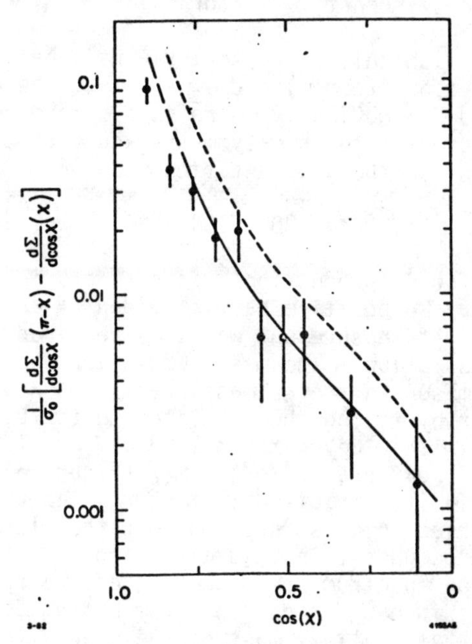

Fig. 5. Asymetry of the energy correlation.

model of ref. 10 the energy correlation cross section can be decomposed as follows:

$$\frac{1}{\sigma}\frac{d\Sigma}{d\cos\chi} = \alpha_s A^{\text{pert.QCD}}_{q\bar{q}g}(\chi) + A^{\text{hadr.}}_{q\bar{q}}(\chi) \tag{2}$$

The first term describes an asymmetric contribution from $q\bar{q}g$ events as calculated in perturbative QCD while the second is symmetric and accounts for the hadronization of $q\bar{q}$ events. At high energies the non-perturbative fragmentation term should be down by a factor of $1/\sqrt{s}$ and the $q\bar{q}g$ term should dominate. Possible contributions from fragmentation of $q\bar{q}g$ events are neglected so far. An attempt of a two parameter fit of eq. 2 in the angular range $40° < \chi < 140°$ yields a bad χ^2 (50 for 22 degrees of freedom) with $\alpha_s = 0.14$. To improve the fit we added a third term for possible $q\bar{q}g$ fragmentation. This term has to be asymmetric since for small angles ($\chi < 90°$) the fragmentation is the same as for $q\bar{q}$ events but the corresponding correlation at 180° vanishes in the 3-jet case. We have approximated this third term as follows:

$$A^{hadr.}_{q\bar{q}g}(\chi) = \alpha_s A^{hadr.}_{q\bar{q}}(\chi) \qquad \text{for } \chi < 90°$$
$$= \alpha_s (1 + \cos\chi) \text{ const.} \qquad \text{for } \chi > 90° \tag{3}$$

A three parameter fit with this extra term yields a better fit (χ^2 = 26 for 21 degrees of freedom) and α_s = 0.19. The result is shown in Figures 4 and 5. Obviously there is a strong contribution from fragmentation processes to the asymmetry and thus the determination of α_s is dependent on the fragmentation model.

IV. D* PRODUCTION

We have searched for D* production[12] in our data in the channel $D^* \to D^\circ\pi^+$, $D \to K^+K^-$. No positive particle identification has been used. The time of flight measurement was only required to be consistent with a π or K assumption. The mass resolution does not allow an observation of the D meson in the $K\pi$ mass spectrum. However, if one does a kinematical fit by fixing the $K\pi$ system to the D mass for all events in the interval 1.810 GeV < $M_{K\pi}$ < 1.93 GeV, a clear D* signal is observed in the mass difference $M_{K\pi\pi} - M_{K\pi}$ (Figure 6). There are 15 D* events at z > 0.4 (z = fractional D* energy) above a background of 1 event. The observed D* cross section is rather large ($\sigma(D^*)$ = 0.36 ± 0.16 nb), but the uncertainty is also large. In Figure 7 the corrected D* production spectrum as a function of the fractional energy, z, is shown. Since D* production from bottom decays is less than 20% and is mainly at small z, most of the events in Figure 7 are from a primary charm quark. Obviously the charm quark fragmentation function is different than the steeply falling light quark fragmentation functions. However, due to the small number of events

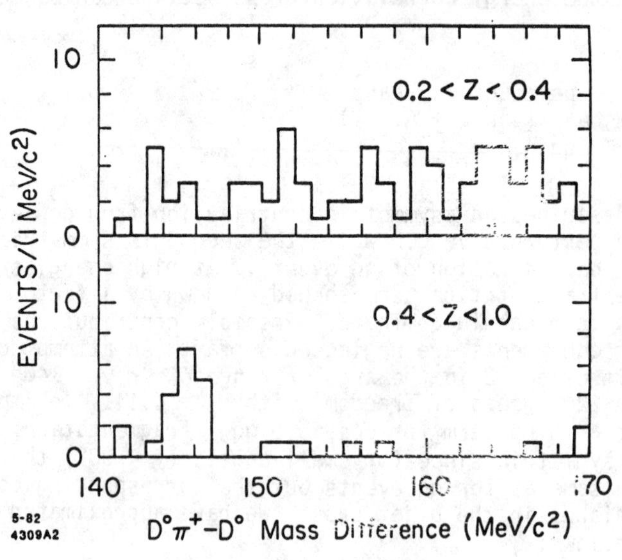

Fig. 6. Mass difference $M_{D\pi} - M_D$.

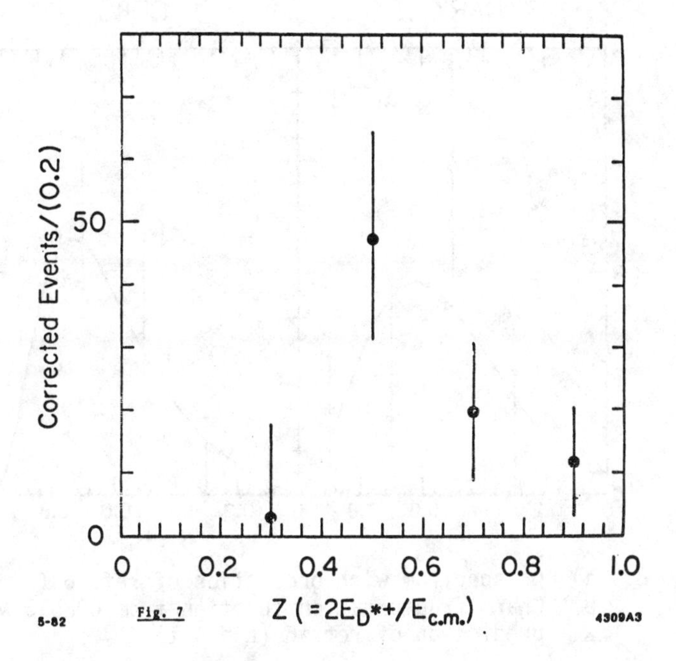

Fig. 7. Number of D* events as a function of z.

a flat fragmentation function cannot be ruled out, but the data would prefer a distribution peaked more at the center. A simple model[6] using kinematical considerations for heavy quark fragmentation gives a reasonable description of the data (fig. 8a). An indirect measurement of a charm fragmentation function has been reported by the CDHS group[13] from $\nu N \to \mu^+\mu^-$ hadrons events. They observe a similar distribution with an average z of 0.7 (fig. 8b).

CONCLUSIONS

The total cross section ratio R has been measured to within 6%, which is still too large to observe deviations from the quark parton model. The inclusive hadron spectrum shows strong scaling violations in the range of 5.2 GeV < $\sqrt{s}$ < 29 GeV. This is in agreement with cascade QCD Monte Carlo models including fragmentation. However, the energy may be still too low to clearly distinguish between perturbative effects of gluon radiation and non-perturbative effects from finite masses. Energy correlation at 29 GeV show an asymmetry as expected from QCD models, but a quantitative result for the strong coupling constant depends on details of the fragmentation model. The observation of D* production allows a first direct measure of the charm fragmentation function. At present small statistics, only steeply falling spectra are ruled out.

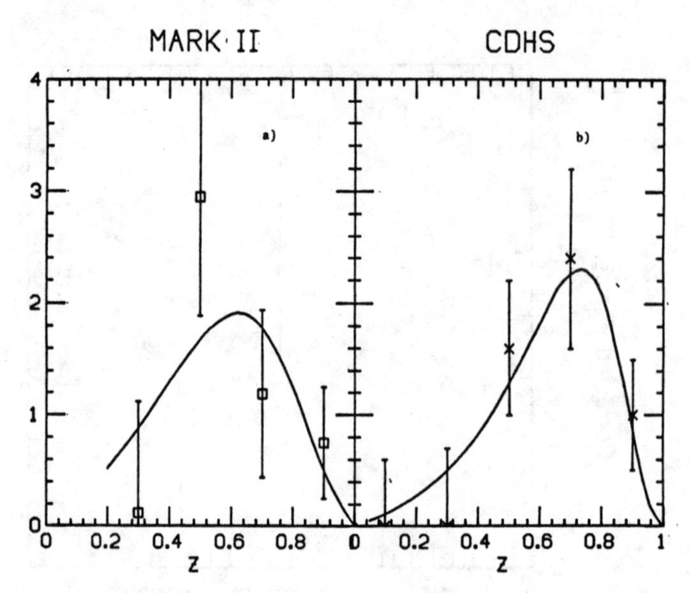

Fig. 8. a) D* spectrum with prediction of ref. 6 (ε = 0.2).
b) Charm fragmentation function from CDHS13 with prediction of ref. 6 (ε = 0.1).

REFERENCES

1. G. S. Abrams, D. Amidei, A. Backer, C. A. Blocker, A. Blondel, Adam M. Boyarski, Martin Breidenbach, D. L. Burke, William Chinowsky, M. W. Coles, G. von Dardel, W. E. Dieterle, J. B. Dillon, J. Dorenbosch, J. Dorfan, M. W. Eaton, G. F. Feldman, M.E.B. Franklin, G. Gidal, L. Gladney, G. Goldhaber, L. Golding, G. Hanson, R. J. Hollebeek, W. R. Innes, J. Jaros, A. Johnson, J. A. Kadyk, A. J. Lankford, Rudolf R. Larsen, B. LeClaire, M. E. Levi, N. Lockyer, B. Lohr, V. Luth, C. Matteuzzi, M. E. Nelson, J. F. Patrick, Martin L. Perl, Burton Richter, A. Roussarie, D. L. Scharre, H. Schellman, D. Schlatter, R. Schwitters, J. Siegrist, J. Strait, G. H. Trilling, R. A. Vidal, I. Videau, Y. Wang, J. Weiss, M. Werlen, C. Zaiser, G. Zhao Stanford Linear Accelerator Center Stanford University, Stanford, California 94305 Lawrence Berkeley Laboratory and Department of Physics University of California, Berkeley, California 94720 Department of Physics Harvard University, Cambridge, Massachusetts 02138.

2. R. H. Schindler et al., Phys. Rev. D24, 78 (1981) and references therein.

3. J. Patrick, Ph.D. thesis, LBL, 1982.

4. J. Patrick et al. SLAC PUB.

5. R. Brandelik et. al., DESY 82-13.

6. C. Peterson et. al., SLAC PUB 2912.

7. G. Altarelli and G. Parisi, Nucl. Phys. B126 (1977) 298 and J. F. Owens, Phys. Lett. 76B, (1978) 85 and T. Uematsu, Phys. Lett. 79B (1978) 97 and E. G. Floratos, C. Kounnas and R. Lacaze, Nucl. Phys. B192 (1981) 417.

8. We used the LUND Monte Carlo, T. Sjoestrand LU TP 80-3.

9. D. Schlatter et al., SLAC PUB 2846.

10. C. Basham, L. Brown, S. Ellis and S. Love, Phys. Rev. D19, 2018 (1979) and Phys. Rev. D24, 2383 (1981).

11. Ch. Berger et al., Phys. Lett. 99B, 292 (1981).

12. J. Yelton et al., SLAC PUB 2926.

13. J. Knobloch, Proc. of International Conference on Neutrino Physics, Maui, Hawaii, 1981.

PROMPT NEUTRINO PRODUCTION BY 400 GEV PROTON INTERACTIONS

R. C. Ball, C. Castoldi, S. Childress, C. T. Coffin, G. Conforto,
M. B. Crisler, M. E. Duffy, G. K. Fanourakis,
H. R. Gustafson, J. S. Hoftun, L. W. Jones, T. Y. Ling,
M. J. Longo, R. J. Loveless, D. D. Reeder, T. J. Roberts,
B. P. Roe, T. A. Romanowski, D. L. Schumann, E. S. Smith,
J. T. Volk, E. Wang.

FIRENZE - MICHIGAN - OHIO STATE
WASHINGTON - WISCONSIN Collaboration

Presented by T. Y. Ling,
Ohio State University, Columbus, OH 43210

ABSTRACT

A sample of 830 neutrino induced 1μ events and 752 muonless events recorded in a prompt neutrino experiment at Fermilab was analyzed to yield the following results: i) 0.65 ± 0.30 for the ratio of prompt $\bar{\nu}_\mu$ to ν_μ flux, ii) 1.0 ± 0.3 for the ratio of prompt ν_e to ν_μ flux for $E_\nu >$ 30 GeV and iii) inclusive $D\bar{D}$ production cross section $\sigma_{D\bar{D}}$ = 16 ± 3 (Stat.) ± 3 (Sys) μb per nucleon assuming $\sigma_{D\bar{D}}$ to be proportional to $(1-X_F)^3\, e^{-2P_\perp}$ and A^1 dependence.

In a beam dump experiment protons are incident on an absorber made of high density material. The beam deposits all its energy in the absorber and in the course of which all varieties of particles are produced. The short absorption length of the absorber causes long-lived particles such as pions and kaons to interact before they decay into neutrinos and muons. However, for particles with shorter lifetimes (< 10^{-10} sec) the reverse is true, i.e. they decay before interacting. The study of the neutrinos produced by these "prompt" decays thus provides a window to look exclusively at the production of new particles such as charm.

Preliminary results from E613, a beam dump experiment carried out at Fermilab, will be reported here. A layout of the experimental set-up is shown in Fig. 1. A 400 GeV/c proton beam is transported onto a tungsten target followed by a charged particle shield consists of 20 Tesla-meters of magnetized iron and 11 meters of passive iron. Strongly interacting particles are absorbed and most muons are ranged out or bent away from the neutrino detector located 56 meters downstream. Two targets, full and one-third density tungsten each 3

0094-243X/82/930225-16$3.00

interaction lengths long, are used in order to extrapolate to infinte density to determine the true prompt neutrino rate.

The employment of magnetized iron to bend muons away represents a unique feature of this experiment. This technique reduces the distance between the detector and the target and hence greatly increases the solid angle for neutrino detection. As a result higher neutrino rate per proton is achieved as compared to previous experiment of this kind. Table 1 lists the comparisons of the essential aspects between this and other beam dump experiments[1-4].

EXPERIMENT	TARGET	SHIELD	DUMP-DETECTOR DISTANCE (m)
CERN (BEBC, CHARM) CDMS	Cu ρ = 1, 1/3	PASSIVE Fe	830
E613-FNAL	W ρ = 1, 1/3 (Cu, Be)	MAGNETIZED + PASSIVE Fe	55

Table 1. Comparison of Beam Dump Experiments.

The detector, shown schematically in Fig. 2, consists of a 150 metric ton calorimeter made up of 30 modules followed by a muon spectrometer. The calorimeter modules are active targets for neutrino interactions. Each module contains 12 teflon-coated lead plates immersed in NE235 liquid scintillator. The lead plates are 6.3 mm thick to give a total of 14.4 radiation lengths, 0.5 hadron absorption length, and 105 gm/cm^2 of material per module. The modules are separated into five horizontal cells. Light from the liquid scintillator is detected by photomultiplier tubes at each end of a cell. Pairs of horizontal and vertical PWC planes are interspaced throughout the calorimeter modules to sample hadronic or electromagnetic shower profiles from neutrino interactions. The muon spectrometer consists of XYU drift chamber planes interspersed in iron toroidal magnets.

A set of six calorimeter modules was calibrated in a test beam to determine the energy scale and resolution. The calorimeter response was found to be quite linear with energy and to have a resolution $\sigma/E = .55/\sqrt{E}$ (GeV) for hadrons and $.27/\sqrt{E}$ (GeV) for electrons.

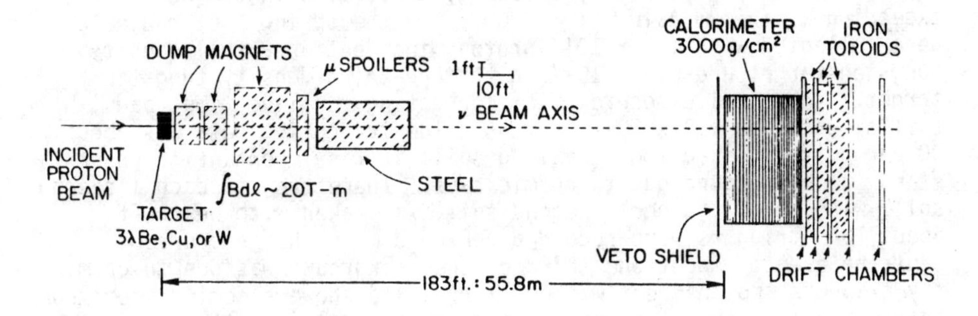

Figure 1. E613 beam dump experimental arrangement.

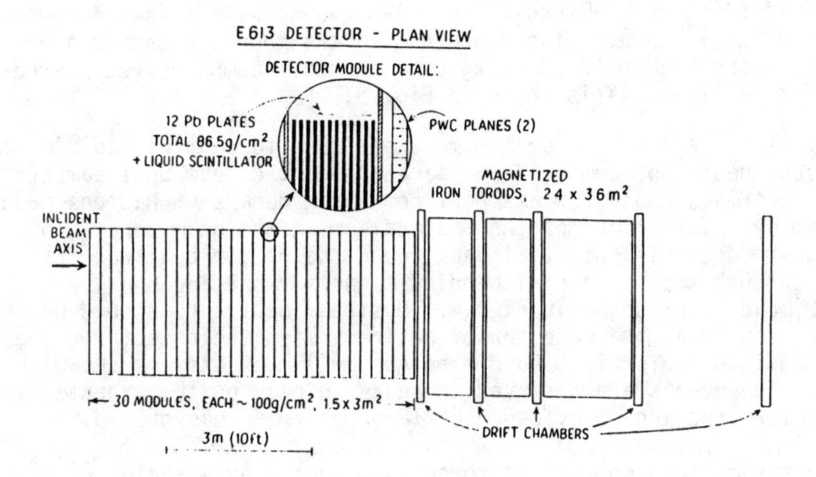

Figure 2. Schematic drawing of the detector.

The spatial resolution of the drift chamber was also measured in a test beam to be σ = 0.25 mm. The muon momentum resolution is mainly dominated by multiple scattering in the toroid. The $\Delta p/p$ for a 40 GeV muon traversing the entire spectrometer is about 15%.

Events are triggered by an energy deposition in the calorimeter exceeding a minimum value of $\sim$ 7 GeV. In the spring 1981 run data were collected with 1.3 x 10^{17} protons incident on a full density tungsten target and 0.3 x 10^{17} on the one-third density tungsten target. The total exposure after correcting for live time, bad spills, etc. was 1.0 x 10^{17} protons. The triggering rate was about 30 per pulse. Because of the long spill it was important to understand the background due to cosmic rays. Therefore for each 1 second spill with beam, another 1 second spill was taken with beam off. About 1000 triggers were recorded per good muon charged current neutrino event. About one third of the background was due to cosmic ray triggers and the rest was due largely to showers coming from muon interactions in the floor or concrete roof shielding blocks.

Neutrino induced events were separated into two categories: i) events with a clearly identified final state muon and ii) events without a muon. The events in the first category are candidates for charged-current ν_μ and $\bar{\nu}_\mu$ interactions-$(\nu_\mu + \bar{\nu}_\mu)^{CC}$. Those in the second category are a mixture of neutral current events from all neutrino types $(\nu_e + \bar{\nu}_e + \nu_\mu + \bar{\nu}_\mu)^{NC}$, charged current ν_e and $\bar{\nu}_e$ interactions-$(\nu_e + \bar{\nu}_e)^{CC}$ and $(\nu_\mu + \bar{\nu}_\mu)^{CC}$ events with a low energy or a large angle muon. Also contained in the muonless sample are cosmic ray and beam related background events. Computer reconstruction of typical events is shown in Fig. 3.

It is essential in a beam dump experiment to control and minimize background neutrinos coming from places along the beam upstream of the dump. Neutrinos could for example, come from decays of hadrons made upstream by interactions of the beam halo with the beam pipe. In order to reduce this potential background the proton beam was transported through two horizontal bending magnets resulting bents of $\sim$ 30 mrad each, shown in Fig. 4. Upstream neutrinos should be off the beam center at the detector of entirely miss the detector. The vertex positions distribution of the $(\nu_\mu, \bar{\nu}_\mu)^{CC}$ events are shown in Fig. 5. The events are observed to center nicely on the expected beam center, showing no evidence of large upstream background.

Upstream scraping was monitored continually by a series of monitors around the beam pipe. These were calibrated by introducing known amounts of material into the beam. The background was measured from these monitors to be less than 0.3% of the non-prompt neutrinos from our high density tungsten target for most of the run. Material just upstream of the target (SWICS, SEM, etc.) introduced a larger but calculable background of 11.6% of the non-prompt neutrino flux from full density tungsten target.

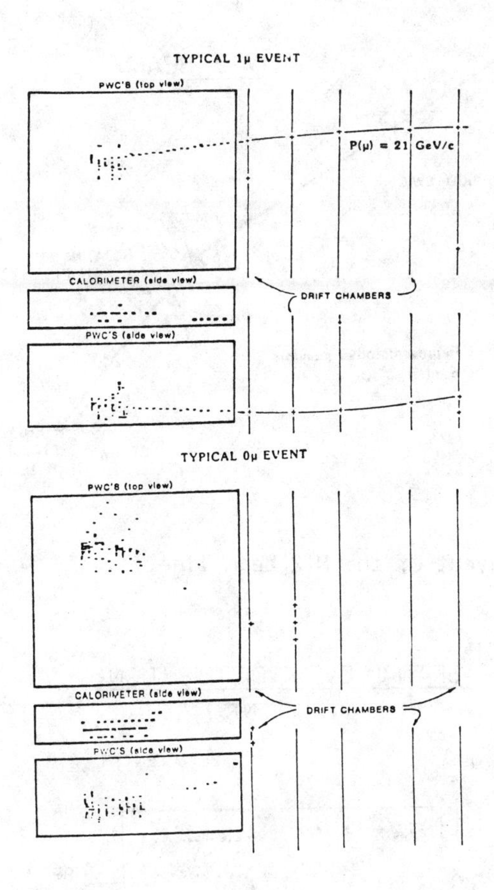

Figure 3. Computer reconstruction of a ν_μ^{CC} event and a muonless event.

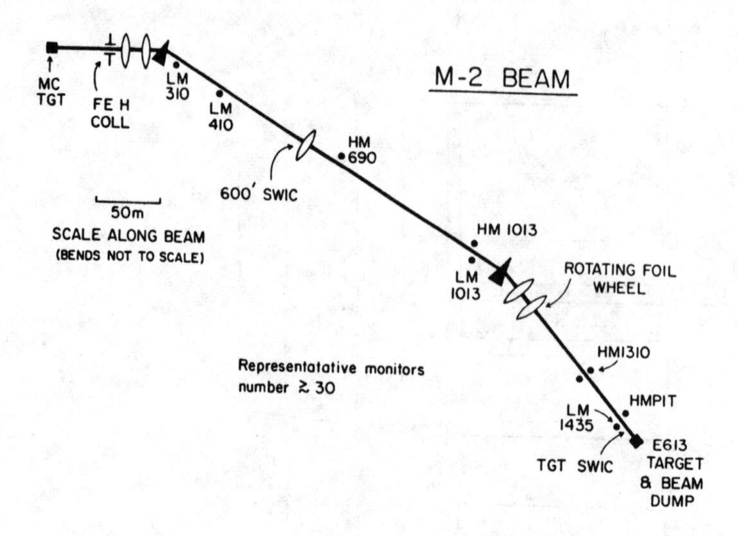

Figure 4. Layout of the M-2 beam line.

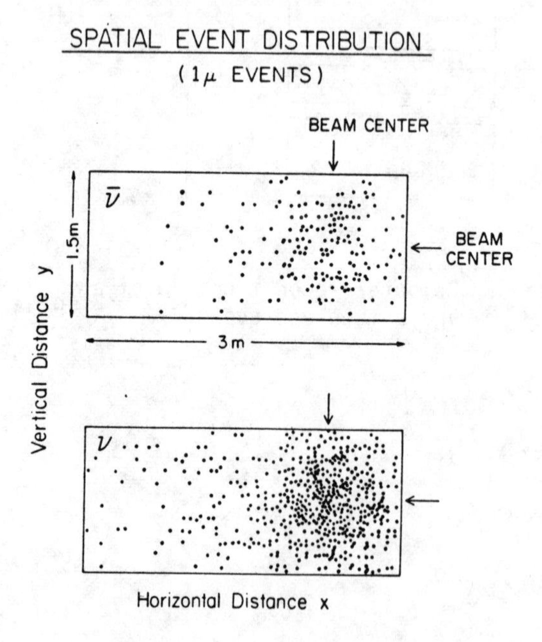

Figure 5. Distribution of transverse vertex position for $\nu_\mu CC$ and $\bar{\nu}_\mu CC$ events.

It was necessary to correct the extrapolation for the upstream background described above. In addition, a correction was required for hadrons which punched through the three absorption length targets into the 11 m iron dump downstream of the target. The net effect of these corrections on the extrapolation is to reduce the effective density ratio of the two targets from 3 to about 2.6.

i) One Muon Data

There are 830 fully reconstructed $(\nu_\mu, \bar{\nu}_\mu)^{CC}$ events within the fiducial volume defined as calorimeter modules 3 through 25 and by a boundary 15 cm in from the lead plates in the transverse dimensions. The combined efficiency of scanning and momentum reconstruction is 80%.

The data had to be corrected for trigger inefficiencies and for muon acceptance through the spectrometer. These effects distort the observed Bjorken x and y distributions. However, the distributions are known and the distortions are calculable. The data are in good agreement with the predicted shape of these x and y distributions, as shown in Figs. 6 and 7.

The raw and corrected numbers of ν_μ and $\bar{\nu}_\mu$ CC events are listed in Table II separately for the runs with two target densities. In Fig. 8, the corrected numbers of events are plotted as a function of $(\text{Density})^{-1}$. By extrapolation to infinite density we obtain 63 ± 13 prompt ν_μ events and 19.6 ± 7.7 prompt ν_μ events per 10^{16} protons on target.

	$\rho = 1$; W	$\rho = 1/2.7$; W	
NO. LIVE PROTONS	7.9×10^{16}	2.1×10^{16}	
ν_μ^{RAW}	415	197	
$\bar{\nu}_\mu^{RAW}$	145	73	PROMPT
$\nu_\mu^{CORR.}/10^{16}p$	117 ± 6	209 ± 15	63 ± 13
$\bar{\nu}_\mu^{CORR.}/10^{16}p$	42 ± 3.5	80 ± 9.4	19.6 ± 7.7

Table II. Numbers of $\nu_\mu^{CC} + \bar{\nu}_\mu^{CC}$ Events ($E_\nu > 20$ GeV; 6" Fid. Vol. Cut).

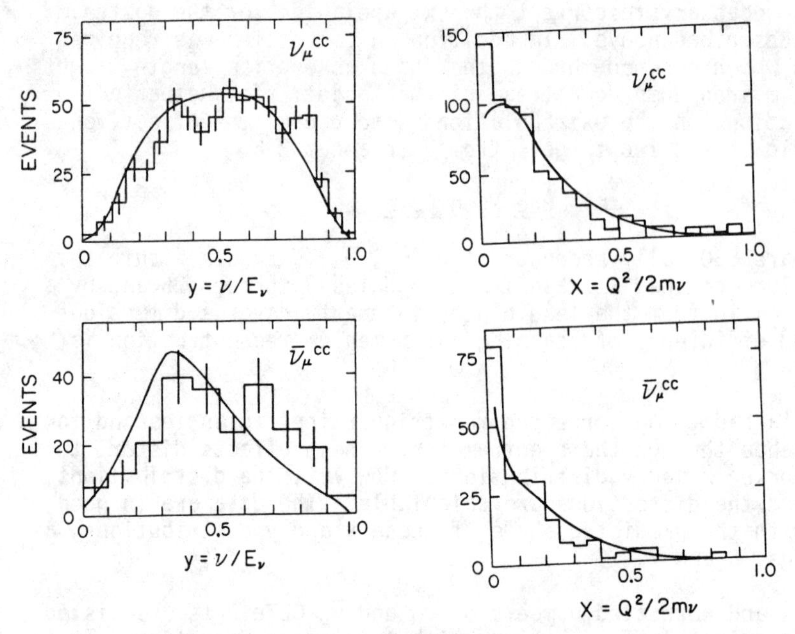

Figure 6,7. Bjorken x and y distributions from ν_μ CC and $\bar{\nu}_\mu$ CC events. Curves represent Monte Carlo predictions.

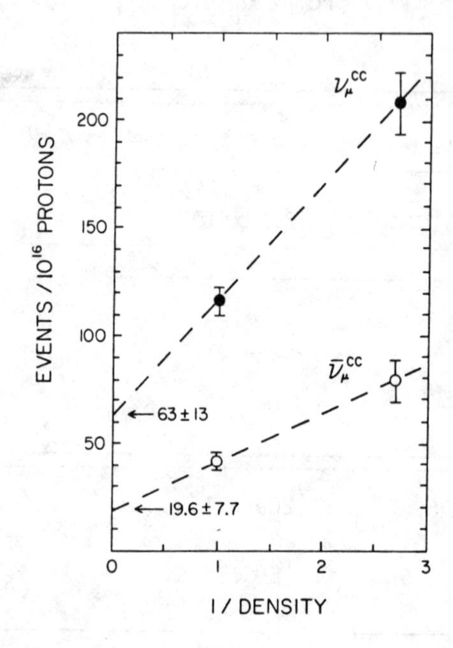

Figure 8. Extrapolation to infinte target density for ν_μ CC $\bar{\nu}_\mu$ CC events.

Taking the world average ratio of 0.48 for $\sigma(\bar{\nu}_\mu^{CC})/\sigma(\nu_\mu^{CC})$ we obtain a flux ratio of prompt $\bar{\nu}_\mu$ to prompt ν_μ of 0.65 ± 0.3. This is consistent with the naive expectation that the charmed particles are produced in pairs, e.g. $D\bar{D}$ so that the ratio $\bar{\nu}_\mu/\nu_\mu$ would be 1. Table III lists the comparison of this ratio between this and other experiments.

EXPERIMENT	$\bar{\nu}_\mu/\nu_\mu$
BEBC	0.79 ± 0.62
CHARM	$1.3 {}^{+0.6}_{-0.5}$
CDHS	$0.46 {}^{+0.21}_{-0.16}$
THIS EXPT	0.65 ± 0.30

Table III. Comparison of $\bar{\nu}_\mu/\nu_\mu$ Flux Ratio

The inclusive $D\bar{D}$ production cross section could be inferred from the prompt ν_μ and $\bar{\nu}_\mu$ rate if we assume all prompt neutrinos to be from decays of $D\bar{D}$ particles. We parameterize the $D\bar{D}$ cross section as

$$E\frac{d^3\sigma}{d_p^3} \alpha(1 - X_F)^n e^{-ap_\perp}$$

where X_F is Feymann X and $p_\perp$ is the transverse momentum of the D's.

In Figures 9 and 10 the energy and transverse momentum distributions of the prompt 1μ events corrected for muon and hadron acceptances are shown. It is shown that the above model with n=3 and a=2 give reasonable fit to the data.

It is also assumed that the protons cascade in the target with an average elasticity (ε) of 0.3 and that the charm production cross section varies as s^k with k = 1.3. (The CERN BEBC group[2] used ε = .67, k = 0.5.) The total $D\bar{D}$ production cross section is then calculated to be 16 μb ± 3 (statistical error) ± 3 (systematic error) for n=3, a = 2.

The BEBC group has obtained a value of $\sigma_{D\bar{D}}$ (n=3, a=2, ε=0.3, k=0.5) of 30 ± 10 μb from $\nu_\mu + \bar{\nu}_\mu$ and 17 ± 10 μb from $\nu_e + \bar{\nu}_e$ events[2]. The CHARM group reported (n=4, a=2, no cascading)

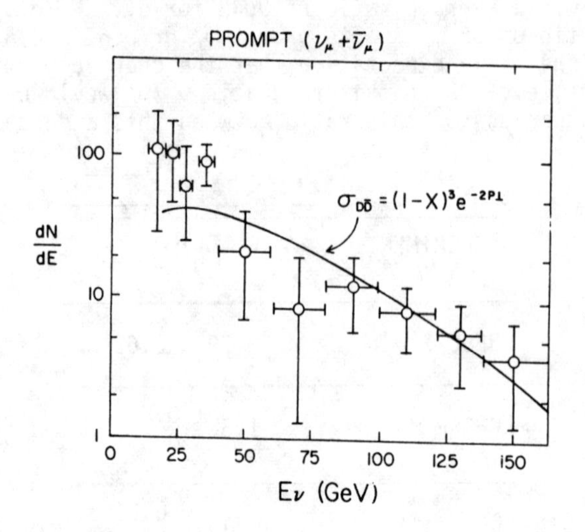

Figure 9. Energy distribution of prompt $\nu_\mu^{CC} + \bar{\nu}_\mu^{CC}$ events.

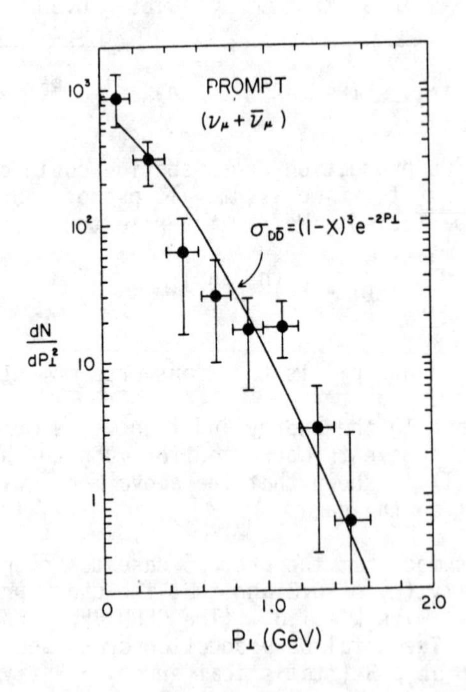

Figure 10. Transverse momentum distribution of prompt ν_μ^{CC} + $\bar{\nu}_\mu^{CC}$ events. The curves in this and the previous figure are the predictions of the n=3, a=2 model.

$19 \pm \mu b^3$. When these results are translated to n=3, a=2, ε=.3, k=1.3, the $D\overline{D}$ production cross sections are calculated to be $45 \pm 15\mu b$ (BEBC $\nu_\mu + \overline{\nu}_\mu$), $25 \pm 15\mu b$ (BEBC $\nu_e + \overline{\nu}_e$, $15 \pm 5\mu b$ (CHARM compared to $16 \pm 3 \pm 3\mu b$ from the present experiment.

ii) Zero Muon Data

The 0μ event sample was obtained by a combination of computer and human scan. The computer analysis required a "good" neutrino event with $E_{vis} > 20$ GeV reduced the trigger rate from $225000/10^{16}$ protons to $1000/10^{16}$. The remaining candidates were visually double scanned (see Fig. 3) for 0μ events (typical ν_e or NC interactions), 1μ events (ν_μCC interactions), and 1μ (miss) events (ν interactions with a muon outside the spectrometer acceptance). The double scan efficiency for 0μ candidates was 94%.

The efficiency of the computer selection for 0μ events was determined from the 1μ data by deleting the muon from the PWC and calorimeter data to create "pseudo 0μ" events which were typical of ν-induced NC interactions. These "pseudo 0μ" events were randomly interlaced among normal triggers with a signal/noise ratio of 1/17 compared with the normal 1/17 in the scan. The "pseudo 0μ" data was then processed in the normal manner; the computer analysis found $88 \pm 2\%$ of the hidden "pseudo 0μ". The scanners found all "pseudo 0μ" events found by the computer giving efficiency of $100 ^{+0}_{-5}\%$.

We found 101 1μ events for which muon misses the spectrometer in this scan. From the number of ν_μ CC events and the geometry of the detector we estimate 94 ± 11, a good agreement. These events are excluded from the 0μ sample.

The remaining sample of 0μ data contains $(\nu_e + \overline{\nu}_e)^{CC}$ events, $(\nu_\mu + \overline{\nu}_\mu + \nu_e + \overline{\nu}_e)^{NC}$ events, and cosmic ray events which have not been identified as such. Throughout the analysis the beam off data has not been distinguished from the beam on data. Therefore, the number of events in the beam off data is indicative of the number of cosmic ray events in the beam on data and can be subtracted. The number of $(\nu_\mu + \overline{\nu}_\mu)^{NC}$ events can be estimated by making an $E_{had} > 20$ GeV cut on the $(\nu_\mu + \overline{\nu}_\mu)^{CC}$ events. Table IV shows the raw data, beam off data, $(\nu_\mu + \overline{\nu}_\mu)^{NC}$ estimates, and the corrected true 0μ results.

Subtracting the $(\nu_\mu + \overline{\nu}_\mu)^{NC}$ events from the true muonless events we obtain the $(\nu_e + \overline{\nu}_e)^{CC+NC}$ events. The extrapolation of both the $(\nu_e + \overline{\nu}_e)^{CC+NC}$ and $(\nu_\mu + \overline{\nu}_\mu)^{CC+NC}$ event rate to infinite density is shown in Fig. 11. It is interesting to observe that the $(\nu_\mu + \overline{\nu}_\mu)$ slope to be much steeper than that of the $(\nu_e + \overline{\nu}_e)$ events, indicating that the non-prompt background for ν_e's to be much smaller than for ν_μ's. This agrees with the expection since non-prompt ν_e's can only arise from Ke3 decays while ν_μ's come from dominant decay modes of both π's and K's.

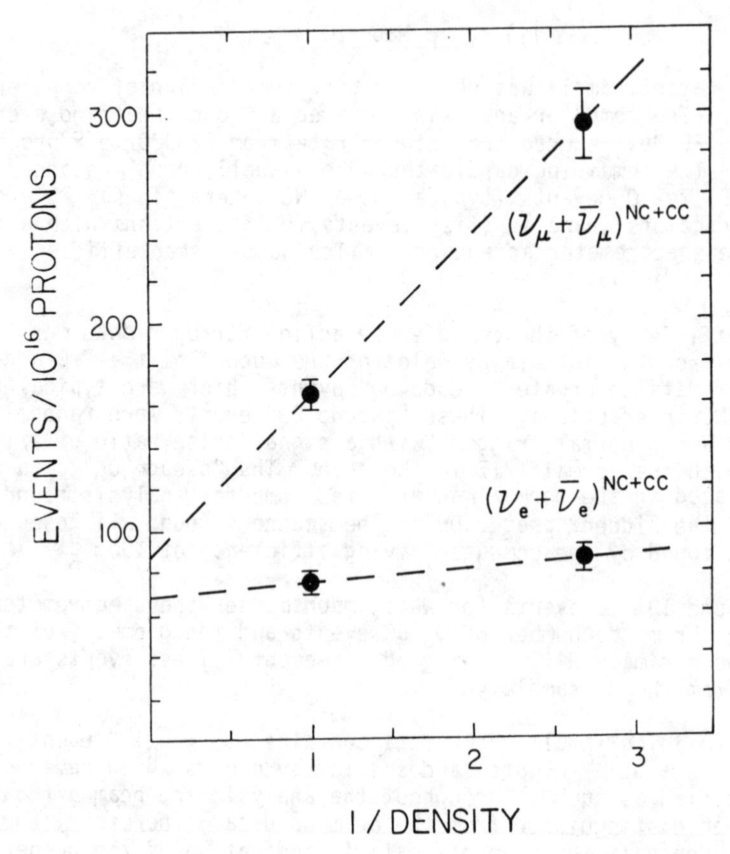

Figure 11. Extrapolation to infinite target density for $(\nu_e + \bar{\nu}_e)^{CC+NC}$ events. Data for $(\nu_\mu + \bar{\nu}_\mu)^{CC+NC}$ are also shown for comparison.

	Full ρ-W Tgt	Partial ρ-W Tgt	Remarks
Total 0μ	117	165	
C.R. BKGD	3	4	Est. From Beam Off Data
$(1\mu)_{E_\mu < 5 GeV}$	18	36	Calculate From 1μ Data
True 0μ	96 ± 4	125 ± 9	
$(\nu_\mu + \bar{\nu}_\mu)^{NC}$	21 ± 1	39 ± 3	Calculate From 1μ Data with the Known Ratio of $(\frac{NC}{CC})$
$(\nu_e + \bar{\nu}_e)^{NC+CC}$	75 ± 4	86 ± 9	
$(\nu_\mu + \bar{\nu}_\mu)^{NC+CC}$	165 ± 7	295 ± 17	From 1μ Analysis (8" F.V. Cut)

Table IV. Corrected Numbers of 0μ Events per 10^{16} Protons. $E_{vis} > 20$ GeV; 8" Fiducial Volume Cut.

We obtain, by extrapolation, 69 ± 8 prompt $(\nu_e + \bar{\nu}_e)^{CC+NC}$ and 88 ± 16 prompt $(\nu_\mu + \bar{\nu}_\mu)^{CC+NC}$ events per 10^{16} protons. The prompt ν_e to ν_μ event ratio is then 0.78 ± 0.17 for E > 20 GeV. This ratio can be interpreted as the prompt ν_e to ν_μ flux ratio if $\sigma(\nu_e)^{CC} = \sigma(\nu_\mu)^{CC}$ and $\sigma(\nu_e{}^{NC})/\sigma(\nu_e{}^{CC}) = \sigma(\nu_\mu{}^{NC})/\sigma(\nu_\mu{}^{CC})$ are assumed.

Fig. 12 shows the energy dependence of this ratio. Most of the systematic effects (acceptance, energy calibration, scan efficiency, etc.) are largest at the low energy limit. At this stage of the analysis the acceptance for $\nu_\mu{}^{CC}$ events from $20 < E_\nu < 30$ GeV is .27 whereas it becomes .77 above 60 GeV. Above 30 GeV (40 GeV) we find $\nu_e/\nu_\mu = 1.0 \pm 0.3$ (1.1 ± 0.4) and we conclude that this experiment finds no evidence that ν_e/ν_μ is not equal to 1.0. In Table V we show

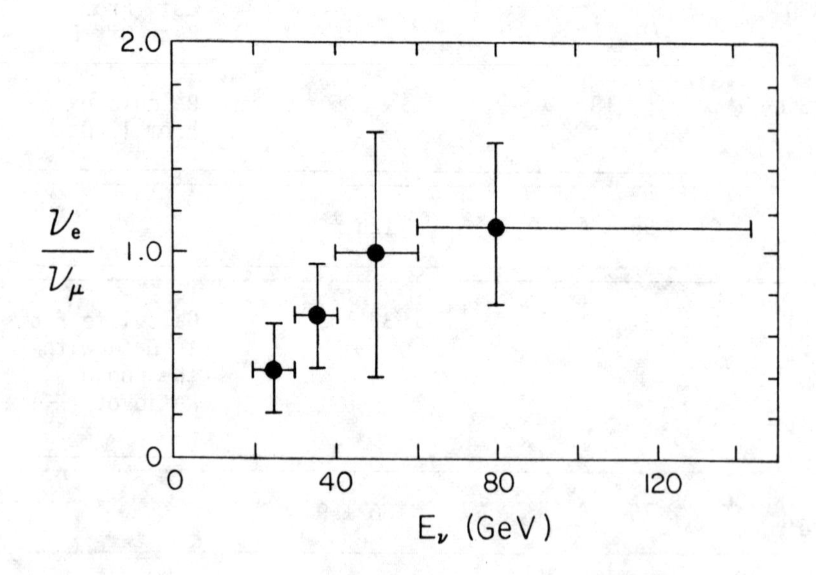

Figure 12. Energy variation of the ratio of ν_e events to ν_μ events.

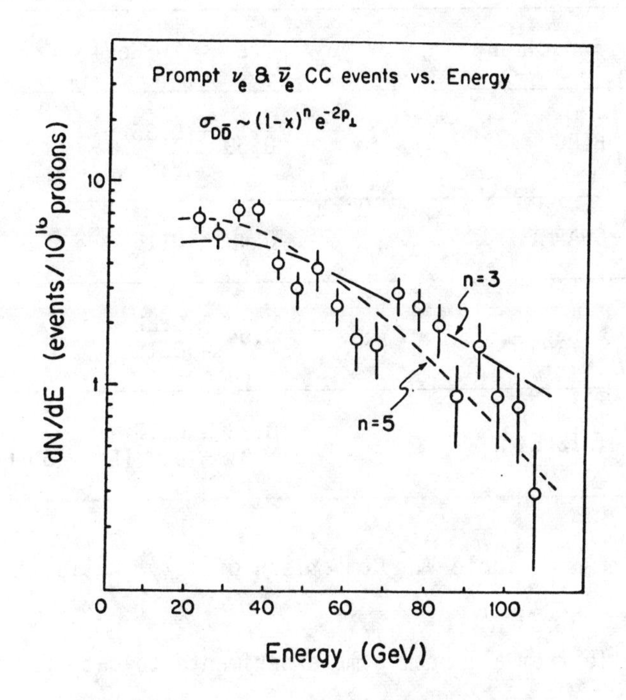

Figure 13. Energy spectrum of prompt ν_e events.

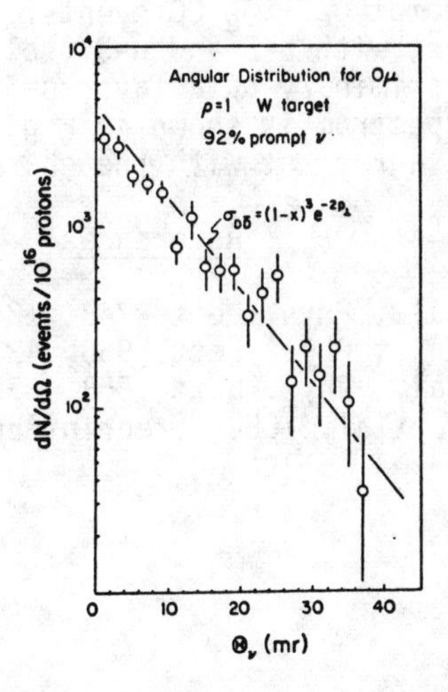

Figure 14. Angular distribution of prompt ν_e events.

Experiment	ν_e/ν_μ (E_ν > 20 GeV)
BEBC	$0.59 ^{+0.35}_{-0.21}$
CHARM	0.48 ± 0.16
CDHS	$0.64 ^{+0.22}_{-0.15}$
THIS EXPT	0.78 ± 0.17 1.0 ± 0.3 (E_ν > 30 GeV)

Table V. Comparison of ν_e/ν_μ Ratios.

this ratio from all beam dump experiments to date for comparison.

The ν_e data require no muon acceptance correction which is particularly useful for investigating the model dependent features of charm production. In Fig. 13 the normalized NC energy distribution is subtracted from the prompt 0μ data resulting in the energy dependence for prompt $\nu_e + \overline{\nu}_e$ CC events. The curves come from the model calculations, with b=2 and n=3 (solids) or n=5 (dashed). At this point the preliminary data favor n=5. The angular dependence of the neutrino spectrum is shown in Fig. 14 along with the model calculation. The agreement with the charm production model is good.

References

1. P. Aliban et al., Phys. Lett. 74B, 134 (1978).
2. P. Fritze et al., Phys. Lett. 96B, 427 (1980).
3. M. Jonker et al., Phys. Lett. 96B, 435 (1980).
4. H. Abramowicz et al., CERN preprint EP 82-17.

CHARM PRODUCTION MEASUREMENTS USING PROMPT MUONS

F.S. Merritt
University of Chicago, Enrico Fermi Institute, Chicago, IL 60637, USA

B.C. Barish, R.L. Messner, M.H. Shaevitz, E.J. Siskind
California Institute of Technology, Pasadena, CA 91125, USA

H.E. Fisk, Y. Fukushima, P.A. Rapidis
Fermilab, Batavia, IL 60510, USA

A. Bodek, R. Breedon, R.N. Coleman, W. Marsh, S. Olsen, J.L. Ritchie, I. Stockdale
University of Rochester, Rochester, NY 14627, USA

G. Donaldson and S.G. Wojcicki
Stanford University, Stanford, CA 94305, USA

ABSTRACT

Measurements of the hadronic production cross-section and distributions of charmed particles have been extracted from the measurements of prompt muons produced in 350 GeV pN interactions. The data indicate a charmed particle production cross-section of 24.6±3.9 μb/nucleon. The measured energy distributions strongly favor central production over diffractive production, and do not support the hypothesis of a 1% "intrinsic charm" component in the nucleon. In addition, the data give a production ratio of $\mu^-/\mu^+ = 1.10 \pm .22$, in good agreement with the ratio of 1 expected from central production of charm.

INTRODUCTION

We report on the measurement of prompt muons produced by 350 GeV pN interactions in an instrumental steel calorimeter.[1] Prompt muons are produced by the semileptonic decays of charmed particles produced by $pN \rightarrow C\overline{C}$ + hadrons, where the $C\overline{C}$ pair may include diffractively produced charm pairs (e.g., $\Lambda_C\overline{D}$) in addition to centrally produced charm (e.g., $D\overline{D}$). The data consist of single μ^+ and μ^- events originating from the semileptonic decay of one of the charmed particles, as well as $\mu^+\mu^-$ events with missing energy originating from the semileptonic decay of both charmed particles (producing $\mu^+\mu^-\nu_\mu\overline{\nu}_\mu$ in the final state).

The production rates and energy distribution of the μ^+, μ^-, and $\mu^+\mu^-$ events are used to distinguish between diffractive and central production, and to extract the production cross-section of the parent charm states. The data is compared to specific production models,[2] as well as to the similar measurements obtained from neutrino beam dump experiments.[3]

0094-243X/82/930241-09$3.00

In addition to the 350 GeV proton data, additional data was taken using a 278 GeV π^- beam. The single-μ pion data is still under analysis and will be presented elsewhere.[4] The $\mu^+\mu^-$ pion events with missing energy have been analyzed,[5] and will be presented here with the similar proton events.

APPARATUS

The hadron beam was targetted on an instrumented steel calorimeter measuring 30" by 30" in transverse dimension and containing 49 steel plates (20 1.5" plates, 25 2.0" plates, and 4 4.0" plates), each followed by a scintillation counter. The 3 meters of steel in the calorimeter gave very good shower containment and a resolution of 4% in hadron shower energy for a 350 GeV shower. Each of the 49 plates was moveable, so the density of the calorimeter could be easily changed to distinguish prompt muon signal from non-prompt π- and K-decays.

The calorimeter was followed by a muon identifier (see Fig. 1), consisting of 3 target carts from the CFRR neutrino experiment, each containing spark chambers and counters sandwiched between fourteen 10' x 10' x 4" steel plates. These were used to give very good acceptance and unambiguous identification of muons, and to impose minimum energy requirements on triggering muons.

A 10-foot diameter toroidal spectrometer followed the muon identifier. Momentum resolution was 11% for muons above the 20 GeV minimum energy needed to traverse the entire system.

BEAM CONDITIONS AND CUTS

The experiment took data using a 350 GeV proton beam (and 278 GeV π^- beam) from the Fermilab N5 beamline. Intensities were typically 10^5 particles per 1.5 second beam spill. To insure a clean beam, the momentum of each beam particle was measured to ±.5% using a dipole magnet and PWC's immediately upstream of the calorimeter.

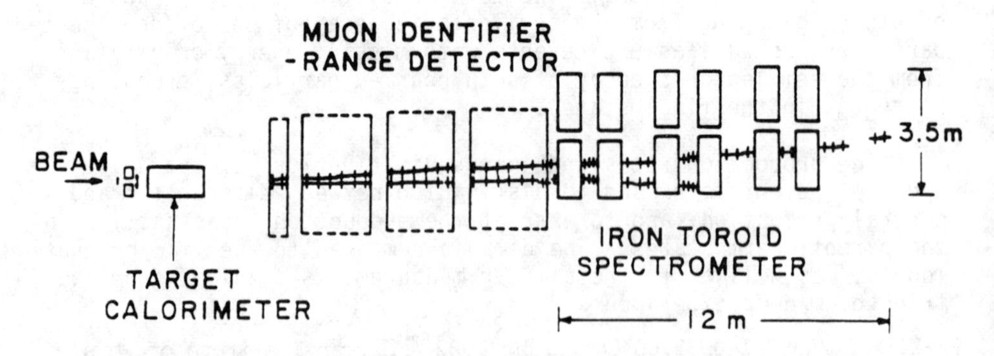

Fig. 1. Plan view of the apparatus.

Each accepted beam particle was also required to pass through a 1" x 1" trigger counter on the front face of the calorimeter. Beam accompanied by any halo particles was rejected by veto counters surrounding the beam counter (a 12" thick "albedo filter" with a hole in the center shielded the veto counters from back-scattered hadrons produced in the calorimeter). In addition, the beam was vetoed if there were any other beam or halo particles within ± 100 ns of the interaction time.

All of these conditions insured that there was no contamination in the beam. As an additional check, special runs were made with material placed upstream in the beam to simulate beam scraping and upstream interaction. No change in trigger rate was observed for these runs.

TRIGGERS

All triggers required a clean beam particle in coincidence with a hadron shower of at least 15 GeV within the first 10 plates of the calorimeter. Interactions satisfying this "clean beam" condition were scaled in order to normalize the muon triggering. A prescaler was used to trigger on one out of every 4096 such interactions through the run in order to give a continuous monitor of beam conditions and calorimeter resolution.

The primary muon trigger required a "clean beam" interaction in coincidence with a muon traversing the entire apparatus (muon identifier plus spectrometer). The minimum muon energy satisfying this trigger was 20 GeV. (A second muon trigger only required a muon in the muon identifier, corresponding to $E_\mu > 8$ GeV. However, this low-energy data is still being studied and only data from the first muon trigger is discussed here.)

Single-muon and two-muon events were taken with the same trigger, and were distinguished in the off-line analysis using the spark chamber and counter information from the muon identifier.

1μ EVENTS (350 GeV PROTONS ONLY).

In order to separate prompt single muons from muons produced by π and K-decay, the experiment was run at three different calorimeter densities: ρ_0, $2/3\ \rho_0$, and $1/2\ \rho_0$. The most compact density corresponded to $\rho_0 = .75\ \rho_{Fe}$. The mean interaction point (plate #5) was kept constant for all densities to insure equal acceptances.

Events were classified as single muon (1μ) or two muon (2μ) using the counters and chambers in the muon identifier. The rates of $(1\mu)^\pm$ and $(2\mu)^\pm$ events are plotted as a function of inverse calorimeter density in Fig. 2. The ± superscript for (2μ) refers to the sign of the triggering muon. Only $\mu^+\mu^-$ events are included in $(2\mu)^+$ and $(2\mu)^-$. The intercepts at $1/\rho = 0$ are the prompt μ^+ and μ^-

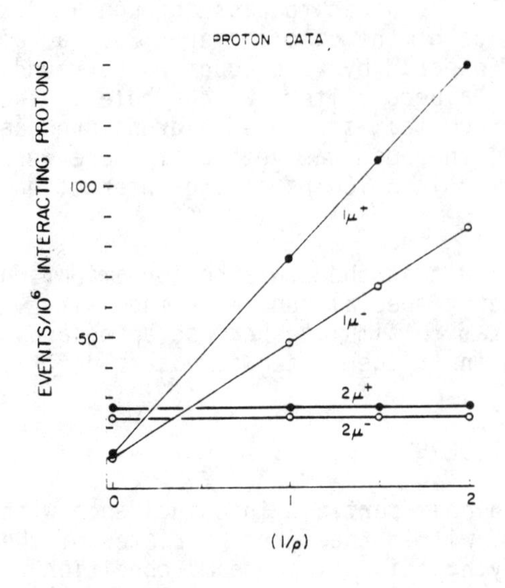

Fig. 2. Measured production rates of 1μ and 2μ events as a function of calorimeter density. (The superscripts refer to the sign of the triggering muon.) The intercepts at $1/\rho = 0$ of the lines through the 1μ data give the extrapolated prompt signals.

signals before corrections. (Note that the 2μ sample here does not have any missing energy requirement and therefore consists mainly of electromagnetic decays, Drell-Yan production, etc.) The uncorrected prompt rates were $11.62 \pm .89/10^6$ protons for $1\mu^+$ and $10.75 \pm .74/10^6$ protons for $1\mu^-$.

The largest background in the 1μ events came from highly asymmetric dimuon events, since muons with less than 5 GeV energy ranged out in the calorimeter and were not identified. This background was subtracted using a Monte-Carlo calculation normalized to the observed number of 2μ events. The calculation included all sources of dimuons including those produced by secondary hadrons and Bethe-Heitler conversion of π° decays.

Errors due to nonlinearities in the extrapolation (due to finite plate thickness and finite lifetimes of π's, K's, and hyperons) were calculated to be negligible. Background from decays occurring downstream of the expanded region of the calorimeter were calculated using a shower Monte-Carlo (propagated through six levels of secondary generation) to be less than 3% of the prompt signal.

After all corrections, the prompt single muon rates were $(6.83 \pm 0.92) \times 10^{-6}$ and $(6.80 \pm 0.77) \times 10^{-6}$ for $1\mu^+$ and $1\mu^-$, respectively. Using the ratio of prompt $(2\mu)^-/(2\mu)^+ = 0.90 \pm 0.01$ as a measure of the acceptance difference for $1\mu^+$ and $1\mu^-$ events, we find a corrected ratio of $1\mu^-/1\mu^+ = 1.10 \pm 0.21$. This agrees with the ratio of 1 expected from symmetric D and $\overline{D}$ production of central production model and does not confirm the asymmetry observed[3] by the CDHS neutrino beam dump experiment, which reported a prompt neutrino flux ratio of $\overline{\nu}_\mu/\nu_\mu = 0.46^{+0.21}_{-0.16}$. However, it should be pointed out that the neutrino experiment is primarily sensitive[3] to decays of charm particle with large values of $x(x_D > 0.8)$, whereas our prompt muon data has good acceptance for $x_D > 0.3$.

The prompt 1μ distributions (Fig. 3) were compared to the predictions of a model in which the D meson cross-section is parameterized as $E\frac{d^3\sigma}{dp^3} = C\ (1-x)^n\ e^{-2.5p_t}$. A Monte-Carlo program was used to generate the muons from the semileptonic decays of these D's and to propagate them through the detector with appropriate multiple scattering and dE/dx effects included. Monte-Carlo data tapes were produced which were then analyzed in the same way as the normal data. D production due to secondary interactions in the calorimeter were included, and amounted to 16% of the total signal. (The dependence of D production on incident energy was taken from a QCD central production model[6])

Comparison of the measured $1\mu^+$ and $1\mu^-$ energy distributions to the model predictions indicate that the invariant cross-sections have a $(1-x)^{6\pm.8}$ dependence (for $x \geq 0.3$). This best fit is compared to the data in Fig. 3.

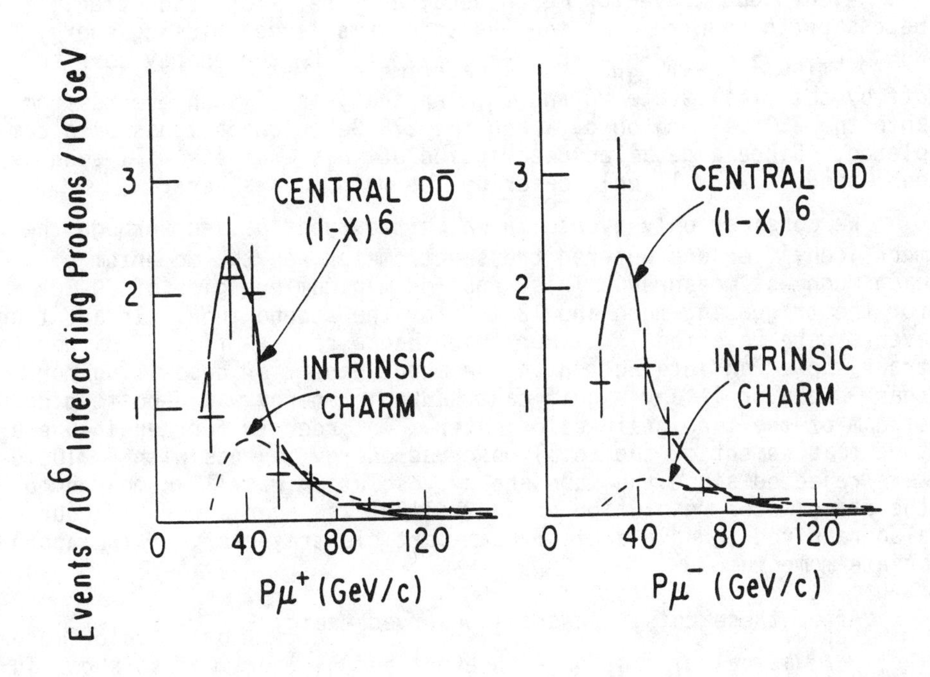

Fig. 3. Comparison of the prompt $1\mu^+$ and $1\mu^-$ proton data to the predictions of central and diffractive (intrinsic charm) production models.

Extrapolation of this fit yields a total $D\bar{D}$ pair production cross-section of (24.6±2.1±3.3) μb/nucleon ($-1<x<1$). The first error is statistical and the second reflects the uncertainty in the fitted exponent n. A linear A dependence and an 8% average semileptonic branching ratio were assumed. This is consistent with previous measurements using only high p_t prompt muons.[7]

We have also compared the data to the distributions expected from diffractive charm production ($pN \to \Lambda_c \bar{D}$ + hadrons) using the intrinsic charm model of Brodsky et al. and assuming a 3% branching ratio for $\Lambda_c \to \mu\nu\Lambda^\circ$.[8] It is not possible to extract a good estimate for the diffractive cross-section from this data alone due to the large backgrounds from central production, even at large E_μ, as shown in Fig. 3. An upper limit of about 10 μb/nucleon (assuming $A^{2/3}$ dependence) can be set from this 1μ data. However, a much better estimate of diffractive charm production can be obtained from the 2μ data as described below.

2μ EVENTS WITH MISSING ENERGY

A second method of studying charm production is to consider only those events in which both charmed particles decay into muonic final states. In order to separate these events from the large number of $\mu^+\mu^-$ events due to vector meson decay and Drell-Yan production, it is necessary to require that the events have a large "missing energy," where $E_{miss} = E_{beam} - E_{\mu^+} - E_{\mu^-} - E_{calorimeter}$ is the energy carried off by the final-state ν_μ and $\bar{\nu}_\mu$. An analysis of such events from both the 350 GeV proton data and the 278 GeV pion data has been completed. Since a detailed description of this analysis will soon be published,[5] we will only describe the main points here.

We consider only events in which two muons passed through the muon identifier and entered the spectrometer, so the momentum of each muon was measured. This required minimum energies of 20 GeV for the triggering muon and 12 GeV for the second muon. In addition, events were rejected if either track had a poor χ^2 fit, if either track showed an interaction in the muon identifier depositing more than 3 GeV, or if any significant hadronic energy was deposited downstream of the fine grain calorimeter. In order to further insure a good measurement of the total observed energy, events with $E_\mu > 100$ GeV were rejected since the muon energy resolution was 11% compared to the calorimeter resolution of 4%. All of the events were of course also required to have a clean beam-particle trajectory of the appropriate momentum.

After these cuts, the total observed energy $E_{tot} = E_{calorimeter} + E_{muons}$ was calculated for each event and histogrammed as shown in Fig. 4. For comparison, the dotted line shows the total energy distribution measured for the unbiased "clean beam" triggers (with no muon requirement) normalized to the number of $\mu^+\mu^-$ events. The measured E_{tot} distributions for $\mu^+\mu^-$ events in both proton and pion data show clear low-energy tails. In the proton data there are 59 $\mu^+\mu^-$ events with $E_{miss} = E_{beam} - E_{tot} > 45$ GeV, and in the pion data there are 154 $\mu^+\mu^-$ events with $E_{miss} > 40$ GeV.

Backgrounds in these regions due to energy resolution are calculated to be 5 events for protons and 10 for pions. Backgrounds due to double π- or K-decay, $\tau^+\tau^-$ production, or $B\bar{B}$ production, are estimated to be less than 5 events in each sample. We conclude that the balance of the events in the region with large E_{miss} is due to double charm decay.

Because of the cuts on minimum missing energy and muon energy, the total leptonic energy in these events must be large ($\gtrsim$ 75 GeV). The events must therefore be produced by charm states of high x ($x_{D\bar{D}}$>0.4). Consequently only a small fraction of centrally produced charm states contribute ($\lesssim$1%), but the acceptance for diffractive charm states is high. For example, using the diffractive intrinsic charm model of Brodsky et al. we calculate an acceptance of 16%.

In Fig. 5 we compare the measured leptonic energy distributions of these events to the predictions of the intrinsic charm model normalized to fit the data. To obtain a cross-section we assume a

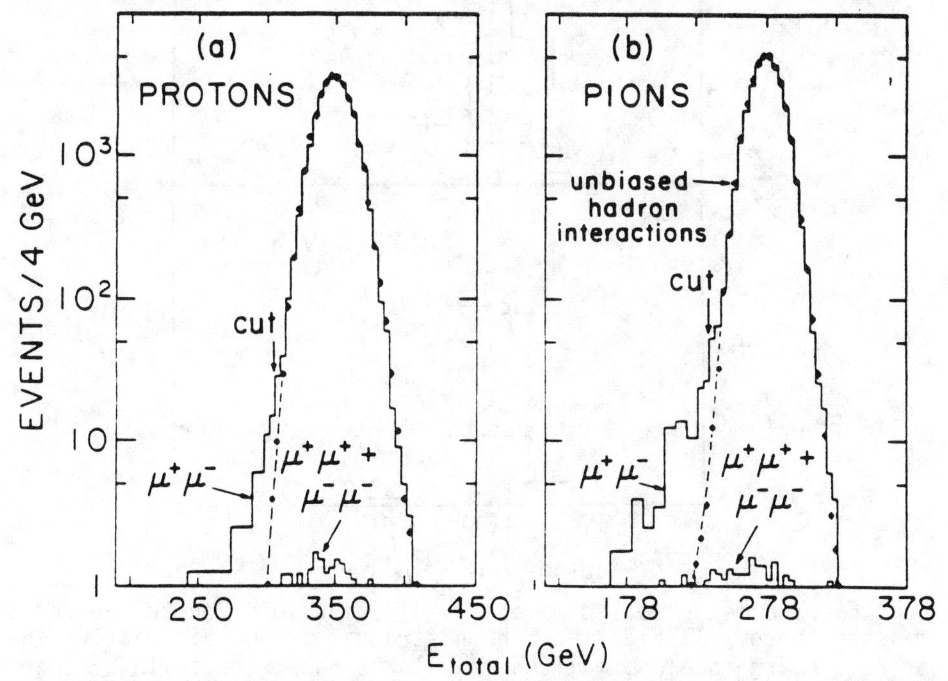

Fig. 4. Total observed energy distribution for opposite-sign and same-sign dimuon events. The solid circles show the observed energy distributions for the unbiased sample of hadron interactions taken simultaneously with the dimuon data.

3% branching ratio for $\Lambda_c \to \mu\nu\Lambda°$, an 8% branching ratio for $D \to \mu\nu K$, and an $A^{2/3}$ dependence on atomic number. If we assume all the events come from diffractive production, we obtain diffractive cross-sections of 5.3 μb/nucleon and 5.4 μb/nucleon for protons and pions, respectively; but these give very poor fits to the data as shown by the dotted-line histogram in Fig. 5. The poor fits are not surprising, since central production is expected to contribute in the lower-energy bins. If we normalize the model only to the data above E_{lepton} = 160 GeV, where central production should be much smaller, we obtain acceptable fits as shown by the solid histrograms. These give cross-sections of 2.5 μb/nucleon and 1.9 μb/nucleon for protons and pions, respectively.

These cross-sections per nucleon are nearly two orders of magnitude smaller than the predictions of 210 μb/nucleon for 350 GeV protons and 90 μb/nucleon for 278 GeV pions, which are given by the hypothesis of a 1% intrinsic charm component in the hadronic wave functions. Note that the measured cross-sections assume an $A^{2/3}$ dependence; if a linear A-dependence were assumed, these cross-sections would be reduced by a factor of 3.8.

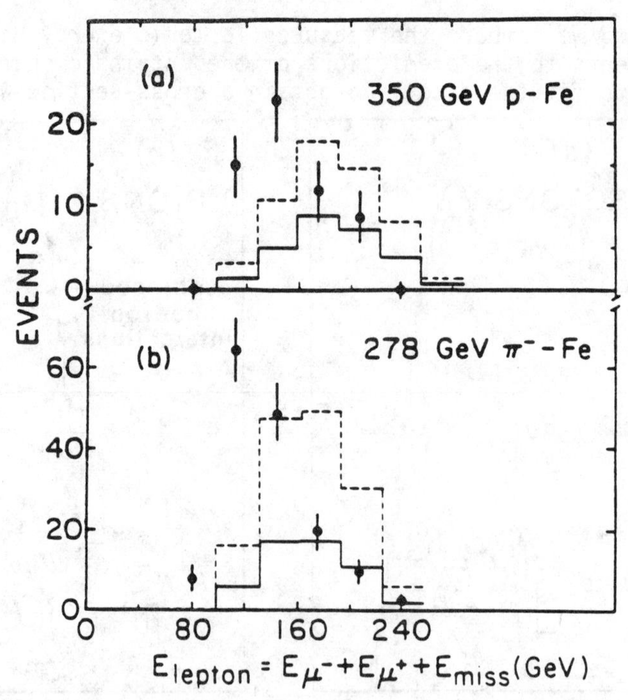

Fig. 5. The measured lepton energy distributions for 2μ events with missing energy. The dotted-line histrogram is the diffractive intrinsic charm model normalized to all the data; the solid histogram is the same model normalized to the data above 160 GeV. Note that the curves are normalized down by two orders of magnitude relative to the predictions of a 1% intrinsic charm component.

CONCLUSIONS

Both the 1μ and 2μ data reported here indicate that the cross-sections for 350 GeV $pN \rightarrow C\overline{C}$ + hadrons is predominantly due to central production rather than diffractive production. The best measurement of central production comes from the 1μ data, which gives a total cross-section of 24.6±3.9 μb/nucleon. The best measurement of diffractive production comes from the 2μ data with missing energy, which gives a model-dependent diffractive cross-section of 2.5 μb/nucleon for protons. The data are not consistent with a 1% intrinsic charm component in the proton wave function. In addition, the ratio $(1\mu^-)/(1\mu^+)$ = 1.10±0.21 from prompt muons is consistent with the ratio of 1 expected from central production of charm (and significantly higher than the analogous flux ratio of $\overline{\nu}/\nu$ reported by the CDHS neutrino beam dump experiment).

REFERENCES

1. A. Bodek et al., University of Rochester Preprints UR-783 and UR-792.
2. S. J. Brodsky et al., Phys. Letters 93B, 451 (1980); Phys. Rev. D23, 2745 (1981).
3. H. Abramowicz et al., Z. Phyzik C13, 179 (1982).
4. A. Bodek et al., University of Rochester Preprint UR-827, presented at "Neutrino 82," Balaton, Hungary, June 1982.
5. A. Bodek et al., Phys. Letters 113B, 77 (1982).
6. C. E. Carlson and R. Suaya, Phys. Letters 81B, 329 (1979).
7. A. Diamant-Berger et al., Phys. Rev. Letters 43, 1774 (1979).
8. Recent measurements indicate a Λ_c semileptonic branching ratio of (4.5±1.7)%. [E. Vella, Phys. Rev. Letters 48, 1515 (1982)].

Charged Current Neutrino Interactions

R.E. Blair[a], A. Bodek[b], Y.K. Chu[c], R.N. Coleman[d], O.D. Fackler[e],

H.E. Fisk[d], Y. Fukushima[f], K.A. Jenkins[e], B.N. Jin[d], Q.A. Kerns[d],

T. Kondo[f], D.B. MacFarlane[a], W.L. Marsh[d], R.L. Messner[g],

D.B. Novikoff[h], M.V. Purohit[a], P.A. Rapidis[d], S.L. Segler[d],

F.J. Sciulli[a], M.H. Shaevitz[a], R.J. Stefanski[d], D.E. Theriot[d]

and D.D. Yovanovitch[d]

Caltech, Columbia, Fermilab, Rochester, Rockefeller Collaboration (CCFRR).

Presented by H.E. Fisk

ABSTRACT

This report reviews our measurement of neutrino and antineutrino total cross sections. We also present new preliminary data on the F_2 and xF_3 structure functions and test the Gross-Llewellyn Smith sum rule $\int xF_3/_x dx$, which, in the quark model, predicts three valence quarks for the nucleon. Muon and neutrino derived $F_2(x)$ are also compared to test the fractional quark charge hypothesis. Finally preliminary results on the logarithmic Q^2 dependence of $F_2(x)$ are given.

[a]Present address: Nevis Laboratories, P.O. Box 137, Irvington-on-Hudson, N.Y. 10533
[b]Present address: Univ. of Rochester, Rochester, N.Y. 14627
[c]Present address: Lauritsen Laboratory, Caltech, Pasadena, CA 91125
[d]Present address: Fermilab, P.O. Box 500, Batavia, IL 60510
[e]Present address: Rockefeller Univ., New York, N.Y. 10021
[f]Present address: Nat'l. Lab. for High Energy Physics, KEK, Oho-machi Tsukuba-gun, Ibaraki-ken, 305 Japan
[g]Present address: SLAC P.O. Box 4349, Stanford, CA 94305
[h]Present address: Hughes Aircraft Co., El Segundo, CA 90245

INTRODUCTION

The structure functions are determined from the scaling variable x,y distributions where:

$$Q^2 = E_\mu E\, \theta^2 ,$$

$$x = Q^2/2ME_H ,$$

$$y = E_H/E = E_H/(E_\mu+E_H).$$

Here θ, E_μ and E_H, are the outgoing muon angle and energy and the hadronic shower energy, respectively. M is the nucleon's mass, and E is the neutrino energy which is E_H+E_μ. Assuming charge symmetry the structure functions F_1, F_2 and F_3 are given by differences and sums of the neutrino and antineutrino cross sections:

$$\frac{d^2(\sigma^\nu - \sigma^{\bar\nu})}{dxdy} = \frac{G^2ME}{\pi}\left[y(1-y/2)\right] xF_3(x,Q^2) \qquad (1)$$

and

$$\frac{d^2(\sigma^\nu + \sigma^{\bar\nu})}{dxdy} = \frac{G^2ME}{\pi}\left[(1-y - \frac{Mxy}{2E}\, F_2(x,Q^2)+\frac{y^2}{2}\, 2xF_1(x,Q^2)\right] . \qquad (2)$$

In terms of the parameter $R\equiv\sigma_L/\sigma_T$, the ratio of longitudinal (transversely aligned boson propagator) to transverse (longitudinally aligned boson propagator) cross sections the structure functions F_1 and F_2 are related by the expression:

$$\frac{2xF_1}{F_2} = \frac{1}{1+R}\left(1 + \frac{4\,M^2\,x^2}{Q^2}\right) . \qquad (3)$$

The cross section sum then becomes:

$$\frac{d^2(\sigma^\nu+\sigma^{\bar\nu})}{dxdy} = \frac{G^2ME}{\pi}\left[1-y +\frac{1}{2}\,\frac{y^2}{1+R}\right] F_2(x,Q^2) . \qquad (4)$$

Recall that in the quark model: $2xF_1 = q+\bar q$, $F_2 = q+\bar q+2K$, and $xF_3 = q-\bar q$, where K represents the non spin 1/2 nucleon constituents.

The quark longitudinal momentum fraction distributions are assumed to contain contributions from the four standard flavors: i.e.

$$q = x[u(x) + d(x) + s(x) + c(x)] \quad .$$

APPARATUS

Figure 1 shows a schematic drawing of the CCFRR detector, which includes a non-magnetic target calorimeter composed of 3m x 3m x 5cm Fe plates sandwiched with liquid scintillation counters and magnetostictive readout spark chambers. The positional accuracy of the chambers is 1/2mm which results in angular resolution for the muon of $68/p_\mu$(GeV/c)mr. The hadron energy obtained from summing, without weighting, pulse height in the counters that sample ionization every 10cm of steel, has a resolution $\Delta E_H/E_H = 0.89/\sqrt{E_H(\text{GeV})}$. The target calorimeter is followed by a 1.8m radius solid iron toroidal muon spectrometer consisting of 24 magnetized discs each 20cm thick.

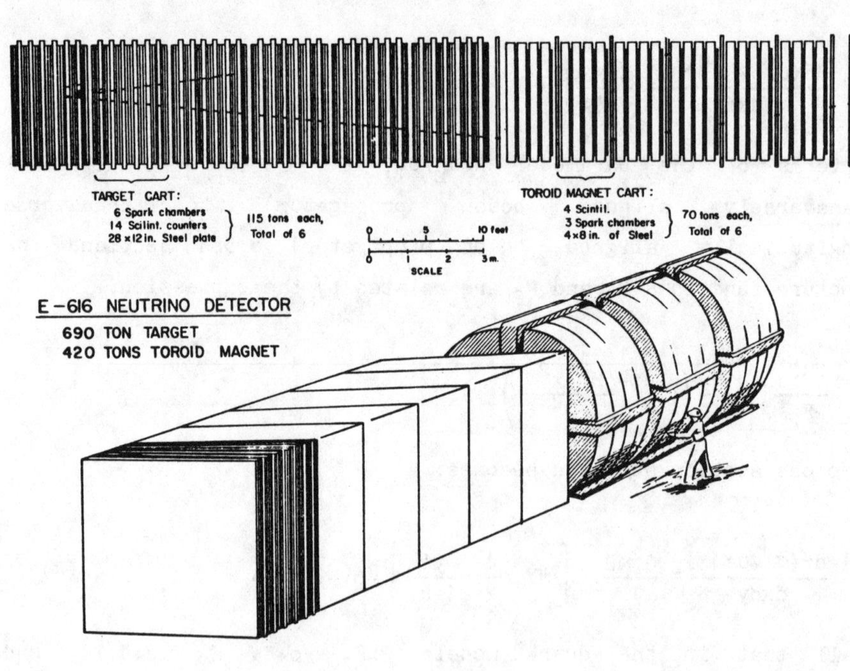

Figure 1. CCFRR neutrino detector.

The toroid is divided into three separate magnets and each magnet is powered by coils that pass through a 12cm hole in the center of the toroids. The spectrometer is instrumented with (a) solid acrylic scintillation counters every 20cm of steel for triggering and for measuring muon energy loss, and (b) spark chambers every 80cm of steel for recording muon track positions. The muon momentum resolution of ±11.5% is dominated by multiple scattering.

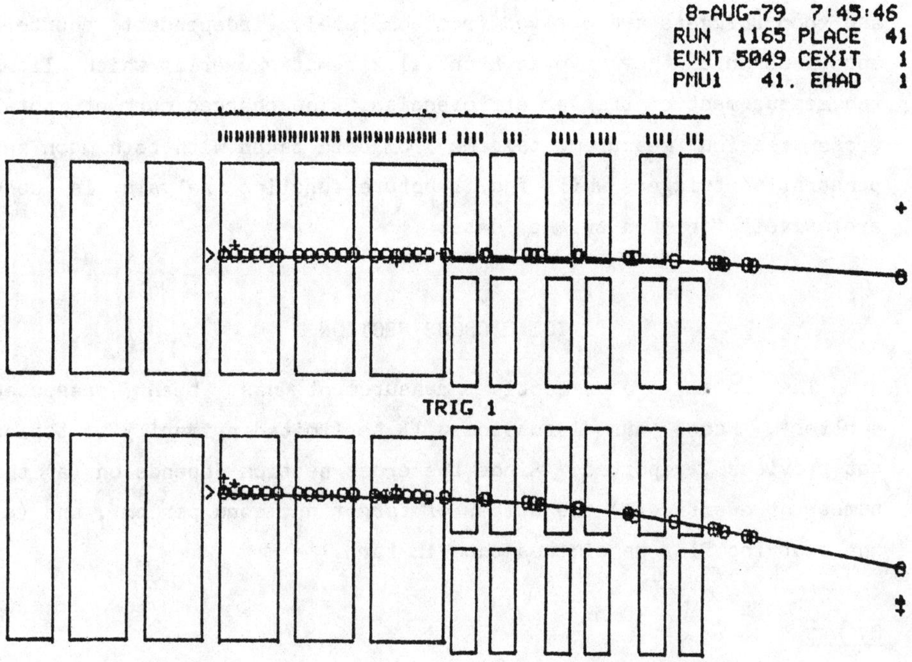

8-AUG-79 7:45:46
RUN 1165 PLACE 41
EVNT 5049 CEXIT 1
PMU1 41. EHAD 1

TRIG 1

Figure 2. Side and top views of a muon trigger event in the CCFRR detector. This event, produced at small y, shows very little energy deposition E_H as indicated by the minimum ionization measured in the counters (top line of the figure).

Between the three toroid magnets there are separate and independent scintillation counters for triggering purposes. The three triggers are designated, "muon", "penetration", and "hadron". The muon trigger demands a muon in the toroid (E_μ> 10 GeV, θ_μ< 100 mrad) while the penetration trigger requires both a hadron energy deposition of 10 GeV in the target and a penetrating muon (E_μ>2.9 GeV, θ_μ<370 mrad). Figure 2 shows an example of a muon trigger event. For neutral current studies, which we do not report here, the hadron trigger is important since it requires only the deposition of hadron energy in the target. Although the penetration and muon triggers are derived from completely independent counters and electronics they have substantial kinematic overlap which allows the measurement of trigger efficiencies. The charged current total cross section measurements come from data taken with both muon and penetration triggers while the structure function analysis is done exclusively for muon trigger data.

TOTAL CROSS SECTION

The total cross section measurement has been presented earlier[1]. Hence our discussion will be limited primarily to topics not previously reported. Since the cross section depends on (a) the number of events, (b) the number of target nucleons per cm^2, and (c) the neutrino flux we discuss them in turn.

Events

The kinematic acceptance of the muon and penetration triggers at E_ν of 100 GeV is shown in Figure 3. Note the significant region of overlap for these two independent triggers. Typically 75% of the observed events satisfied both triggers. For θ_μ<100 mr and p_μ>10 GeV/c the muon trigger was 99.5±0.5% efficient while the penetration trigger, with θ_μ<270 mr and E_H>10 GeV, was 100.0±0.1% efficient. Aside from small corrections for the events missed at large y (see Figure 3), the cross section data were weighted with a geometrical detection efficiency that is obtained by a simple azimuthal rotation, (ϕ) of the event about the neutrino beam direction as shown in Figure 4. Figure 5 shows the efficiency of the muon trigger events after the ϕ rotation is applied to both the muon and penetration trigger data. The graph demonstrates that for θ_μ<100 mr the number of ϕ weighted events is the same for both muon and penetration trigger data. The efficiency decrease beyond θ_μ of 100 mr is due to the fact that large angle muons miss the muon trigger counters in the toroid spectrometer. However, events with θ_μ>100 mr are picked up with the penetration trigger. The average ϕ weight for events is 1.06.

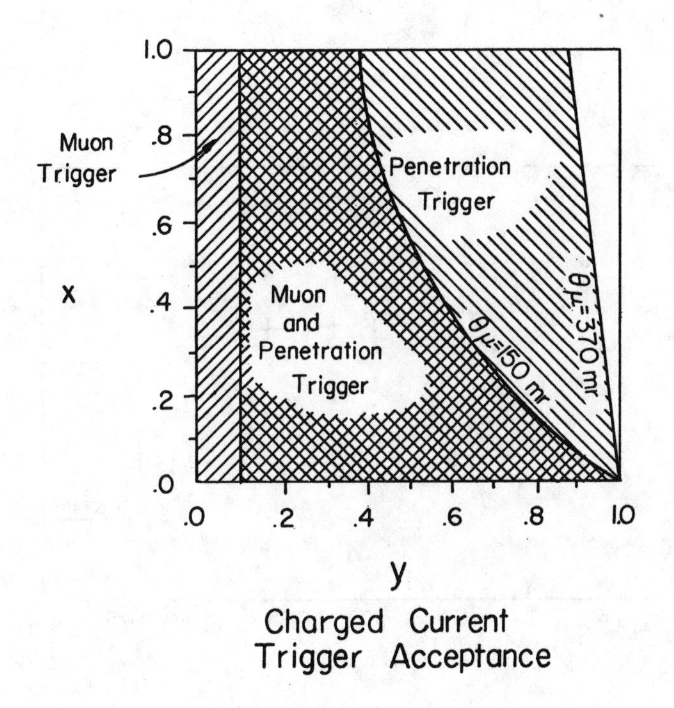

Figure 3. Scaling variable x and y charged current trigger acceptance.

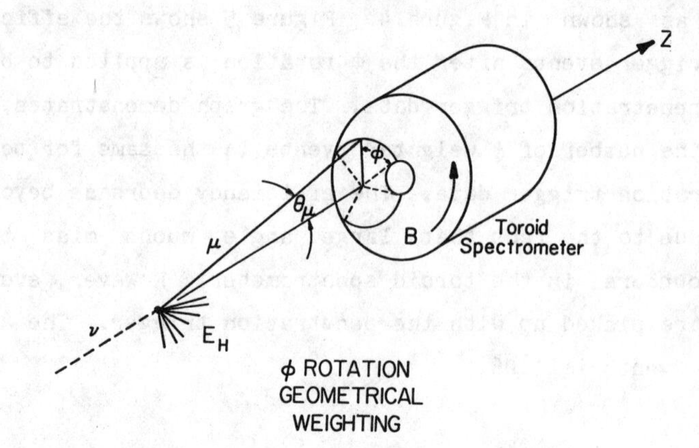

Figure 4. φ Rotation Diagram.

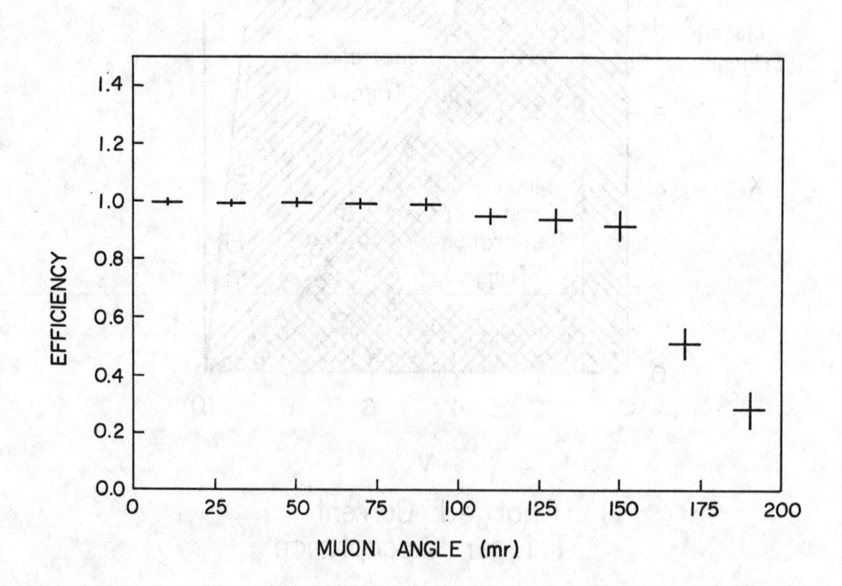

Figure 5. Muon trigger efficiency as a function of muon polar angle, ϕ_{μ}.

To determine the cross section the events need to be identified as coming from either K or π neutrinos. For the muon trigger events this is done by separating the events on an energy versus radius plot as shown in Figure 6. The penetration trigger only events, which do not have a measured muon momentum, are also easily classified as K or π neutrino induced events, since the large angle muon implies a hadron energy which uniquely establishes the category. The small number of charged current events with $\theta_{\mu}>370$ mr, that are not contained in the muon and penetration trigger data, are estimated by extrapolating the differential cross section. This large y loss varies from 1.9% for 250 GeV secondary beam kaon neutrinos to 9.6% for 120 GeV pion neutrinos. Losses for antineutrinos are even smaller because of the $(1-y)^2$ behavior of the differential cross section.

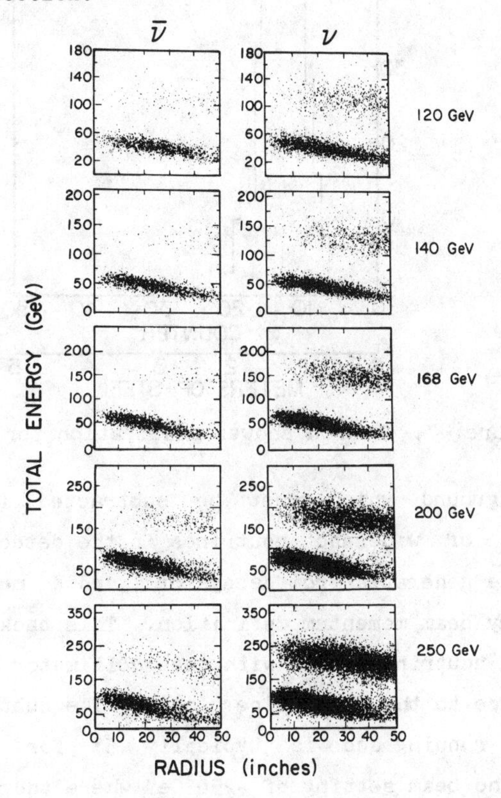

Figure 6. Energy vs. radius for the CCFRR ν, $\bar{\nu}$ data.

There is a small contamination of neutral current events that can satisfy the penetration trigger requirements: i.e. a track that penetrates 2m of Fe. Figure 7 shows an experimentally measured hadron shower penetration distribution for 90 GeV pions. Three per cent of the pions have a shower which extends beyond 2 meters of Fe. When track reconstruction of length 2m is required, the background in the charged current event sample from neutral current events is less than 0.5%.

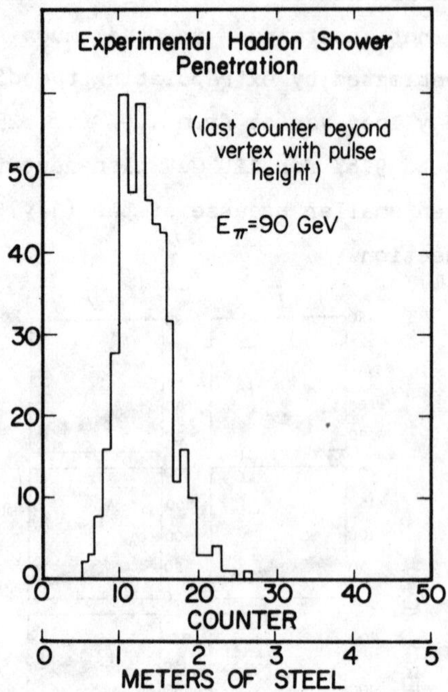

Figure 7. Hadron shower penetration for 90 GeV pions.

A background which must be subtracted is that due to interactions of wideband neutrinos in the detector. The wideband neutrinos are generated from decays of π and K mesons upstream of the secondary beam momentum definition. This background is measured by obtaining neutrino events with the collimator upstream of the beam entrance to the decay pipe closed. The subtraction is 2 to 3% for neutrino running and is typically 5% for antineutrino data except at the beam setting of -250 GeV where there is ~16% wideband background.

Apparatus deadtime has been monitored for data accumulated with both slow (1 s) and fast (1 ms) resonant extraction, from the main accelerator. The data presented here are derived from fast spill except where noted. During the fast spill only one event per cycle could be recorded. The fraction of beam for which the experiment was sensitive averaged ~70% (90%) during neutrino (antineutrino) running. This live fraction was measured two ways: by recording the flux transmitted during the triggerable and non-triggerable times, and by counting all possible triggers during both times. Measurements of the livetime obtained with both methods agreed to about 1%. The livetime derived from the flux and triggers also agreed with the mean number of triggers per pulse as obtained from Poisson statistics. Neutrino data taken during the 1 sec slow spill, where the livetime fraction was 85% yielded cross sections that agreed to 1% with the fast (1 ms) spill data.

Flux

The cross section measurements require precise information about the neutrino flux. Since σ/E_{ν} is reported, the mean neutrino beam energy must also be known. The secondary hadron beam momentum is measured with the Cerenkov counter which, in turn, is calibrated with 200 GeV protons extracted from the main ring. The calculated neutrino beam energy is then checked directly with the mean energy of neutrino events in the detector. The energy calibration of the detector is found by exposing it to hadrons and muons of known energy. The secondary beam angular divergence, which is important for calculating the pion neutrino flux as a function of radius, is measured with the pion Cerenkov counter curves.

Knowledge of the $e/\pi/K/p$ relative secondary beam particle fractions and the absolute magnitude of the secondary beam flux, along with the items discussed above, allow calculation of the neutrino flux at the detector. The particle fractions were measured, as described by Rapidis et al.[1], and Blair[2] with a focussing Cerenkov counter to a typical accuracy of 1 to 4% for pions and 4 to 7% for K mesons.

The magnitude of the secondary flux was measured during data collection with ion chambers located in the decay pipe at distances of 132 m and 284 m, respectively. The ion chambers were calibrated three independent ways: by Cu foil irradiation with the subsequent counting of Na^{24}; by use of an RF cavity in the secondary beam that is sensitive to the 53 MHz beam structure and whose quality factor and consequent calibration are known; and by measuring the response of the ion chamber when it was placed in a secondary beam that could be counted with standard scintillation techniques. The different calibration methods showed systematic variations with energy and beam polarity of 6.5% that are believed to be due to the slight difference in response of the ion chamber to pions and protons whose total cross sections are 24 and 40 mb, respectively. The linearity of all beam monitors was checked throughout the range of operational secondary beam intensities, and beyond, during the 200 GeV machine energy calibration studies. The estimated overall ion chamber calibration error is ±2.5%.

Horizontal and vertical neutrino beam position at the lab E detector were measured by monitoring the secondary beam with split plate ion chambers located in the decay pipe. During the analysis stringent cuts were imposed on the data to insure beam centering to an accuracy of about ±2.5 cm.

Target Density

The number of nucleons per cm^2 has been measured by weighing the steel plates which make up the target. Since the cross sectional area of each plate is known, the number of gm/cm^2 is directly calculated. The liquid counters, spark chambers, and miscellaneous materials make up an additional target mass of 7%. The steel density we calculate, by also measuring the thickness of the Fe plates, is 7.85 gm/cm^3 which agrees very well with the wallet card[3] value of 7.87 gm/cm^3.

Data were recorded with the secondary beam magnets energized to transport and focus 10 different momenta of ±120, ±140, ±168, ±200, and ±250 GeV/c. These conditions provided both neutrino and antineutrino data in the energy range 30 to 250 GeV. The total cross sections were determined from a final event sample of 57,500 ν and 16,600 $\bar{\nu}$ events inside a fiducial radius of 1.27 m. The σ/E results are given in Figure 8. Each data point shown typically contains events from several different secondary beam settings. In combining events from different energy settings the average, and individual data points that make up the averages, show very consistent chi squares which indicate systematic errors are well understood. The high energy data from an earlier run with the secondary beam momentum tuned to 300 GeV/c are also shown[4]. These cross section results should be compared to other published data from the CFRR[5], BEBC[6], CDHS[7], and CHARM[8] groups. The present values for σ_ν/E_ν, $\sigma_{\bar{\nu}}/E_{\bar{\nu}}$, of 0.70±0.04 and 0.35±0.03 are higher by 10% to

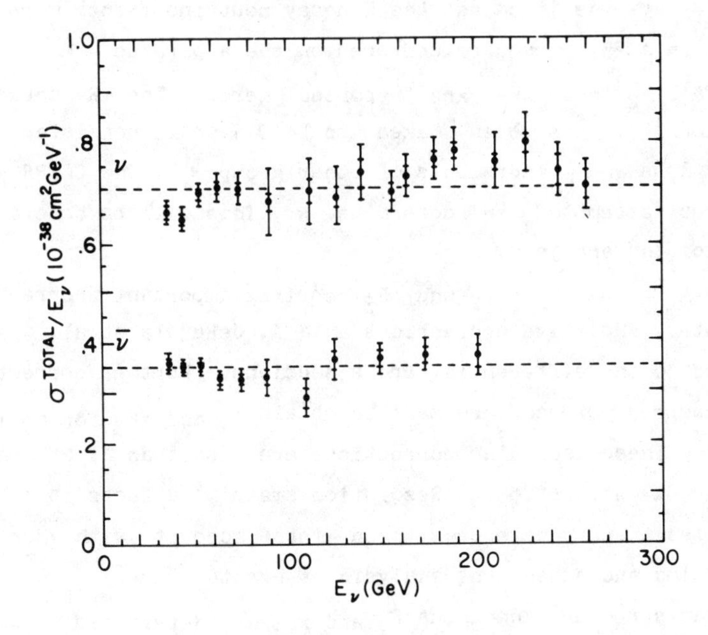

Figure 8. CCFRR neutrino and antineutrino cross section slopes versus energy.

15% than the previously published data. Several of the people involved in the 1977 CFRR measurement have also participated in the present experiment. No specific cause for the different result is known although it should be mentioned that the neutrino detector, dichromatic beam and flux monitoring apparatus are new and as such contain redundancy and sophistication not previously available.

STRUCTURE FUNCTIONS

Structure functions have been extracted from the muon trigger data including events taken during both fast (1ms) and slow (1 sec) spill. From an initial sample of 116,000 ν and 16,600 $\bar{\nu}$ events cuts were applied to the data to obtain 60,000 ν and 7,000 $\bar{\nu}$ events. A large fraction of the eliminated events had a hadron energy, E_H, of less than the 10 GeV as required for good resolution in the x and y scaling variables. The fiducial radius for neutrino events from pion decay was 76 cm and the K decay neutrino events were required to lie in a 254 cm square centered on the apparatus.

Only F_2 and xF_3 are reported here. The R parameter in equation (3), has been taken to be 0.1 as in consistent with the assumptions and measurements of other groups[9,10]. The CCFRR group has not yet attempted to determine R. This will be done at a later stage of the analysis.

Determination of F_2 and xF_3 requires important corrections to the data. Radiative corrections a la A. deRujula et al.[11] have been applied to the differential cross sections. Further corrections for the neutron excess are made to obtain F_2 and xF_3 for an isoscalar target. These isoscalar corrections are less than 1% at small x and reach ~5% at x=0.85. Resolution smearing effects in the scaling variables x and y have been taken into account with Monte Carlo generation and subsequent analysis of events.

The structure functions F_2 and xF_3 as determined from the y distributions of the sum and difference of neutrino and antineutrino data (equations (4) and (1)) are given below in Figures 9 and 10.

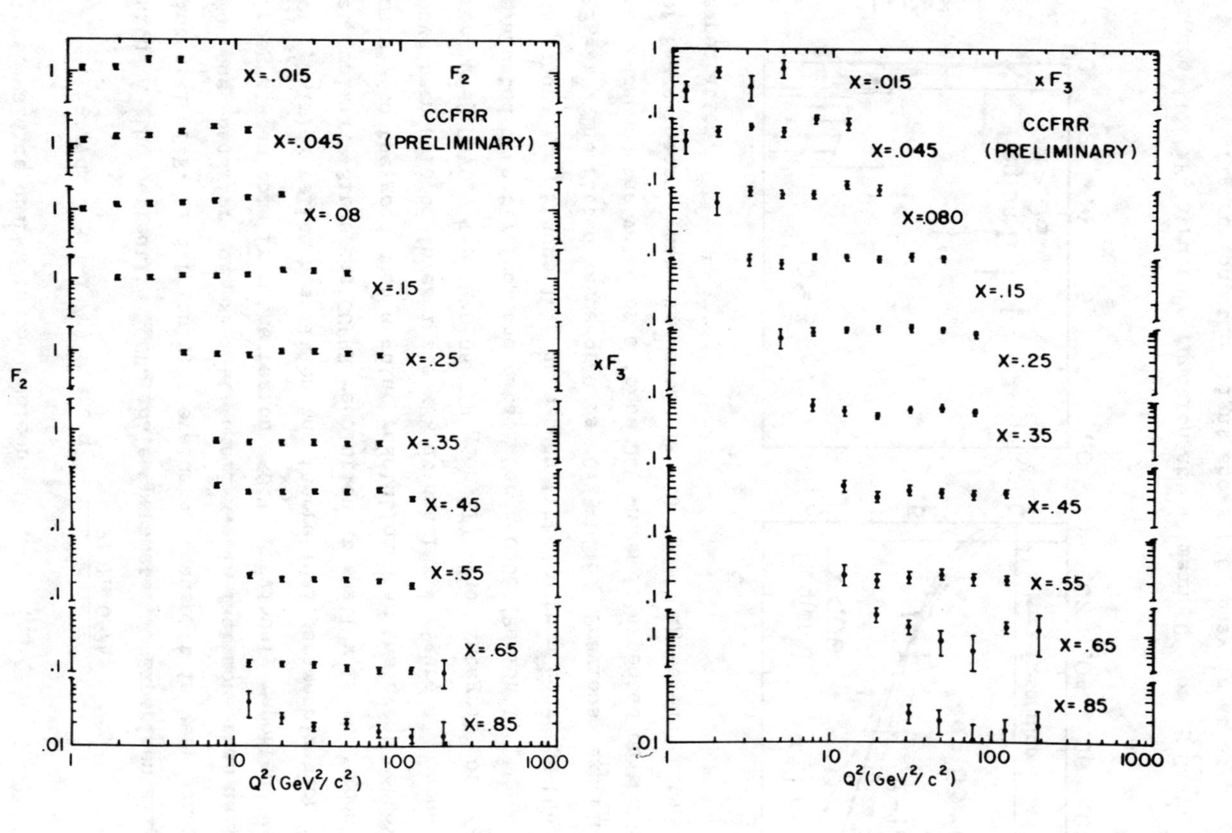

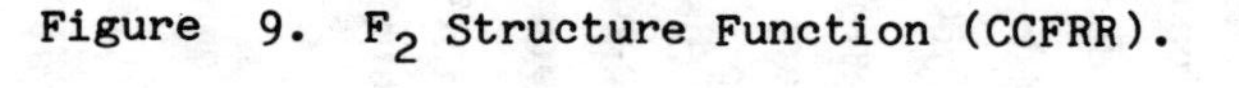

Figure 9. F_2 Structure Function (CCFRR).

Figure 10. xF_3 Structure Function (CCFRR).

THE GROSS-LLEWELLYN SMITH SUM RULE

The Gross-Llewellyn Smith sum rule[12], amended to include higher order QCD corrections[13,14] tells us there should be approximately three valence quarks in the nucleon, i.e. (as $Q^2 \to \infty$)

$$\int_0^1 F_3(x)dx = 3(1-\alpha_s/\pi) \text{ where } \alpha_s = \frac{12\pi}{27 \ln Q^2/\Lambda^2}$$

with Λ the effective QCD scale parameter. Experimentally the integral of F_3 is difficult to evaluate because it is very sensitive to the behavior of xF_3 at small x. In fact more than half the integral can come from the region x<0.1. Previously reported values of $\int F_3(x)dx$ relied heavily on the theoretical parametrization of F_3 at small x. Since the CCFRR resolution at small x is very good, due mainly to the good muon angular resolution near the event vertex, we have been able to investigate xF_3 at small x. Figure 11 shows xF_3 for Q^2=3 GeV^2. For x>0.06 (Region II) the integral of F_3 is evaluated directly from the data. For x<0.06 (Region I) the data, as shown in Figure 11, have been fitted with a curve of the form $xF_3=Ax^{\alpha}$. The fitted value of α is 0.53±0.16. Previous experiments have assumed the value of α to be 0.5 which is expected on the basis of Regge theory[15] and our data is in good agreement with this expectation.

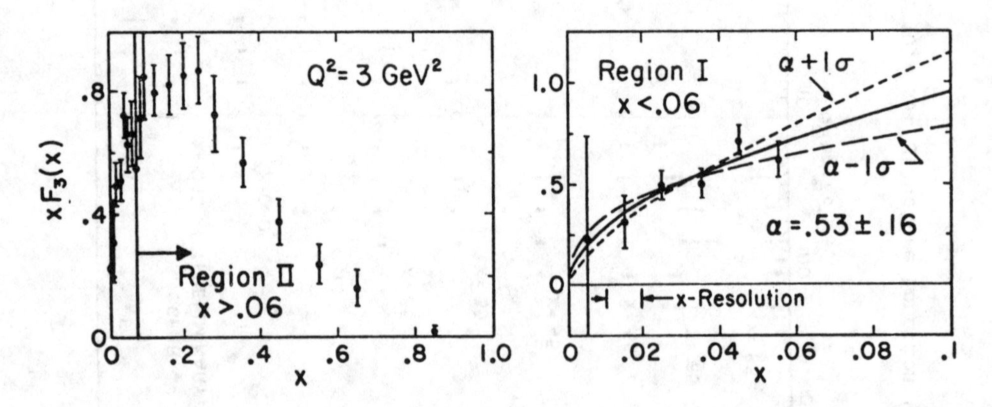

Figure 11. xF_3 structure function at a mean Q^2 of 3 GeV^2. The figure on the right shows the behavoir at small x and the fit to the form Ax^{α}.

The contribution to $\int F_3(x)dx$ in Region I, $x<0.06$, has been evaluated using the fitted value of α. The data yields the following values of $\int F_3(x)dx$ for three different fixed Q^2 regions.

$\langle Q^2 \rangle$	α	$\int F_3 dx$
$2(GeV/c)^2$	0.49 ± 0.20	$2.89 \pm 0.40 \pm 0.15$
$3(GeV/c)^2$	0.53 ± 0.16	$2.95 \pm 0.33 \pm 0.15$
$5(GeV/c)^2$	0.49 ± 0.13	$3.22 \pm 0.46 \pm 0.16$

The first error on the integral is statistical and includes the uncertainty in α while the second error results from assuming a 5% normalization uncertainty. Since the integral is very close to 3 a small value of Λ is implied. This result is consistent with other determinations of the integral[7,16,17] but this is the first experiment with definitive data and good resolution in the low x region.

COMPARISONS OF F_2 DERIVED FROM MUON AND NEUTRINO EXPERIMENTS

Muon and neutrino data provide a direct test of the fractional quark charge hypothesis since F_2 in the two cases is:

$$F_2^{\mu} = \frac{4}{9}x(u+\bar{u}) + \frac{1}{9}x(d+\bar{d}) + \frac{1}{9}x(s+\bar{s}) + \frac{4}{9}x(c+\bar{c}) \qquad \text{and}$$

$F_2^{\nu} = q+\bar{q}$ where q is given earlier. Neglecting any contribution due to charm quarks in the nucleon one finds:

$$F_2^{\nu} = \frac{18}{5} \frac{F_2^{\mu}}{1 - \frac{3}{5}\frac{s+\bar{s}}{q+\bar{q}}} . \qquad (5)$$

The strange sea correction is relatively important at low x where, with the assumption at a half SU(3) symmetric sea, the correction is 5% at x=0.1. The half SU(3) symmetric sea, i.e. $(s+\bar{s})/(\bar{u}+\bar{d})=0.5$, is consistent with neutrino induced production of charm as measured in dimuon data[18].

Figure 12 shows the ratio of the CCFRR and CDHS F_2 iron data divided by the EMC iron data as scaled by the mean square charge correction factor in equation (5). The quark model prediction with fractional charges is satisfied to ~10%. The CCFRR data are ~13% higher than EMC would predict. If the EMC hydrogen and deuterium data are compared to SLAC[19] and CHIO[20] results, the EMC data are on average lower by 10% and 8%, respectively. The comparison of EMC and CDHS shows a difference which is x dependent but averaged over x implies that F_2 for EMC is higher than for CDHS by about 10%.

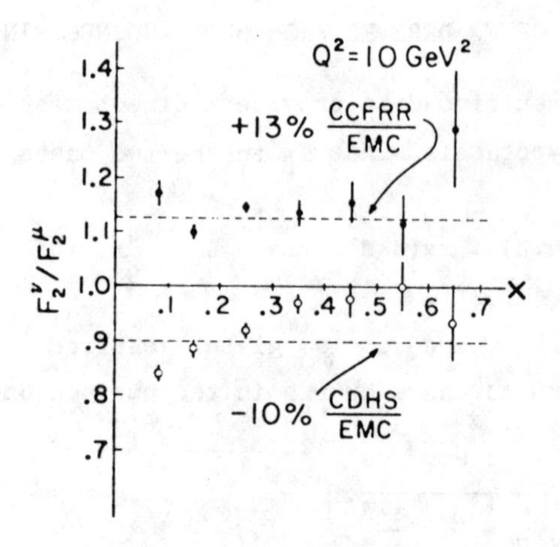

Figure 12. Comparison of muon and neutrino F_2 structure functions.

Disregarding differences in normalization one can ask whether the x and Q^2 scaling violations are consistent among the various sets of data. To answer this question we have fitted the data to the form $F_2(x,Q^2) = A(x)(1+b \log_{10}Q^2/10)$. In this way the normalization is contained in A(x) and the values of b at fixed x show the Q^2 evolution of the data. Figure 13 shows the fitted values of b vs x for the CDHS, EMC, and CCFRR iron data. Although there may be some systematic differences at large x the data agree reasonably well. Differences in Λ are correlated with different logarithmic behavior of F_2 at large x. Thus, according to Figure 13 we may expect variation in Λ from the different experiments. CCFRR has not yet made a fit to Λ since our data are still preliminary.

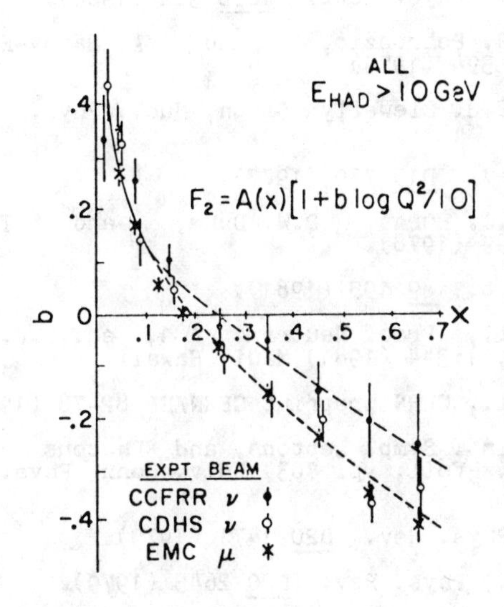

Figure 13. Logarithmic Q^2 dependence of the F_2 structure function for the CCFRR, CDHS, and EMC experiments.

REFERENCES

1. P.A. Rapidis et al., Proc. of the 1981 Summer Inst. Part. Phys., ed. A. Mosher, p. 641(1982) Stanford, Calif. SLAC.
2. R.E. Blair, Ph.D. Thesis, A Total Cross Section and y-Distribution Measurement for Muon-Type Neutrinos and Antineutrinos on Iron (1982), Caltech, Pasadena (unpublished).
3. Particle Data Group, Rev. Mod. Phys. 52, part 2, S1(1980).
4. J.R. Lee, Ph.D. Thesis. Measurements of νN Charged Current Cross Sections from 25 to 260 GeV, (1980) Caltech, Pasadena (unpublished).
5. B.C. Barish et al., Phys. Rev. Lett. <u>39</u> 1595 (1977).
6. P. B^setti et al., Phys. Lett. <u>110B</u> 167 (1982).
7. J.G.H. deGroot et al., Z. Phys. <u>C1</u> 143 (1979).
8. M. Jonker et al., Phys. Lett. <u>99B</u> 265 (1981).
9. H. Abramowicz et al., Phys. Lett. <u>107B</u> 141 (1981).
10. J.J. Aubert et al., Phys. Lett. <u>105B</u> 315 (1981).

    J.J. Aubert et al., Phys. Lett. <u>105B</u> 322 (1981).
11. A. deRujula, R. Petronzio, and A. Savoy-Navarro, Nucl. Phys. <u>B154</u> 394 (1979).
12. D.J. Gross and C.H. Llewellyn Smith, Nucl. Phys. <u>B14</u> 337 (1969).
13. M. Calvo, Phys. Rev. <u>D15</u> 730 (1977).
14. W.A. Bardeen, A.J. Buras, D.W. Duke, and T. Muta, Phys. Rev. <u>D18</u> 399 (1978).
15. E. Reya, Phys. Rept. <u>69</u> 195 (1981).
16. P. Fritze et al., Proc. Neutrino '81, ed. R.J. Cence, E. Ma, A. Roberts, 1:344 (1981) Maui, Hawaii.
17. T. Bolognese et al., CERN preprint CERN/EP 82-78 (1982).
18. H.E. Fisk, Proc. Int. Symp. Leptons and Photons at High Energies, ed. W. Pfeil, p. 703 Univ. Bonn Phys. Inst. (1981).
19. A. Bodek et al., Phys. Rev. <u>D20</u> 1471 (1979).
20. B.A. Gordon et al., Phys. Rev. <u>D20</u> 2645 (1979).

RECENT RESULTS ON THE ν_μ and $\tilde{\nu}_\mu$ CHARGED CURRENT CROSS SECTIONS*

P. Igo-Kemenes**
Columbia University, New York, NY 10027

1. INTRODUCTION

The cross section of the charged current processes,

$$\nu_\mu + N \rightarrow \mu^- + X$$
$$\tilde{\nu}_\mu + N \rightarrow \mu^+ + X \quad ,$$

have been previously measured in many high energy ($E_\nu \leqslant$ 10 GeV) narrow band experiments.[1] The cross sections rise linearly with neutrino energy. However, the agreement between the experiments on the value of the slope parameter, σ/E, is not good. In particular, the Caltech -Columbia-Fermilab-Rochester-Rockefeller (CCFRR) collaboration results[2] are higher than the values from the CERN experiments. A possible source of this discrepancy is the very different techniques used in the determination of the neutrino flux: most of the early experiments, in particular those of the CDHS and CHARM collaborations at the CERN SPS narrow band beam, relied on counting the muons from the decays $\pi \rightarrow \mu\nu_\mu$ and $K \rightarrow \mu\nu_\mu$ at different depths in the absorbing shield; the CCFRR collaboration measured the neutrino flux by monitoring the flux of the parent pions and kaons along the decay pipe.

The new results presented here come from experiments using the FNAL 15-ft bubble chamber. Both the neutrino and anti-neutrino data were taken simultaneously with the CCFRR collaboration. The neutrino data were obtained and analyzed by the BNL-Columbia-Rutgers collaboration[3] and the determination of the cross section slope will be discussed below in detail. The anti-neutrino data were analyzed by the Berkeley-Fermilab-Hawaii (BFH) collaboration and the result was also submitted to this conference.[4]

* Research supported in part by the National Science Foundation and the Department of Energy.

** On leave from Heidelberg University, Heidelberg, F.R. Germany.

It is important to note that the precision of all past counter experiments is limited by systematic uncertainties. For this reason bubble chamber experiments, subject to completely different systematic errors, have an important role in the cross section determination, in spite of their comparatively low statistical accuracy. It should also be emphasized that the bubble chamber experiments used the neutrino flux monitoring of the CCFRR collaboration; consequently, in comparing results, the common uncertainty related to the neutrino flux measurement cancels.

2. THE EXPERIMENTAL ARRANGEMENT

The layout of the Fermilab narrow band neutrino beam is shown in Fig. 1. 400 GeV protons hit a beryllium target and produce secondary particles which are momentum and sign selected by the N-30 magnetic train. The source of neutrinos is the decay of secondary pions and kaons along the 340 m decay pipe. All particles except neutrinos are absorbed in the beam dump and the 910 m earth shielding which precedes the neutrino detectors. In the figure, the positions of the CCFRR counter experiment and of the 15-ft bubble chamber are indicated.

2.1 Magnetic Train and Neutrino Flux Monitoring

Data were taken with the momentum of the secondary particle beam selected at five different values: 120, 140, 165, 200, and 250 GeV/c. Neutrinos from pion decays covered the energy range $0 \leq E_{\nu} \leqslant 80$ GeV and those from kaon decays extended the energy range to 240 GeV.

The neutrino flux at the detectors was obtained from counting the charged secondary particles in ionization chambers situated at the expansion port and at the target manhole (see Fig. 1). Segmented wire ionization chambers measured the beam divergence and allowed to discard, off line, beam bursts with unsatisfactory steering. The proportion of pions and kaons in the secondary beam was measured by a differential Cherenkov counter, situated at the expansion port, which was made to intercept the secondary beam at regular time intervals. The monitoring of the secondary beam, the calibration of the ionization chambers, and the analysis of the Cherenkov counter data were performed by the CCFRR collaboration.* Details may be found in

---

* We are grateful to the CCFRR collaboration for supplying us with all the relevant information concerning the neutrino flux and for many critical discussions.

Ref. 5. To obtain the neutrino flux at the bubble chamber on a burst-by-burst basis, that is, the correct normalization to the bubble chamber exposures, the relevant monitor information (flux of secondaries, pion to proton, and kaon to proton ratios) was combined with a beam transport and decay Monte Carlo. In Fig. 2, the energy-weighted neutrino flux at the bubble chamber, as a function of radial distance from the beam center, is compared to a similar distribution obtained by the CCFRR collaboration. Comparisons of this kind were performed for all momentum settings of the secondary beam and the agreement in shape and in absolute scale was found to be within 1%. Thus, possible differences in the cross sections of the two experiments, due to the flux normalization, are smaller than 1%.

2.2 The 15-ft Bubble Chamber

Neutrino interactions were recorded in the Fermilab 15-ft bubble chamber, filled with a heavy neon hydrogen mix and operated in a magnetic field of 30 kG. The liquid contained 59 atomic % neon; this mixture is an almost perfect "isoscalar" target with a slight proton excess of 3.4%. The liquid density was monitored during data taking by the bubble chamber support group: liquid samples were submitted to a chromatographic analysis. The average density determined by this method was $\langle\rho\rangle = 0.71 \pm 0.04$ g cm^{-3}. An independent determination of $\langle\rho\rangle$ was obtained from the data: using the range of protons stopping in the liquid and their momentum from the curvature in the magnetic field, the value $\langle\rho\rangle = 0.715 \pm 0.020$ g cm^{-3} was obtained, in good agreement with the chromatographic analysis. The error includes systematic uncertainties of 2%.

The liquid had a radiation length of $\approx$ 40 cm and an interaction length of $\approx$ 125 cm; thus, for a typical charged current neutrino interaction, the muon left the bubble chamber, pions, protons, and neutrons were likely to interact and gamma rays had a high probability to be converted.

3. ANALYSIS OF THE NEUTRINO DATA

3.1 Selection of Charged Current Events

At the five settings of the secondary beam momentum quoted in Section 2.1, a total of 98,000 bubble chamber frames were recorded. All frames were scanned for possible neutrino interactions but for the cross section determination only 78,000 frames, with a valid flux monitor information, were retained. A fiducial volume requirement assured a high detection efficiency for charged current interactions.

The scanning efficiency was obtained from double scanning 60% of the frames. It was found to be (93±4)%.

Neutrino interactions were selected by requiring the total visible energy of the event to be more than 10 GeV and the angle of the total visible momentum vector with the neutrino direction to be less than 45°. These requirements greatly reduced backgrounds from cosmic radiation and from neutral particles produced before the bubble chamber and faking a neutrino interaction. The loss of genuine charged current events was only about 1%. At this stage the number of events was 1043.

To select charged current interactions, the fastest negative particle leaving the bubble chamber was designated to be a muon; its momentum was required to be higher than 2 GeV/c. The charged current sample obtained in this way contained 830 events.

3.2 Backgrounds and Losses

3.2.1 Neutral Current Background

The principal background to the above charged current sample came from neutral current interactions. Occasionally a π^- from such an interaction leaves the bubble chamber without interacting in the liquid. If the π^- satisfies the > 2 GeV/c momentum requirement, the event is accepted as a charged current interaction. The method to determine this "hadronic punchthrough" background is described below.

Because all positive tracks in the data sample are hadrons, the ratio L^+/I^+ of the number of leaving to interacting positive tracks is a measure of the hadronic punchthrough; it is a property of the bubble chamber liquid and depends on the track momentum.* This ratio, multiplied by the number of interacting negative tracks, I_p, of a given momentum and summed over all momenta** gives L_h^-, the number of leaving negative hadrons:

$$L_h^- = \sum_p (L^+/I^+)_p \cdot I_p^- \quad .$$

---

* The fact that positive hadrons contain pions and protons, whereas negative hadrons are mostly pions, was taken into account by studying the L/I ratios for pions of both signs from identified K^0 decays.

** One has to include all events, neutral and charged current, in the track counting.

For a given event, only the fastest of the leaving negative hadrons is simulating a muon; thus to obtain the number of fake charged current events, one has to subtract from L_h^- the number of successive leaving negative tracks with $P > 2$ GeV/c. In this way, a neutral current background of $(12 \pm 4)\%$ was obtained. The method has the additional merit to include also the background from neutral particles other than neutrinos.

An alternative method consists of disregarding, for each event in the charged current sample, the fastest leaving negative track which is supposed to be a muon. The truncated events simulate the hadronic part of neutral current events; thus asking for a second leaving negative track with a momentum above 2 GeV/c determines the hadronic punchthrough probability. Using this probability in an iterative way, one obtains a punchthrough background of $(9 \pm 2)\%$. In what follows, the combined result of the two methods, $(10 \pm 2)\%$, is used.

3.2.2 Wide Band Background

For the cross section determination, one has to disregard events for which the neutrino originated upstream of the momentum selecting slit (see Fig. 1) because the corresponding parent pions or kaons are usually not detected by the ionization chambers, thus they are not considered in the neutrino flux measurement. This wide band background was directly measured by the CCFRR collaboration by closing the momentum selecting slit of the narrow band beam.* Their result of $(2.5 \pm 0.5)\%$ was used in the present analysis.[5]

3.2.3 Cosmic Radiation Background

Interactions produced by cosmic radiation were largely suppressed by the total energy and angle requirements of Section 3.1. The residual background was found to be $(0.4 \pm 0.4)\%$ by counting events with the momentum vector pointing into the backward direction with respect to the neutrinos.

3.2.4 Losses Due to Kinematic Cuts

Corrections due to the kinematic cuts were small. The visible energy cut at 10 GeV implied a loss of $(1 \pm 1)\%$. To correct for the cut at 2 GeV/c on the muon momentum, each event was weighted according to the unsampled part of the scaling-y distribution.** The

---

* The corresponding bubble chamber data were statistically insignificant.

** For the y distribution $(d\sigma/dy) \sim [(1+B)+(1-B)(1-y)^2]$ was used with $B = 0.8$.

total correction, averaged over all events, amounted to $(3.6\pm1)\%$.

The possibility of losing events with low track multiplicity at the scanning stage was investigated. Events with a μ^- and one positive track are expected from the quasi-elastic reaction, $\nu_\mu + n \rightarrow \mu^- + p$, and from some of the single pion channels; their contribution to the total cross section was estimated to be of $(3\pm1)\%$. By counting these low topology events in the sample, this loss was estimated to be $(1\pm1)\%$.

Table 1 summarizes all corrections to the charged current data sample. The number on the bottom line is the corrected number of charged current events which enters into the cross section determination.

3.3 Determination of the Cross Section Slope

The cross section slope averaged over all neutrino energies (E_ν) is given by :

$$\sigma_\nu/E_\nu = N_{cc}/W \quad .$$

N_{cc} is the corrected number of charged current events in Table 1, and

$$W = [\text{nucleons/fiducial volume}] * \sum_{i=1}^{N_\nu} E_\nu^{\,i} \cdot \ell_i$$

is the corresponding normalization obtained from the beam transport and decay Monte Carlo. The summation runs over all generated neutrinos; $E_\nu^{\,i}$ is the energy of the ith neutrino and ℓ_i the potential path length of the ith neutrino within the fiducial volume of the bubble chamber. The three sources of neutrinos, $\pi^+ \rightarrow \mu^+\nu_\mu$, $K^+ \rightarrow \mu^+\nu_\mu$, and $K^+ \rightarrow \pi^o\mu^+\nu_\mu$, have been considered. Normalizing to the total number of parent pions and kaons detected by the ionization chambers during the active time of the experiment, one obtains:

$$W = (1298 \pm 66)\ 10^{38}\ \text{GeV/cm}^2 \quad .$$

The error is due to uncertainties in the flux monitoring and beam composition measurement (4%), to uncertainties of the Monte Carlo parameters (1%) and to the error on the bubble chamber liquid density (3%). Using the above number, one obtains $\sigma_\nu/E_\nu = 0.635 * 10^{-38}$ cm^2/GeV. To express the cross section slope for any isoscalar target, one has to correct for the slight proton excess of the neon-hydrogen mixture. Using $2\sigma_p = \sigma_n$, one obtains the final result:

$$\sigma_\nu/E_\nu = (0.64 \pm 0.02 \pm 0.04)\ 10^{-38}\ \text{cm}^2/\text{GeV}\ .$$

The first error is statistical, coming essentially from the number of observed charged current events. The second error includes systematic uncertainties related to the corrections of Table 1, to the flux monitoring and to the knowledge of the liquid density. The systematic error reduces to 0.02, if the uncertainties of the flux monitoring, common to this experiment and the CCFRR measurement, are not considered.

The following consistency checks and searches for systematic biases have been performed:

a) The five settings of the secondary beam momentum constitute five independent experiments. In Fig. 3, the cross section slope is plotted against the secondary beam momentum and shows the consistency of the five results.

b) The data taking continued over a period of several months. In order to test the long term stability of the whole experiment, we plot in Fig. 4 the cross section slope obtained from the 55 individual rolls of film. The distribution does not exhibit any significant trend with time and the spread of individual points is consistent with the statistical expectation.

c) In order to detect possible biases related to specific parts of the bubble chamber, we compared the distribution of events along the beam direction to what is expected from neutrino interactions in the fiducial volume; no significant deviation was found. We also calculated the cross section slope separately for the upper and lower half of the fiducial volume and in four radial bins perpendicular to the beam direction. The results are given in Table 2; they show no significant effect.

d) The scanning of the film was performed by the three institutions of the collaboration. In order to check the homogeneity of the scanning, we determined the cross section slope separately for the three institutions. The results are also presented in Table 2.

In Table 3 the result of the BNL-Columbia-Rutgers experiment is compared to previous measurements. Although not inconsistent with the CCFRR result, it favors the lower value obtained by earlier experiments.

3.4 Cross Section as a Function of Neutrino Energy

To determine σ_ν/E_ν as a function of E_ν, one has to reconstruct the energy of the incoming neutrino for each event. One approach is to correct the visible energy for the undetected energy carried by neutral particles; techniques based on transverse momentum balance between the muon and the hadronic shower become less accurate as the neutrino energy increases. The resulting energy resolution is typically $\Delta E_\nu/E_\nu \approx 0.15$ for $E_\nu \leq 100$ GeV and higher for $E_\nu > 100$ GeV, with asymmetric tails.

A more reliable approach is based on the dichromatic property of the narrow band beam which permits one to determine the neutrino energy from the radial position of the interaction vertex. The neutrino energy is related to the radial distance, R, from the center of the bubble chamber by the expression:

$$E_\nu \approx \frac{2\gamma\, E_\nu^*}{1+\gamma^2(R/L)^2} \quad ,$$

where E_ν^* is the neutrino energy in the parent's center of mass system, L is the mean separation between the parent's decay point and the neutrino interaction point, and γ is the Lorentz boost factor of the parent particle. This method implies a decision for each event regarding the parent (pion or kaon) of the neutrino. Because for the predominant decay modes, $\pi \to \mu\nu$ and $K \to \mu\nu$, the c.m.s. neutrino energies are very different, [$E_\nu^*(\pi) = 0.03$ GeV, whereas $E_\nu^*(K) = 0.236$ GeV], a crude determination of the event energy is sufficient to reach this decision.

A convenient way of separating from each other the events from the two sources of neutrinos is based on the quantity,

$$S = \frac{E_{vis} - E_\pi^R}{E_K^R - E_\pi^R} \quad ,$$

where E_{vis} is the visible energy, increased by 5% to take into account the energy carried by the undetected neutral particles; E_π^R and E_K^R are the energies predicted from the radial position, assuming the parent of the neutrino to be a pion or kaon, respectively. The distribution of charged current events as a function of S is shown in Fig. 5. The clustering of the events around the expected values of 0 for $\pi^+ \to \mu^+\nu_\mu$ and 1 for $K^+ \to \mu^+\nu_\mu$ is clearly visible.

The cross section slope as a function of neutrino energy is shown on Fig. 6. The error bars include the uncertainty related to the use of the S-distribution to separate the two sources of neutrinos. The data do not indicate any deviation from a linear rise of the cross section.

4. DISCUSSION

The recent CCFRR counter experiment results for both σ_ν/E_ν and $\sigma_{\bar\nu}/E_{\bar\nu}$ have been consistently higher than values obtained in other high energy narrow band beam experiments. The bubble chamber results presented here were obtained from data taken in the same beam and at the same time as the CCFRR data. The basic data on the secondary particle rates and on the beam composition were recorded by the CCFRR group and formed the basis of a common neutrino flux monitoring.

Based on the observation of 830 bubble chamber events, the BNL-Columbia-Rutgers collaboration found a linear rise of the ν_μ charged current cross section with neutrino energy; the slope was found to be:

$$\sigma_\nu/E_\nu = (0.64 \pm 0.05)\ 10^{-38}\ \mathrm{cm}^2\ \mathrm{GeV}^{-1} \quad .$$

This analysis, while not in disagreement with the CCFRR result (see Table 3), yields a lower value of the slope parameter.

A similar conclusion is reached by the Berkeley-Fermilab-Hawaii collaboration from the analysis of the $\tilde\nu_\mu$ data. Their value,

$$\sigma_{\tilde\nu}/E_{\tilde\nu} = (0.30 \pm 0.02)\ 10^{-38}\ \mathrm{cm}^2\ \mathrm{GeV}^{-1} \quad ,$$

is discussed in a separate contribution to this conference.[4]

All high energy narrow band beam measurements of σ_ν/E_ν and $\sigma_{\tilde\nu}/E_{\tilde\nu}$ are compiled in Fig. 7 and Table 3.

References

1 Review of Particle Properties, Rev. Mod. Phys. 52, #2, 1980. For cross sections and related refs., see p. 554.

2 R. Blair et al, Proc. of the 1981 Int. Conf. on Neutrino Physics and Astrophysics, Maui, Hawaii, July 1-8, 1981. Vol. I, p. 311.
R. Blair et al, to be published in Phys. Rev. Lett.

3 N.J. Baker, P.L. Connolly, S.A. Kahn, M.J. Murtagh, M. Tanaka, BNL; C. Baltay, M. Bregman, D. Caroumbalis, L.D. Chen, H. French, M. Hibbs, R. Hylton, P. Igo-Kemenes, J.T. Liu, J. Okamitsu, G. Ormazabal, A.C. Schaffer, K. Shastri, J. Spitzer, Columbia; E.B. Brucker, P.F. Jacques, M. Kalelkar, E.L. Koller, R.J. Plano, P.E. Stamer, A. Vogel, Rutgers.

4 Contribution to this conference by H. Bingham.

5 R. Blair, Ph.D. Thesis, Caltech (1982).

Table 1: Corrections to the Charged Current Event Sample

	Correction	Events
Uncorrected sample:		830 ± 29
Scanning efficiency:	93±4%	
Low-multiplicity losses:	1±1%	
Neutral current background:	10±2%	
Wide band background:	2.5±0.5%	
Cosmic radiation:	0.4±0.4%	
Visible energy cut:	1±1%	
Muon momentum cut:	3.6±1%	
Corrected number of charged current events:		823 ± 50

Table 2: σ_ν/E_ν for Specific Parts of the Data

Upper half of fiducial volume:		0.65±0.03 (stat)
Lower half of fiducial volume:		0.63±0.03
Radial bins:	0 - 50 cm :	0.62±0.04
	50 - 70 :	0.66±0.04
	70 - 100 :	0.63±0.05
	> 100 :	0.62±0.05
Institutions:	1 :	0.64±0.03
	2 :	0.58±0.04
	3 :	0.67±0.04

Table 3: Summary of High Energy Neutrino and Anti-Neutrino Charged Current Cross Sections from Narrow Band Experiments

Experiment	$\sigma_{\bar{\nu}}/E_{\bar{\nu}}$ $[10^{-38}\ cm^2 GeV^{-1}]$	σ_{ν}/E_{ν} $[10^{-38}\ cm^2 GeV^{-1}]$
CDHS [a]	0.30 ± 0.02	0.62 ± 0.03
CHARM [b]	$0.301\pm0.008\pm0.016$	$0.604\pm0.009\pm0.031$
BEBC [c]	$0.303\pm0.010\pm0.011$	$0.663\pm0.012\pm0.025$
CCFRR [d]	$0.350\pm0.004\pm0.022$	$0.701\pm0.004\pm0.025$
BFH [e]	0.30 ± 0.02	
BNL-Columbia-Rutgers		$0.64\pm0.02\pm0.04$

a J.G.H. deGroot et al, Z. Phys. C1, 143 (1978).
b M. Jonker et al, Phys. Lett. 99B, 265 (1981).
c P. Fritze, Proc. of the 1981 Int. Conf. on Neutrino Physics and Astrophysics, Maui, Hawaii, Vol. I, p.344.
d Ref. 2.
e Ref. 4.

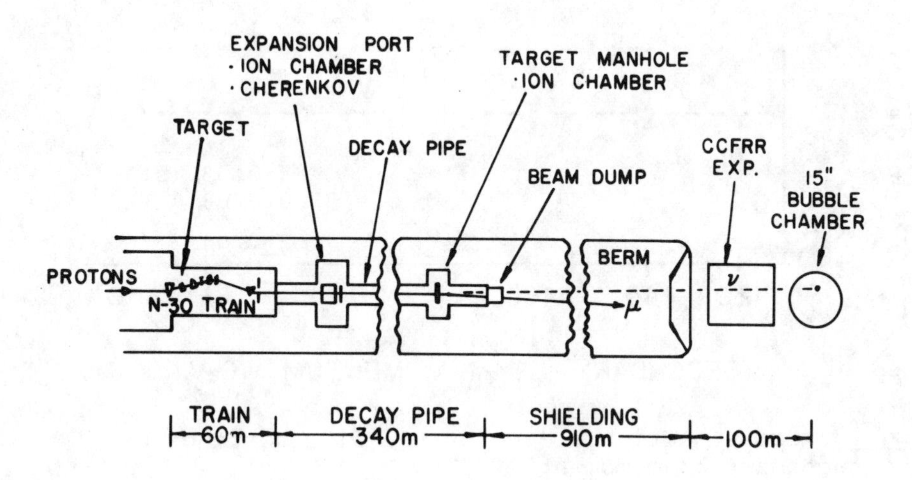

Fig. 1 Layout of the Fermilab narrow band neutrino/anti-neutrino beam. The approximate position of the 15-ft bubble chamber and of the CCFRR counter experiment are indicated.

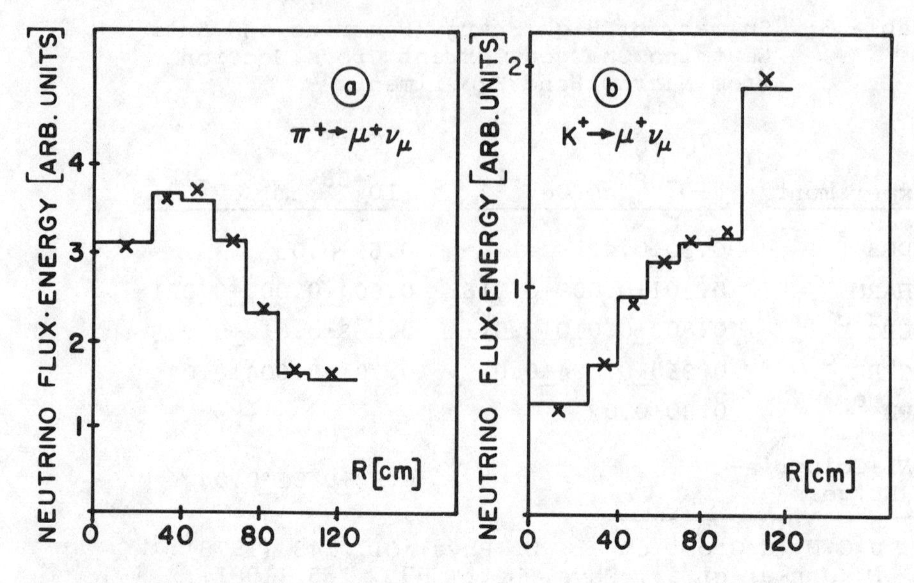

Fig. 2 Energy-weighted neutrino flux at the bubble chamber in radial bins around the beam center. Secondary beam momentum 200 GeV/c. Neutrinos from a) $\pi^+ \to \mu^+ \nu_\mu$ and b) $K^+ \to \mu^+ \nu_\mu$. Full histogram: BNL-Col-Rutgers expt.; crosses: CCFRR experiment.

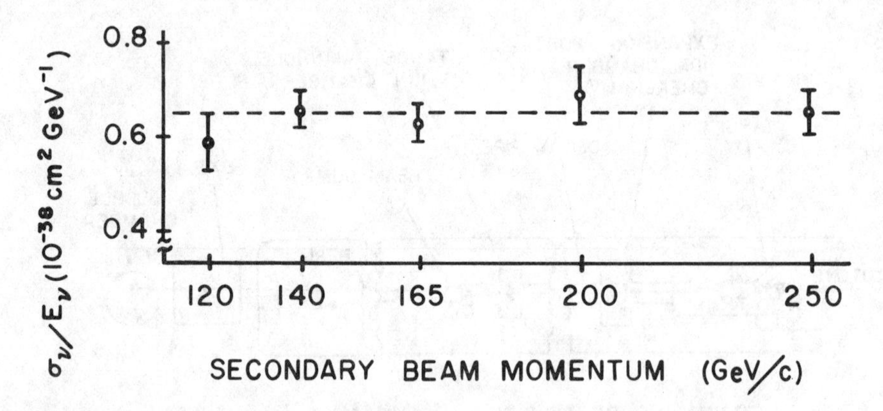

Fig. 3 Cross section slope for the five settings of the secondary beam momentum.

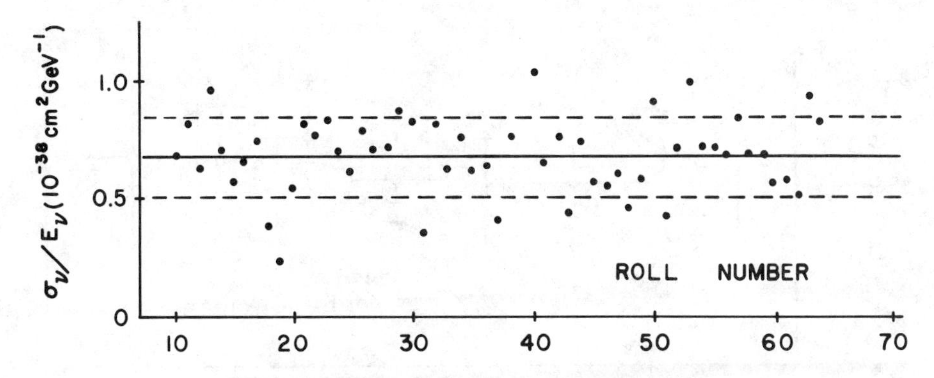

Fig. 4 Cross section slope as determined for individual rolls of bubble chamber film. The mean value of σ_ν/E_ν = 0.64 and the $\pm$ 1 standard deviation lines, corresponding to an average of 15 events/roll, are indicated.

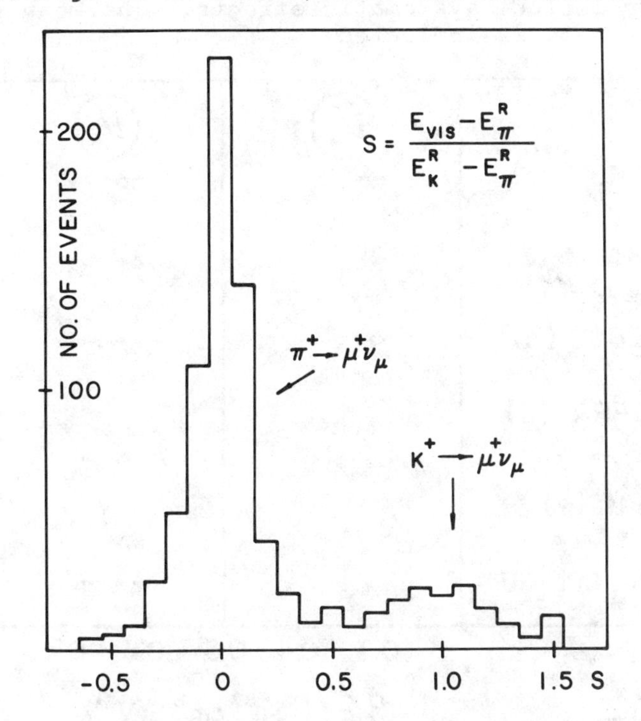

Fig. 5 Distribution of the charged current events as a function of S, defined in the text. Events from $\pi^+ \rightarrow \mu^+ \nu_\mu$ and $K^+ \rightarrow \mu^+ \nu_\mu$ are seen to cluster at the values 0 and 1, respectively.

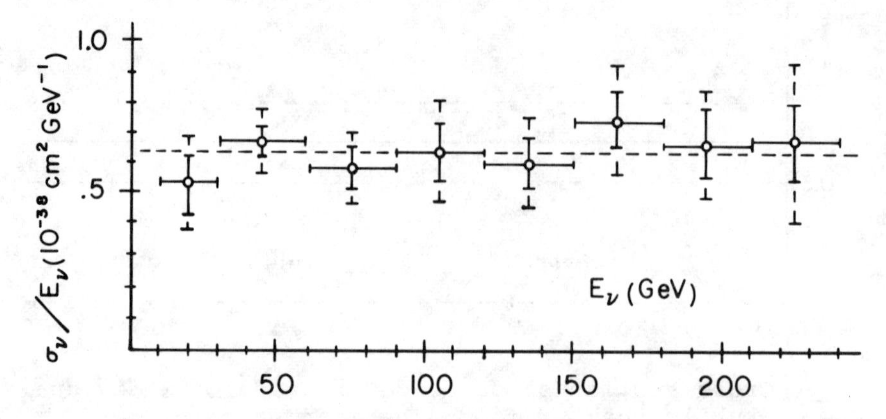

Fig. 6 Cross section slope as a function of neutrino energy. Inner error bars: statistical only; outer error bars include systematic effects. The mean value of σ_ν/E_ν = 0.64 is indicated.

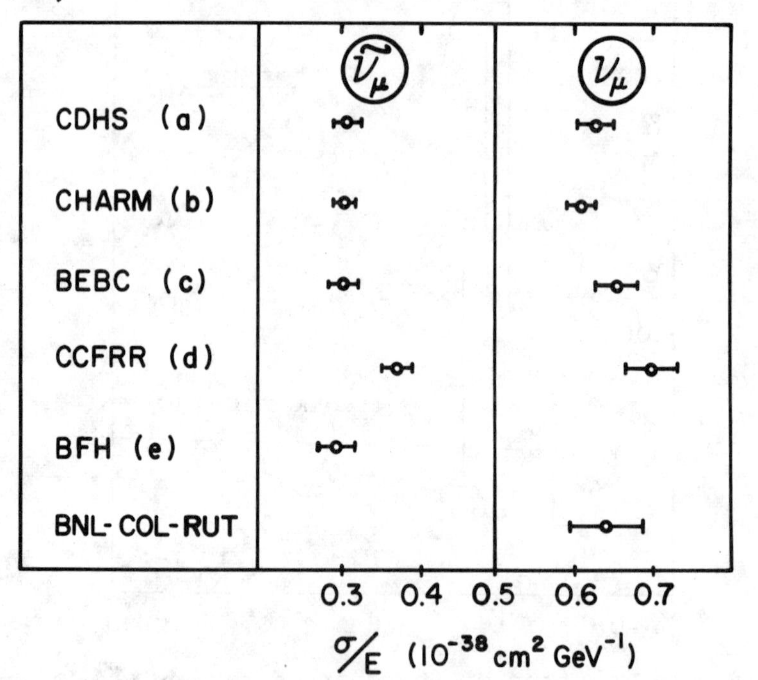

Fig. 7 Neutrino and anti-neutrino cross section slopes obtained by different experiments. The letters in parentheses refer to Table 3.

Anti-neutrino-nucleon Charged Current Interaction Cross Sections Measured by the Fermilab 15′ Ne-H_2 Bubble Chamber in a Dichromatic Beam†

G.N. Taylor , R.J. Cence, F.A. Harris, M.D. Jones , S.I. Parker, M.W. Peters, V.Z. Peterson, V.J. Stenger
University of Hawaii at Manoa, Honolulu, Hawaii 96822

H.C. Ballagh, H.H. Bingham , T.J. Lawry, G.R. Lynch, J. Lys , M.L. Stevenson, G.P. Yost
University of California and Lawrence Berkeley Lab
Berkeley, CA 94720

D. Gee, F.R. Huson, E. Schmidt, W. Smart, E. Treadwell
Fermi National Accelerator Lab, Batavia, IL 60510

Presented at the 1982 Vanderbilt Conference on Particle Physics by H.H. Bingham, May 25, 1982, Revised Sept. 17, 1982

ABSTRACT: Fermilab experiment E388 measurements of antineutrino-nucleon total charged-current interaction cross sections are reported for 20 $< E_{\bar{\nu}} <$ 200 GeV. E388 used the Fermilab 15′ Neon (56.5% atomic)-hydrogen bubble chamber (plus external muon identifier), exposed to the N30 dichromatic beam. E388 ran downstream of the CCFRR experiment E616 in this beam and used the same flux monitoring information. No evidence is seen for any variation of $\sigma_{\bar{\nu}}/E_{\bar{\nu}}$ in this $E_{\bar{\nu}}$ range. The *preliminary* average $\sigma_{\bar{\nu}}/E_{\bar{\nu}} = [0.30 \pm 0.02(\text{stat.}) \pm 0.02(\text{sys.}) \pm 0.02(\text{flux})] \times 10^{-38}$ cm^2/GeV/nucleon, is in good agreement with most other measurements, but 20% below the E616 result presented at this conference.

† Work supported in part by the U.S. National Science Foundation and Department of Energy.

Introduction:

The total cross sections for the charged current (CC) neutrino (ν) and antineutrino ($\bar{\nu}$) reactions $^{(}\bar{\nu}^{)}+N\rightarrow\mu^{(\mp)}+X$ are of fundamental importance for understanding the weak interactions, the structure of nucleons, possible violations of scaling laws, etc. Any departure from a constant value of $\sigma_{\bar{\nu}}/E_{\bar{\nu}}$ could be an indication of the finite mass of the intermediate vector boson, or of new physics. The previous speakers [1,2] have reported a disagreement between the average values of σ_{ν}/E_{ν} obtained by two experiments, E616 and E380, running together in the same ν beam. We report here a similar disagreement between our (E388) results for $\sigma_{\bar{\nu}}/E_{\bar{\nu}}$ and E616's. We ran together with E616 in the same N30 dichromatic beam, tuned for antineutrino parents this time, and used E616's flux monitoring and calibration.

As the previous two speakers have stressed, the disagreement over σ_{ν}/E_{ν} between E616 and E380 is about three times the statistical uncertainty of E380 and the flux uncertainties should cancel out. Thus some other systematic error of the counter experiment, E616, or the bubble chamber experiment, E380, is probably to blame. Our analysis offers not only a similar contrast of bubble chamber (E388) and counter (E616) systematics, but also some important differences between our (E388) and E380's bubble chamber analyses. We discuss these differences below in hopes that our results may provide further insight into where the key systematic errors may lie. Geoff Taylor's thesis[3] has further information about our experiment.

Measurement of $\sigma_{\bar{\nu}}$:

The cross section, $\sigma_{\bar{\nu}}$, for the CC reaction $\bar{\nu}_{\mu}+N\rightarrow\mu^{+}+X$, as a function of the antineutrino beam energy, $E_{\bar{\nu}}$, is measured here essentially by the average over the bubble chamber (BC) fiducial volume and apparatus acceptance of the expression

$$\sigma_{\bar{\nu}}(E_{\bar{\nu}})=\frac{N_{evt}-N_{BG}}{\epsilon\, F_{\bar{\nu}}\Delta E_{\bar{\nu}} n\Delta V}.$$

where N_{evt} is the number of $\bar{\nu}$ -induced (CC) events observed in this element of BC volume, ΔV, and this $E_{\bar{\nu}}$ bin, $\Delta E_{\bar{\nu}}$; N_{BG} is the number of background events among them; ϵ supplies corrections for various event losses and for smearing from one $E_{\bar{\nu}}$ bin to another due to experimental resolution; $F_{\bar{\nu}}$ is the flux of beam antineutrinos per unit area per unit $E_{\bar{\nu}}$ through this BC volume element and in this $\Delta E_{\bar{\nu}}$; n is the number of target nucleons per unit volume. Let us discuss each of these factors in turn and take note of where we use the same, or different, information and/or procedures than does E616, or E380.

The determination of the ν flux, F_ν, from the N30 dichromatic beam, and its monitoring and calibrations have been described by the previous two speakers [1,2]. For the $\bar{\nu}$ flux, $F_{\bar{\nu}}$, the monitoring and calibration was similar to that for F_ν. We used E616's K^-/π^- ratios, which agree within a few percent with those measured at CERN. Because we use E616's flux monitoring and calibration, the 5% uncertainty in the flux should cancel out in a comparison of E616 and E388. After accounting for decays in the train (parent beam transport system), within $\pm 3\%$ we predict the same $F_{\bar{\nu}}$ (as a function of $E_{\bar{\nu}}$ and R) per beam π or K decay as E616 does. We find no evidence for any systematic difference between the respective predictions. (R is the radial distance from the beam center line in the BC.) To compare with E616, we restrict our data sample to those frames where there is good E616 monitor information, the beam is properly steered (according to E616) and the slit is open. We are making further checks of the bookkeeping, etc., but it seems unlikely that any discrepancy between E616 and E388 is due to errors in estimating $F_{\bar{\nu}}$.

The density of target nucleons, n, depends upon the Ne/H composition of the BC liquid and its pressure and temperature. Pressure and temperature were monitored continuously during the run and the composition was measured from time to time at both the top and the bottom of the BC. The resulting n varied by a few percent as the run progressed, and, at a given time, varied by a few percent with depth in the BC. Wes Smart (of Fermilab) has analyzed these data and estimates that the average composition (over time and depth) was 72.2 Molar % neon (56.5 atomic %), the average temperature was 29.38 ± 0.06° K and the average density was 0.666 ± 0.014 gm/cm^3. (The corresponding radiation length was 44 cm and the absorption distance for high energy pions was ~ 165 cm, making the 380 cm diameter chamber (in a 30 KG magnetic field) an excellent calorimeter for the $\bar{\nu}$ energy in CC events. The fiducial volume was 17.7 m^3 containing 11.8 tons of liquid.) E380 found that Wes Smart's estimates of n were confirmed to a percent or so by measurements of range vs momentum of stopping protons, etc. We are not that far along yet but our fits to $\Lambda^0 \rightarrow p\pi^-$ with stopping protons are compatible with the above density. The liquid density (and BC geometry) is unlikely to be off by more than a percent or so.

What about $(N_{evt} - N_{BG})/\epsilon$? We **scan** the film for essentially anything that happens in the BC and we **measure** each neutral-induced event with total $(\Sigma p_\|)$ above $\sim$ 1 GeV, where $\Sigma p_\|$ is the total visible momentum component parallel to the $\bar{\nu}$ beam direction. (Tracks coming into the BC are usually so signed by δ rays.) The E388 film was even cleaner than E380's and real events are unmistakable. We find $\epsilon_{scan} = 97^{+3}_{-4}\%$ for events passing the cuts below. At this stage ϵ_{meas} = 96 ± 2% and climbing. After measurement, we **require** for a good CC event, i) $\Sigma p_\| > 5$ GeV, ii) vertex in fiducial volume ($|Z| < 130$ cm, potential length to downstream wall > 50 cm), iii) good μ^+ with $p_{\mu^+} > 4$ GeV/c identified in the external muon identifier (two plane CL > 10^{-4}).

We differ importantly from E380 in muon identification proceedure. We use the external muon identifier, EMI [4], to select the good μ^+ while E380 uses kinematic criteria. The correction for neutral current (NC) background thus introduced is E380's largest (~ 10%). Our correction for NC BG is negligible, but on the other hand, we rely on our calibration of the EMI acceptance (ϵ^{EMI}_{Geom} = 93 ± 3%) and electronic efficiency ($\epsilon^{EMI}_{elec2plane}$ = 91 ± 3%), about which more below. In practice, we correct for the EMI acceptance and efficiency and for the cuts on p_μ and $\Sigma p_\|$, as a function of position in the BC, and of the kinematic variables of the event, using a Monte Carlo program which takes our resolution smearing, etc, into account.

There are two other small corrections in ϵ: i) ϵ_{misc} = 99 ± 1% for partial loss of 1 prong events; (We expect 10 quasi-elastic events; we see 8-- these look like through muons without a big hadronic shower (maybe a short proton or two) and are the hardest events to find in scanning.) ii) ϵ_{I-0} = 1.01. The Ne/H mixture has a 2.8% excess of protons over neutrons so σ^{I-0} = 99.07% σ^{obs}.

We get rid of non-antineutrino-induced events with our cuts but some of the good $\bar{\nu} + N \rightarrow \mu^+ + x$ events we observe are due to $\bar{\nu}$ coming from upstream of the narrow-band-beam (NBB) parent π^- and K^- we (i.e., E616) monitor. These background (BG) events are of two types, "**Wide Band Background**" (WBBG) and "**prompt background**" (PBG). E616 and E380 use the measured event rate obtained with the NBB closed off with a slit to estimate these BG's. We have the EMI, however, and in contrast to E616 our acceptance for μ^- is about the same as for μ^+. There should be no μ^- from the NBB (focussing $\pi^-(K^-) \rightarrow \mu^- \bar{\nu}_\mu$; $\bar{\nu}_\mu \rightarrow \mu^+$), thus the μ^- are a measure of WBBG and PBG. Using measured ratios of μ^-/μ^+ in unfocussed wide band and prompt $\bar{\nu}$ beams we estimate 9 wide band and 1 prompt background events should be among our 279 μ^+ events passing cuts, i.e., ~ 3%. After correction by ϵ, 279-10 become 324 signal events.

Within our limited closed slit statistics (and correcting for train decays) we would get the same answer if we used E616's procedure. For us our method is more precise. E380 has an additional complication in doing the closed slit subtraction and not using the EMI. The μ^-/μ^+ ratio for them is much larger in the NBB than in the closed slit data.

Results:

Our **preliminary** result for the slope with energy of the total antineutrino charged current cross section is

$$\sigma_{\bar{\nu}}/E_{\bar{\nu}} = [0.30 \pm 0.02\,(\text{stat}) \pm 0.02\,(\text{sys.}) \pm 0.02\,(\text{flux})] \times 10^{-38}\text{cm}^2/\text{Gev/nucleon}.$$

We see no significant variation of σ/E with E between 20 and 200 GeV. See Fig. 1. This average value of $\sigma_{\bar{\nu}}/E_{\bar{\nu}}$ is to be compared with E616's value of 0.36 ± 0.02 and on the other hand [5] with CDHS 0.30 ± 0.02, CHARM 0.30 ± 0.02, BEBC 0.30 ± 0.02. .Thus we agree with the rest of the world and are 20% below E616, even though we use E616 flux monitoring and calibration.

Discussion -- what could be wrong?

-- Did our scanners miss some class of event systematically? (Not likely if they found 8 of the expected 10 one-prong events-- certainly the hardest to find--and if we measure $\epsilon_{scan} = 97^{+3}_{-4}\%$ averaged over all event classes.)

-- Did our scanners miss events in some hard-to-see part of the BC? (Not likely, for $0 < R < 10$ cm, $\sigma/E = 0.29 \pm 0.06$ (stat), for $10 < R < 15$ cm, $\sigma/E = 0.29 \pm 0.07$, for $15 < R < 60$ cm, $\sigma/E = 0.30 \pm 0.05$. (R is the radial distance from the center of the beam line in the BC.) This checks the flux vs E_ν and R, too. The distribution of events in x, y, and z is also in good agreement with the Monte Carlo predictions. We have included in the systematic error an estimate of possible small loss of events deep in the BC. The three participating labs each separately got the same value of σ/E, within errors, as well.)

-- Was the BC insensitive during the arrival time of part of the flux? (Not likely--we require a minimum beam flux at the right time to trigger the BC flash. The BC is sensitive for 10's of millisec before and after the ~ 1 ms beam pulse and during this time **the BC has no dead time.** We can tell from the track quality if an event occurs more than a few ms early or late and we have seen no such events.)

-- Was the BC sometimes expanded at the wrong time and thus insensitive during the entire beam flux? (Not likely--the BC timing and its sensitivity and track quality were routinely checked by the crew and by the physicists during the run. For not more than a fraction of a percent of the normal running time could such a gross maladjustment have gone undetected. There was no evidence of jitter or of erratic pulsing. Further evidence against periodic or occasional mistiming is given by a plot vs. time of the event rate per beam particle (or of the rate of "beam" muons traversing the BC, which has better statistics.) This plot is flat enough for each tune to rule out more than a few percent of occasional or periodic insensitivity during part of a run. A further, related check has been done to the ~ 5% level and is being improved currently: the EMI is used to signal those frames which should have an in-time muon and the film is reexamined to find it if it has not already been measured. There is no evidence of BC insensitivity here, once correction is made for muons produced in the material between the BC liquid and the EMI counters.)

-- Was the EMI insensitive as a whole sometimes, causing CC events to be interpreted as NC? (Not likely--muons measured in the BC, making an angle with the beam direction of less than a few degrees, are detected in the EMI with efficiency consistent with the individual EMI chamber electronic efficiencies. (These individual efficiencies are measured by comparing the response, to a given muon, of one chamber with that of another in line with it.) With 95% confidence, the EMI as a whole was insensitive less than ½ % of the time that the BC was sensitive and good beam was present.)

Conclusions:

See "Results" above. We have no explanation, other than statistics, for the disagreement between E616's and our result. We hope that our discussion of systematics and procedures will help shed light on possible reasons for the disagreement.

References

1) CCFRR (E616) collaboration, paper presented at this conference by E. Fisk.

2) CBR (E380) collaboration, paper presented at this conference by P. Igo-Kemenes.

3) G.N. Taylor, et al., B.A.P.S. *27*,1, (1982) Abstract BE 7; G.N. Taylor, Ph.D. Thesis, University of Hawaii, 1982.

4) M.L. Stevenson, in Proc. Top. Conf. on Neutrino Physics at Accel., ed. by A.G. Michette and P.B. Renton (Rutherford Lab, England, 1978), p. 392.

5) CDHS: J.G.H. de Groot, et al., Z. Phys, *C1* (1979)143;
CHARM: M. Jonker, et al., Phys. Lett. *99B* (1981) 265;
BEBC: P Bosetti, et al., Phys. Lett. *110B* (1982) 167;
and references cited in these papers.

Figure Caption

1) E388 results for $\sigma_{\bar{\nu}}/E_{\bar{\nu}}$ vs. $E_{\bar{\nu}}$. (Taken from G. Taylor's thesis [3].)

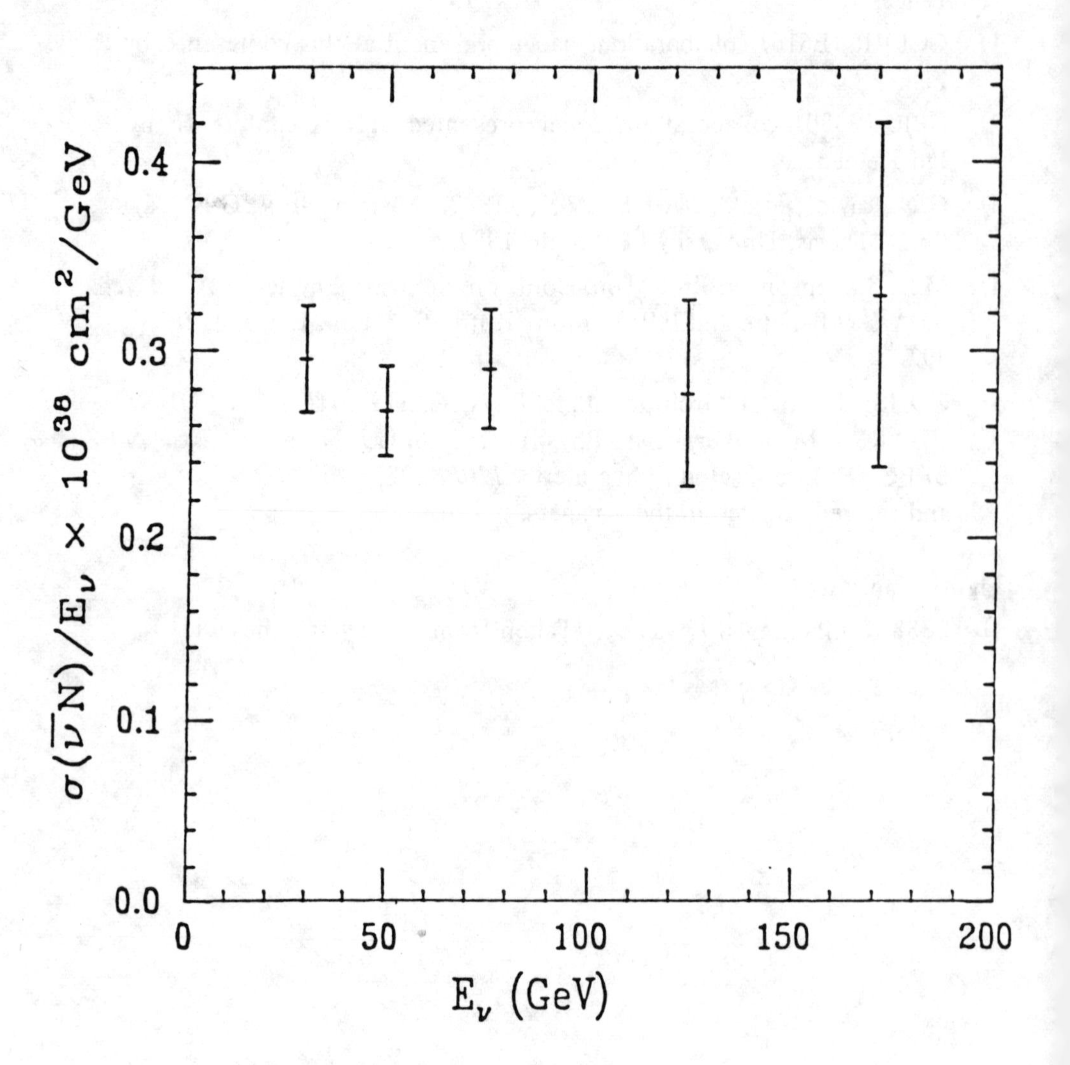
$\sigma(\overline{\nu}N)/E_\nu \times 10^{38}$ cm^2/GeV
0.0
0.1
0.2
0.3
0.4
0
50
100
150
200
E_ν (GeV)

Figure 1

GUTs, SUSY GUTs AND SUPER GUTs

Mary K. Gaillard
Lawrence Berkeley Laboratory and Department of Physics
University of California, Berkeley, California 94720

ABSTRACT

We review the motivations for extending grand unified theories with particular emphasis on supersymmetry and its phenomenological and cosmological fallout, and comment on the relevance of quantum gravity.

INTRODUCTION

The notion that the elementary forces of nature should ultimately reveal themselves as part and parcel of a single unified force has been around for some time.[1] The prototype SU(5) model[2] for unification of the now "standard" $SU(3)_C \otimes SU(2)_L \otimes U(1)$ theory of strong, electromagnetic and weak interactions was proposed at a time when experimentalists were still uncertain as to even the existance of neutral currents. However the original[3] and simplest version of a renormalizable electroweak theory withstood the test of time and survived experimental scrutany to the point where deviationists from theoretical orthodoxy have been forced to artifically adjust their more complex scenerios to mimic the straight-forward predictions of the standard model. The present explosion of variations on grand unified models (GUTSs): technicolored[4] (TC), supersymmetric (SUSY)[5-9] supercolored,[10] superunified,[11,12] weak-confining,[13] compactified[14,15]..., is reminiscent of the earlier proliferation of electroweak models, with the unfortunate difference that there is very little data to stem the flow of speculation.

In fact experimental support for the idea of grand unification rests essentially on a single piece of data: the value[16,17] of the parameter $\sin^2\theta_W$ which characterizes the strength and structure of weakly coupled neutral currents. There has been a steady convergence between the radiatively corrected[18,19] experimental value

$$\sin^2\theta_W = 0.215 \pm 0.002.$$

and the value predicted from the simplest version of SU(5), most recently evaluated at[18]

$$\sin^2\theta_W = 0.214 \pm 0.002.$$

This model also relates some quark and lepton masses. A recent comparison[21] between estimates of quark masses from analyses of low

0094-243X/82/930291-14$3.00

energy data and from SU(5) calculations indicates agreement within 20% for the b-quark to τ mass ratio and about a factor two for the s-quark to μ mass ratio. Perhaps one can't expect better: application of perturbative QCD techniques to low energy becomes increasingly unreliable especially since thresholds are involved. It is well known that the d-quark to electron mass ratio is incorrectly predicted but it has been argued[22] that this discrepancy can be accounted for by effects of quantum gravity in that these masses are so tiny that effects of order m_{GUT}/m_P relative to the overall fermion mass scale are not negligible.

So why all the fuss? The objections to the minimal GUT are largely aesthetic. While GUTs unify the three independent coupling constants of the low energy theory, there remains a large number of parameters which must be put in by hand: the Yukawa coupling of scalars to fermions which determine the fermion mass spectrum and the Cabbibo-like angles which govern weak decays; the scalar self couplings which govern the pattern of symmetry breaking and hence the vector boson mass spectrum; the θ-parameter of QCD which characterizes the strength of P and CP violation in "strong" interactions. Two of these "fine tuning" problems are particularly acute: the ratio of mass scales characteristic of SU(5) breaking and of $SU(2)_L \otimes U(1)$ breaking which differ by 14 orders of magnitude, and the θ-parameter which must be adjusted to a very tiny value. The first problem is the notorious gauge hierarchy problem: it is this problem which forms the focal point for most of the attempts at generalizing the minimal GUT. The "strong CP" problem which has been discussed by Dine[23] may in fact also be brushed under the gravitational rug,[24] since introducing a cut-off $\Lambda \simeq m_P$ in the radiative corrections to θ yields an acceptably small value for the neutron dipole moment. On the other hand if we are using quantum gravity as a garbage pail for our lack of understanding we must ultimately address the second major failing of GUTs: it makes extrapolations from present day laboratory energies to energy scales only four orders of magnitude below the Plank scale, but cannot include gravitational interactions in the unified picture. This is because the renormalizable gauge theories based on local (i.e. space-time dependent) internal symmetries cannot accommodate fields of spin greater then one, whereas quantum gravity requires a spin-2 graviton.

The two major failings of GUTs - arbitrariness of parameters and the failure to include gravity - has led theorists to take seriously the possibility that supersymmetry[25] may have something to do with nature. Supersymmetry goes beyond ordinary internal symmetries in that it relates fields of different spin. This means that gauge couplings can be related by supersymmetry to Yukawa couplings and to scalar self-couplings, thus promising to remove the arbitrariness aluded to above. In addition, the higher degree of symmetry provides extra cancellations among divergent contributions to radiative corrections, and it is hoped that supergravity (SUGRA),[26] the supersymmetric version quantum gravity, will provide a tractable theory of gravity as well as its unification with gauge theories. Unfortunately, these ambitious programs require what is called _extended_ supersymmetry which embeds internal symmetries with supersymmetries. On the other hand, the data forces us to describe our particle world by _chiral_ gauge theories, in which left and right handed fermions couple with different strengths to gauge bosons. A chiral gauge theory, it turns out, can be embedded only in a _simply_ supersymmetric theory, which does not lead to a finite theory of gravity, nor does it remove any arbitrariness as applied to our present unified gauge theories. All it does is to double the number of particle species.

SIMPLE SUPERSYMMETRY

Nevertheless, simply supersymmetric grand unified theories, or SUSY GUTs have become very popular as a line of attack[27] on the gauge hierarchy problem, which can be formulated as follows. If the scalar sector of the unified gauge theory can be described using convergent perturbation theory, then phenomenology requires[28]

$$m_H \leqslant 1 \text{ TeV} \quad (1)$$

in the standard electroweak model.[3] However the natural scale[2,16] of the strong and electroweak unified model is 10^{15} GeV. Fermion masses can be much smaller than this scale because they are protected by chiral symmetry, and vector boson masses are similarly protected by gauge symmetries. The problem for scalars is that their masses are unprotected in an ordinary gauge theory; that is, one expects them to be governed by the largest mass scale around since all the interacting scalars communicate with one another through radiative corrections.

Supersymmetry offers a simple solution: protect the mass of the standard model Higgs doublet by tying it to the mass of a chiral fermion superpartner. But in a realistic GUT, things are not so simple. For example, in the minimal SU(5) model[2] the Higgs doublet is part of an SU(5) 5-plet which also contains scalars which transform like a triplet under color SU(3). These scalars can mediate proton decay and are therefore constrained to be very heavy:

$$M_{\text{triplet}} \geqslant 10^{11-12} \text{ GeV} \quad (2)$$

as opposed to (1). These phenomenological requirements can be simultaneously satisfied,[17] but this requires an artificial adjustment of parameters which in an ordiniary gauge theory is highly unstable against radiative corrections. One of the (somewhat mysterious) properties of supersymmetric theories is that they allow[29] such a parameter adjustment to be _stable_ against radiative corrections. In minimal SU(5) it remains arbitrary,[5] but this can be cured by appealing to a higher symmetry.[6]

If we wish to protect the electroweak Higgs mass using

supersymmetry, then the above argument suggests that the scale m_S of supersymmetry breaking should be no more than a TeV. This would imply[30] that quarks (q) and leptons (ℓ) have scalar supersymmetric partners called squarks ($\tilde{q}$) and sleptons ($\tilde{\ell}$) with masses

$$m_{\tilde{q}}, m_{\tilde{\ell}} \leqslant m_S \leqslant 1 \text{ TeV}. \tag{3}$$

In addition the gauge bosons have supersymmetric fermionic partners (inos); those associated with the massless photon and gluons aquire masses only through radiative corrections, so we expect for the photino ($\tilde{\gamma}$) and gluinos ($\tilde{g}$):

$$m_{\tilde{\gamma}} \leqslant \frac{\alpha}{\pi} m_S \leqslant \text{a few GeV}$$

$$m_{\tilde{g}} \leqslant \frac{\alpha_S}{\pi} m_S \leqslant 30 \text{ GeV}. \tag{4}$$

Experimental evidence against the existance of these objects is meager; for example, analyses[32] give only $m_{\tilde{g}} \geqslant 2$ GeV, with almost no limit on $m_{\tilde{\gamma}}$. Groups[33] at PETRA have looked for the process

$$e^+e^- \to \tilde{\ell}^+\tilde{\ell}^-, \quad \tilde{\ell}^\pm \to \tilde{\gamma} + \tilde{\ell}^\pm.$$

giving limits

$$m_{\tilde{\ell}} \geqslant 16 \text{ GeV} \tag{5}$$

if the photino does not decay to produce a photon shower in the detector. A similar limit[34] for the smuon ($\tilde{\mu}$) mass follows from the experimental and theoretical errors on the muon anomalous magnetic moment.[35] It should be possible[36] to push the limit on the selection mass beyond the beam energy in e^+e^- reactions by looking for single selectron production via the quasi-real photon process

$$\gamma + e \to \tilde{\gamma} + \tilde{e}, \quad \tilde{e} \to \tilde{\gamma} + e \tag{6}$$

signed by an energetic, large angle electron from the selectron decay with the spectator electron emerging at a very small angle with respect to the beam direction; if the photino has a short enough lifetime the high $p_\perp$ of the decay electron would be partically balanced by photons from the photino decays. The total cross section corresponds to 5% of a unit of R for $m_{\tilde{e}} = E_{beam}$ and drops to 0.15% for $m_{\tilde{e}} = 1.5\ E_{beam}$. At present "low energy" supersymmetry seems to be phenomenologically acceptable. But is it not without difficulties. Aside from the fact that it has proven quite difficult[37-40] to write down a realistic SUSY model of just the electroweak and strong gauge theories, the upper limit (1) is itself somewhat artificial and in fact rather generous. It corresponds to allowing the self coupling constant λ of the Higgs scalar to reach its unitarity limit

$$\lambda = 4\pi \gg 1. \tag{7}$$

The more plausible value $\lambda = O(g^2)$, where g is the weak gauge coupling constant, would reduce the limit on m_H by an order of magnitude and correspondingly reduce the limits (3) and (4) on squark, slepton and ino masses. As usual, an eventual conflict with experiment can be avoided by enlarging the theory. If one puts[40,7,8] the supersymmetry breaking into a sector of the theory which doesn't couple directly to the Higgs scalars, one can get

$$m_H \sim (\frac{\alpha}{\pi})^n \, m_S \times \text{(GIM type suppression factors)} \tag{8}$$

where the power n can be made arbitrarity high, or the mass suppression factors arbitrarily small, allowing m_S to become arbitrarily large-perhaps even as large as the Planck mass m_P[7] - depending on the extent to which one is willing to complexity the theory.

Having dispensed with the immediate phenomenological problems, we should address aesthetic problems. We started out in the hope of limiting the number of independent mass scales. Let's see what has been acheived. In the standard GUT-less strong and electroweak theory we had two scales: m_W and the parameter Λ of QCD which measures the energy scale at which the strong coupling constant becomes strong. Incorporating this theory into a GUT, we introduced a third scale, m_X (the mass of superheavy gauge bosons), but also eliminated one: the value of Λ can be related[16] to the value of m_X through the effects of radiative corrections. Going now to a SUSY GUT, we end up a priori with three mass scales: m_W, m_X and m_S. In all cases we still have the Planck mass m_P unrelated to anything (except that if $m_S \sim 1$ TeV, one finds that m_X actually approachs m_P in magnitude), as well as all the fermion masses which one tends to ignore in discussions of scales on the grounds that they are really Yukawa couplings - but this makes them no less arbitrary. In short, supersymmetry has forced us to introduce more parameters and removed none. Can we remedy this?

In fact, in models[7,8,40] of the type leading to Eq. (8), the breaking of $SU(2)_L \otimes U(1)$ and consequently the ratio m_W/m_S is determined by radiative corrections (which, however, depend on the arbitrary Yukawa couplings). Can we also get the ratio m_X/m_S from radiative corrections? As emphasized by Witten,[41] spontaneously broken supersymmetric theories have a large vacuum degeneracy which is lifted by radiative corrections where one encounters a dependence on scales of the form $\alpha/\pi\ \ln(m_1/m_2)$ which could lead, under certain conditions, to a ratio of scalar field vacuum expectation values $m_1/m_2 \sim \exp(\pi/\alpha)$. This is similar to the effect of radiative corrections in determining[16] the ratio Λ/m_X and also to previous attempts to determine the ratio m_W/m_X by introducing technicolor[4] (for which a viable GUT model was never found and which has serious phenomenological difficulties with strangeness changing neutral currents) or by radiative corrections[42] in the standard model with m_H = 10 GeV (which involves an ad hoc assumption). It seems fair to say that while the SUSY context appears a more promising framework for realizing this scenario, a quantitative and realistic example of its implementation has not yet been formulated although interesting work[8] in that direction is going on. Unfortunately, specific calculations[43] without fine tuning yield scale hierarchies which are only of order $m_1/m_2 \sim \exp(1/2\pi\alpha)$, which is not very large in a GUT model.

GRAVITINO AND GRAVITY

Another feature of supersymmetric theories is that in addition to doubling the number of known particle species at least one new species must be added to the zoo. A spontaneously broken internal symmetry gives rise to a massless scalar particle called a Goldstone boson; in spontaneously broken gauge theories the Goldstone bosons are eaten by the massive vector mesons. A spontaneously broken supersymmetry gives rise to a massless fermion called a goldstino. This particle is eliminated in a similar way[44] if we include gravity in its supersymmetric form. Then we must introduce a spin-3/2 superpartner for the graviton (gravitino, $\tilde{G}$) which acquires a mass and eats the Goldstino when supersymmetry is broken.

This of course brings in gravity again and we wonder to what extent it can be ignored in the construction of models and in their phenomenological implications. For simple supergravity coupled to a simply supersymmetric matter theory, the observation that the cosmological constant as measured today is essentially zero leads to the prediction:[44]

$$m_{\tilde{G}} = \frac{m_S^2}{m_P} \equiv \kappa\, m_S^2 \tag{9}$$

in the tree approximation, where $\kappa = m_P^{-1}$ is the gravitational coupling constant. The longitudinal (helicity ± 1/2) components of the gravitino have enhanced couplings at high energy

$$g_{eff} = \kappa \frac{E^2}{m_{\tilde{G}}^2} = \frac{E^2}{m_S^2} \tag{10}$$

which will not necessarily be negligible at low energy if m_S is not too large. What about quantum[45] corrections? Since simple supergravity is not a renormalizable theory, we must introduce a cut-off, and unless a unified theory including gravity becomes effective below the Planck mass, the only available cut-off is the Planck mass itself. Then we expect that those particles (scalars, inos) whose masses are not protected by low energy gauge or chiral symmetries will get mass contributions of order

$$m_{ino} \sim m_{\tilde{G}} \frac{\kappa^2\Lambda^2}{\pi} \sim \frac{m_{\tilde{G}}}{\pi}$$

$$m^2_{scalar} \sim \pm m^2_{\tilde{G}} \frac{\kappa^2\Lambda^2}{\pi} \sim \frac{1}{\pi} m^2_{\tilde{G}} \qquad (11)$$

The bound (1) on m_H would then require $m_{\tilde{G}} \leqslant 1$ TeV or $m_S \leqslant 10^{11}$GeV. For the scalar fields ψ one should really consider the full effective potential

$$V_{eff}(\psi) = m^2_{\tilde{G}}\, \kappa^{-4} \frac{\Lambda^2}{\pi} f(\kappa^2\psi^2) + \text{less divergent terms} \qquad (12)$$

where f is a polynomial function of the dimensionless fields $\kappa\psi$. An amusing possibility is the case where V_{eff} has its minimum away from the origin. This would lead to a vacuum expectation value $\langle\psi\rangle \sim m_P$, independent of the value of m_S for the leading term. Since gravity sees no internal symmetries, the potential (12) depends only on $|\psi|^2 = \sum_{i=1}^{n} |\psi_i|^2$ where the sum is over all (complex) scalar particles in the theory. The vacuum is degenerate under SU(n) and leaves 2n - 1 massless Goldstone scalars. This degeneracy will be removed, and the remaining scalars acquire masses, when the gauge and other interactions are included. From the interplay of various radiative corrections one could imagine a scenario where the desired hierarchy of mass scales does arise, but in which the inclusion of gravity is an essential element.

SUPERUNIFICATION

Up to now we have concentrated on simple supersymmetry and abandoned our initial goals of removing arbitrary parameters and achieving the unification of gauge forces with gravity. For this program we turn to extended supersymmetry. The maximum number of supersymmetries we allow is 8; otherwise we are led[46] to an elementary particle spectrum including spins greater than 2, for which we cannot even write down a field theory. N = 8 extended supergravity is an ideal candidate for a truly unified theory: it has a unique particle spectrum[46] whose couplings are completely specified and there is hope that it may have finite S-matrix elements.[47] The trouble is that it fails to reproduce the "observed" particle spectrum, in spite of its richness. Among the 28 vector fields in the elementary N = 8 supermultiplet, there are none that can be identified with the $W^{\pm}$, let alone the X and Y of the minimal SU(5) GUT. Among its 56 fermion fields there are none that can be identified with the muon or with the (τ, ν_τ, b, t) generation of fermions. Interest in N = 8 supergravity was renewed when Cremmer and Julia[48] discovered that this theory has a local SU(8) invariance associated with 63 gauge fields which are not elementary but are composites of the elementary "preon" fields of the N = 8 supergravity multiplet. This led to the conjecture[11] that, aside from the graviton, the preons are all confined and the "observed" spectrum, i.e. those particles which make up our unified gauge theories (neutrinos, leptons, quarks, photon, gluons, $W^{\pm}$, Z, X, Y and Higgs scalars) are all bound states. The N = 8 supersymmetry is dynamically broken in such a way that those states which survive in the "low energy" theory ($E \ll m_P$) are such as to allow a renormalizable effective field theory (SUPERGUT). This set includes vectors in the adjoint of the surviving gauge group, an anomaly free set of spin-1/2 fermions, and scalar particles. To date this speculation has not yielded much predictive power. It does restrict the simple group unifying the strong, weak and electromagnetic interactions to be no larger than SU(5), although it could allow for some extra U(1)'s such as have been found necessary[37] to introduce in constructing realistic supersymmetric gauge theories. It further restricts the type of representations to which scalars and fermions may be assigned. These constraints become tighter if one assumes[11] that the particle content of the effective renormalizable gauge theory arises from a single N = 8 supermultiplet. Then the maximal fermion content compatible with a viable SU(5) GUT is

$$3(\bar{5} + 10) + 9(1) + 3(5 + \bar{5}) + 9(10 + \overline{10}) + 4(24) + 45 + \overline{45}$$

for left handed fermions, along with their CPT conjugate right handed fermions. Under the hypothesis[11] that one of the original 8 supersymmetries remains unbroken at "low" energy leaving us with a simple SUSY GUT, the number of allowed singlets (1) is reduced to 6 and the $(10 + \overline{10})$ states to 3. In either case we can have the usual three generations of $\bar{5} + 10$, plus a number of additional fermions which are presumably super heavy because they can aquire SU(5) invariant masses, which are consequently "unprotected". Thus in addition to accommodating the spectrum of fermions, gauge bosons and scalars of the minimal SU(5) model a single supermultiplet of N = 8 supergravity bound states also accounts easily for the additional fermions and scalars (in a simple SUSY effective gauge theory one 24-plet of fermions is associated with the gauge vectors; each of the remaining fermions has a complex scalar superpartner) which appear to be necessary for the construction[5-9] of a realistic SUSY GUT.

An alternative approach to superunification is based on a generalization[14] to supergravity of the old Kaluza-Klein approach[49] to the unification of gravity with electromagnetism. One starts with simple supergravity in a space of dimension greater than four. Upon "compactification" or the curling up of the extra dimensions into circles of infinitesimally small radius, their associated degrees of freedom appear as internal symmetry degrees of freedom of fields of lower spin. The difficulty with this approach is that it appears to generate non-chiral gauge theories.[14] Recently it has been shown[15] that chiral theories can be generated by compactification of initially non-chiral gauge theories in higher dimension, but these examples have no obvious relevance to gravity.

COSMOLOGICAL PROBES

Whatever the underlying theory, supersymmetric models tend to generate new stable or long-lived objects such as the gravitino, photino, selectron, or random Goldstone-like objects associated with either the spontaneous breaking of global chiral symmetries which are characteristic of SUSY models, or with the large vacuum degeneracy. Observational cosmology permits such an object if it is light enough to contribute negligibly to the cosmological mass density:[50,51]

$$m \leqslant O(\text{KeV}) \tag{13}$$

which from (1) implies[51]

$$m_S \leqslant 10^6 \text{GeV} \tag{14}$$

for a stable gravitino. In models which rely on the Witten[41] mechanism one gets a pseudo-goldstone scalar of mass

$$m \simeq m_S^2/m_X \tag{15}$$

which would correspondingly require[52]

$$m_S \leqslant 10^3 \text{GeV}. \tag{16}$$

Alternatively such a stable object would be acceptable if it is heavy enough[53] to decay or annihilate very early in the expansion of the universe. This does not give a very strong constraint for the photino,[50] but it is relevant to more exotic objects which decouple at very high temperatures. Depending on whether one is considering a gravitino,[53] or a Wittino[52] one gets mass bounds in the range

$$m \geqslant 10^4 - 10^{11} \text{ GeV} \tag{17}$$

and SUSY breaking scales in the range

$$m_S \geqslant 10^{11} - 10^{16} \text{GeV}. \tag{18}$$

Thus models with $10^6 \leqslant m_S \leqslant 10^{11}$ appear to be ruled out and a wider range of scales is excluded for Witten-type models.[41] However it should be remembered that the gravitino analyses are based on the

N = 1 SUGRA tree graph relation (9) which could be violated if the underlying theory is an extended SUGRA - in fact the only explicit example[54] of a broken N = 8 SUGRA (which is not a realistic model) provides a counter example to (9). In addition quantum corrections, which we have argued can give significant contributions to photino, gluino and scalar masses, may also invalidate the relation (9). We remark in passing that some of the above mentioned random goldstone particles are candidates for the invisible axion discussed by Dine,[23] so that SUSY theories may at least provide a neat, albeit nearly untestable, solution to the strong CP problem.[10,55]

LOW ENERGY PROBES

As a concluding remark, I would like to emphasize the importance of precision low energy experiments for probing the very high energy sector of our theory which may be out of reach of even the next generation of accelerators. The most exciting example is proton decay - a clear signal for any decay mode is of prime importance in itself. If supersymmetry is valid down to mass scales of 1 TeV or less, then the mass of the superheavy X of the GUT gets pushed up[56] to a value much higher than 10^{15} GeV, and proton decay is no longer dominated by X-exchange. In some models the most important contributions[38,57] arise from diagrams involving superheavy fermions. These are higher order in the coupling constant but lower order in the inverse superheavy mass. If they dominate one expects[58] the dominant modes for nucleon decay to be $N \to K + \nu$. On the other hand if the dominant mechanism is the exchange[8] of the usual color triplet Higgs of SU(5), the dominant mode will be $K\mu$. But beware of drawing conclusions. By adding scalars in 10-plets of SU(5) one can recover[59] the minimal SU(5) prediction that the πe mode is dominant. A second example is the neutron electric dipole moment. If the resolution of the strong CP problem lies in the existance of an axion, visible or not, most theories predict a neutron dipole moment much smaller than the present experimental limit,[60] whereas in the absence of an axion,[61] the observed baryon to photon cosmological density ratio suggests that the neutron dipole moment should be within the reach of future experiments.[62]

Rare K-decays can continue to play an important role in constraining theorists' fantasies. As an example an experiment[63] which could reach the level of 10^{-10} in branching ratio for the process $K^+ \to \pi^+$ + nothing would contain a considerable amount of physics. Firstly, a null result is not expected, except in the advent of a perverse cancellation. The standard model alone predicts[64]

$$\text{B.R.}(K^+ \to \pi^+ + \nu\nu) \sim 10^{-10} + \text{top quark correction} \quad (19)$$

and this is undoubtedly, among the various K-decay processes, the theoretically cleanest test of weak radiative corrections. The same experiment would probe the mass of a mediator of generation-changing neutral processes like $K^+ \to \pi^+ \nu_e \bar{\nu}_\mu$ to a scale of 25 TeV. Under the hypothesis that photinos are quasi massless and quasi stable it would probe the squark mass to some fraction[65] of m_W. General flavor changing neutral current processes which provide a severe headache for technicolor theories[4] are more limited in their ability to restrict[66] SUSY model building - tending to yield limits on, e.g., slepton and squark mass differences rather than slepton or squark masses. Mass differences among squarks of a given flavor appear to be more strongly constrained by measurements of parity violating nuclear transitions.[67]

To conclude: theorists are off on a binge of unrestrained speculation; we badly need guidance from experiment.

ACKNOWLEDGMENTS

This paper is dedicated to Leon M. Lederman: may future birthdays bring future discoveries. I have enjoyed benifical discussions with Larry Hall, Joe Polchinski, Mahiko Suzuki, Bruno Zumino, and especially Ian Hinchliffe whose critical comments on the manuscript are gratefully acknowledged. This work was supported by the Director, Office of Energy Research, Office of High Energy and Nuclear Physics, Division of High Energy Physics of the U.S. Department of Energy under Contract DE-AC03-76SF00098.

REFERENCES

1. J. C. Pati and A. Salam, Phys. Rev. D8, 1240 (1973).
2. H. Georgi and S. L. Glashow, Phys. Rev. Lett. 32, 438 (1974).
3. S. L. Glashow, Nucl. Phys. 22, 579 (1961);
S. Weinberg, Phys. Rev. Lett. 19, 1264 (1967);
A. Salam, Proc. 8th Nobel Symposium, ed. N. Svarthholm, (Amquist and Wiksell, Stockholm 1968) p. 367.
4. S. Weinberg, Phys. Rev. D13, 974 (1975)
L. Susskind, Phys. Rev. D20, 2619 (1979).
5. S. Dimopoulos and H. Georgi, Nucl. Phys. B193, 150 (1981).
N. Sakai, Z. Phys. C11, 153 (1982).
6. R. N. Cahn, I. Hinchliffe and L. J. Hall, Phys. Lett. 109B, 426 (182).
S. Dimopoulos and F. Wilczek, "Incomplete multiplets in supersymmetric unified models", Santa Barbara ITP preprint (1981).
7. J. Ellis, L. Ibanez and G. G. Ross, Rutherford preprint RL-82-024 (1982).
8. M. Dine and W. Fischler, "A supersymmetric G.U.T.", Inst. of Advanced Study, Princteon preprint (1982).
S. Dimopoulos and S. Raby, Los Alamos preprint LA-UR-82-1282 (1982).

9. T. N. Sherry, ICTP Trieste preprint 1C/79/105 (1979).
H. Georgi, Harvard preprint HUTP-81/A041.
A. Masiero, D. V. Nanopoulos, K. Tarvakis, and T. Yanaside, CERN-TH-3298 (1982).
10. M. Dine, W. Fischler and M. Srednicki, Nucl. Phys. B189, 575 (1981).
S. Raby and S. Dimopoulos, Nucl. Phys. B192, 353 (1981).
11. J. Ellis, M. K. Gaillard, L. Maiani and B. Zumino, in "Unification of the Fundamental Particle Interactions", Ed. S. Ferrara, J. Ellis and P. van Nieuwenhuizen (Plenum Press 1980) p. 343.
J. Ellis, M.K.Gaillard and B. Zumino. Phys. Lett. 94B, 343 (1980).
J. Ellis, M. K. Gaillard and B. Zumino, Annecy preprint LAPP-TH-44, TH.3152-CERN (1981). To be published in Acta Polmica Physica.
12. T. Curtright and P. G. O. Freund, in "Supergravity", ed. P. van Nieuwenheusen, and D. Z. Freedman (North-Holland, Amsterdam, 1979);
P. H. Frampton, Phys. Rev. Lett. 46, 881 (1981);
J. P. Derendinger, S. Ferrara and C. A. Savoy, Nucl. Phys. B188. 77 (1981).
J. E. Kim and H. S. Song, Seoul National University preprint (1981).
G. Altarelli, N. Cabibbo, and L. Maiani, Istituto di Fisica "G. Marconi", Universita degli Studi, Rome, preprint n. 282 (1982).
13. L. F. Abbott and E. Farhi, Nucl. Phys. B180, 547 (1981)
Y.-P. Kuano and S.-H. H. Tye, Cornell preprint CLNS-82/599 (1982).
14. E. Witten, Nucl. Phys. B186, 412 (1981).
15. G. Chapline and R. Slansky, Los Alamos preprint LA-UR-82-1076 (1982). Submitted to Nucl. Phys.
16. H. Georgi, H. R. Quinn and S. Weinberg, Phys. Rev. Lett. 33, 451 (1974).
17. A. J. Buras, J. Ellis, M. K. Gaillard and D. V. Nanopoulos, Nucl. Phys. B135, 66 (1978).
18. W. Marciano and A. Sirlin, Phys. Rev. Lett. 46, 163 (1981).
19. A. Sirlin and W. Marciano, Nucl. Phys. B189, 442 (1981).
C. H. Llewellyn Smith and J. F. Wheater, Phys. Lett. 105B, 146 (1981).
20. M. S. Chanowitz, J. Ellis and M. K. Gaillard, Nucl. Phys. B128, 506 (1977).
21. J. Gasser and H. Leutwyler, Bern Univ. preprint BUTP-6/1982-BERN (1982).
22. J. Ellis and M. K. Gaillard, Phys. Lett. 88B, 315 (1979).
23. M. Dine, this volume.
24. J. Ellis and M. K. Gaillard, Nucl. Phys. B150, 141 (1979).
25. D. V. Volkov and V. P. Akulov, Phys. Lett. 46B, 109 (1973);
J. Wess and B. Zumino, Nucl. Phys. B70, 39 (1974).
26. S. Ferrara, D. Z. Freedman and P. van Nieuwenheusen, Phys. Rev. D13, 3214 (1976);
S. Deser and B. Zumino, Phys. Lett. 62B, 335 (1976).
27. E. Witten, Nucl. Phys. B188, 513 (1981).

28. M. Veltman, Acta Phys. Polon. B8, 475 (1977).
B. W. Lee, C. Quigg and H. B. Thacker, Phys. Rev. D16, 1519 (1977).
29. J. Wess and B. Zumino, Phys. Lett. 49B, 52 (1974);
J. Iliopoulos and B. Zumino, Nucl. Phys. B76, 310 (1974).
30. P. Fayet, XVI Rencontre de Moriond, First Session.
G. Farrar in "Supersymmetry in Nature". Erice Subnuclear Physics 1978, p. 59.
31. G. Barbiellini et al., Desy report DESY 79/67 (1979).
32. G. R. Farrar and P. Fayet, Phys. Lett. 76B,575 (1978).
G. L. Kane and J. P. Leveille, Univ. of Mich. preprint UMHE 81-68 (1982).
33. See J. Burger, Desy preprint 81-074, to be published in Proc. Int'l Symposium on Lepton and Photon interactions (Bonn, 1981).
34. P. Fayet, Phys. Lett. 84B, 416 (1979).
35. S. Ferrara and E. Remiddi, Phys. Lett. 53B, 347 (1974).
36. M. K. Gaillard, L. Hall and I. Hinchliffe, LBL preprint 14521 (1982), to be published in Phys. Lett.
37. P. Fayet, Phys. Lett. 69B, 489 (1977); 70B, 461 (1977).
38. S. Weinberg, Harvard preprint HUTP-81/A047 (1981).
39. L. Ibanez and G. Ross, Phys. Lett. 110B, 215 (1982);
M. Dine and W. Fischler, Phys. Lett. 110B, 227 (1982);
C. Nappi and B. Ovrut, "Supersymmetric Extension of the $SU(3) \otimes SU(2) \otimes U(1)$ Model",IAS, Princeton preprint (1982);
L. Hall and I. Hinchliffe, Phys. Lett. 112B, 351 (1982);
R. Barbieri, S. Ferrara and D. Nanopoulos, TH.3226-CERN (1982).
40. L. Alvarez-Gaumé, M. Claudson and M. B. Wise, Harvard preprint, HUTP-81/A063 (1981).
41. E. Witten, Phys. Lett. 105B 267 (1981).
42. E. Gildener and S. Weinberg, Phys. Rev. D13, 3333 (1976).
J. Ellis, M. K. Gaillard, D. V. Nanopoulos and C. T. Sachrajda, Phys. Rev. Lett. 83B, 339 (1979).
43. H. Yamagishi, Renormalization Group analysis of supersymmetric mass hierarchies, Princeton preprint (1982).
44. B. Zumino, in Lectures at the 1976 Scottish Universities Summer School, Aug. 1976 (Edinburgh, 1976).
S. Deser and B. Zumino. Phys. Rev. Lett 38, 1433 (1977).
45. The importance of quantum gravity effects for SUSY models was first emphasized to me by J. Polchinski, private communication.
46. D. Z. Freedman in "Recent Results in Gravitation", ed. M. Levy and S. Deser (Plenum Press, NY, 1978) p. 549.
47. M. Gell-Mann, Talk at the 1977 meeting of the American Physical Society (unpublished).
48. E. Cremmer and B. Julia, Nucl. Phys. B159, 141 (1979).
49. T. Kaluza, Sitzungsber Preuss Akad. Wiss. Berlin, Math. Phys. Kl. 1 966 (1921);
O. Klein, Z. Physik 37, 895 (1926); Arkiv Mat.Astron. Fys. B 34A (1946).
50. N. Cabibbo, G. R. Farrar and L. Maiani, Phys. Lett. 105B, 155 (1981).
51. J. Primack and H. Pagels, Phys. Rev. Lett. 48, 223 (1982).

52. P. Q. Hung and M. Suzuki, U.C. Berkeley preprint (1982).
53. S. Weinberg, Phys. Rev. Lett. 48, 1303 (1982).
54. E. Cremmer, J. Scherk and J. Schwartz, Phys. Lett. 84B, 83 (1979).
55. M. Wise, S. L. Glashow and H. Georgi, Phys. Rev. Lett 47, 402 (1981);
    H. P. Nilles and S. Raby, Nucl. Phys. B198, 102 (1982).
56. S. Dimopoulos, S. Raby and F. Wilczek, Phys. Lett. 112B, 133 (1982).
57. N. Sakai and T. Yanagida, Nucl. Phys. B197, 533 (1982).
58. J. Ellis, D. V. Nanopoulos and S. Rudaz, CERN preprint CERN-TH 3199 (1981).
59. Y. Igarashi, J. Kubo and S. Sakakibara, Dortmund preprint DO-TH 82109 (1982).
60. I. S. Altarev et al., Preprint 636, Leningrad Nucl. Phys. Inst. (1981).
61. M. K. Gaillard, in High Energy Physics -1980 (Proc. XX Int'l. Conf., Madison, Wis., 1980);
    J. Ellis, M. K. Gaillard, D. V. Nanopoulos and S. Rudaz, Phys. Lett. 99B, 101 (1980).
62. N. Ramsey, private communication.
63. M. Ferro-Luzzi, H. Steiner, private communication.
64. M. K. Gaillard and B. W. Lee, Phys. Rev. D10, 897 (1974).
65. This calculation is being performed by Y.-C. Kao and I.-H. Lee.
66. J. Ellis and D. V. Nanopoulos, Phys. Lett. 110B (1982), 44 (1982).
    M. Suzuki, U. C. Berkeley preprint, UCB-PTH-82/8 (1982).
67. M. Suzuki, U.C. Berkeley preprint, UCB-PTH-82/7 (1982).

THE EXPERIMENTAL ASPECTS OF NUCLEON DECAY*

Marvin L. Marshak
School of Physics, University of Minnesota, Minneapolis, MN 55455

ABSTRACT

The prediction of nucleon decay provides an experimental test of many grand unified models of the strong, weak and electromagnetic interactions. Past experiments indicate a lifetime longer than 10^{30} years. Current experiments seek to control background in order to extend their sensitivity to longer lifetimes. Specific, negative results from the Soudan 1 detector are presented.

THE CURRENT STATUS OF NUCLEON DECAY

The dominant theme of recent years in elementary particle physics has been the development and verification of the now standard model, a theory of a unified electroweak interaction. The success of this approach has motivated an attempt at so-called grand unification, the inclusion in this framework of the strong interaction. The most commonly proposed verification of these grand unification models is that the proton and bound neutron are very slightly unstable, that they should spontaneously decay with a lifetime of order 10^{31} years. These predictions have generated intense interest in nucleon decay experiments. However, the question of whether protons and bound neutrons, indeed more than 99 percent of all matter, are stable is ultimately an experimental one. It is also a question which has not been entirely neglected over the years.

The first order argument for stability is simply that we and the universe around us exist and that objects around us do not spontaneously disappear or decrease in mass. A more sensitive indication of proton stability is the non-observation of a gamma ray background from the e^{+}-e^{-} annihilations which must follow proton decays in order to preserve the apparent charge neutrality of the universe. More sensitive and more quantitative observations and experiments have also been made. The non-observation of spontaneous fission in ^{232}Th was interpreted as evidence for a minimum lifetime of 2×10^{22} years, relatively independent of decay modes [1]. The failure to find ^{129}Xe in samples of tellurium ore was taken by Evans and Steinberg as evidence for a lifetime longer than 1.6×10^{25} years [2]. This historical approach was further refined by Bennett [3], who searched for etched spallation tracks in mica samples removed from very deep mines. Although limited by background from spontaneous ^{238}U fission, this investigation implied a lower bound of about 2×10^{27} years.

* Work supported by the U.S. Department of Energy.
0094-243X/82/930305-09$3.00

In addition to the historical experiments, there have been several dedicated experiments which have attempted to measure nucleon lifetimes. Fireman and Evans and Steinberg [4] used the techniques developed by Davis for the solar neutrino experiment to search a deep underground sample of ^{39}K for ^{37}Ar. In this case, the failure to find a signal was interpreted as a lower bound of 2.2×10^{26} years. A Case-Witwatersrand-Irvine collaboration measured deep underground stopping muons. An exposure of 67 ton-years yielded a limit of about 10^{30} years for decays into muons, either directly or indirectly [5]. Another experiment used a large array of water Cerenkov detectors, also to search for stopping muons. Using average values for SU(5) branching modes, the Pennsylvania group operating this experiment deduced a lower bound of $1\text{-}3 \times 10^{30}$ years for the nucleon lifetime [6].

The experiments performed in the past can be divided into two groups on each of two orthogonal axes, as shown in Fig. 1. One can differentiate between experiments which use historical samples vs. ones which are dedicated to the purpose of nucleon decay. The former have an exposure time advantage of at least one million, but the latter have better control over background. The second point of difference is between detectors which search for an integrated effect vs. those which report on an event-by-event basis. Again, the latter give better information on decay-simulating events. Indeed, it is background control which is now viewed as the key to making any further progress in this field. For that reason, all experiments which are now taking data, under construction or proposed are of the dedicated, event-by-event type. It seems that only by the use of these techniques can lifetimes be measured, which are of the magnitude predicted by grand unification models.

	Integrated	Event-by-event
Historic	We exist Thorium decay Tellurium decay	Mica tracks
Dedicated	Potassium decay	Stopping muons All new experiments

Figure 1: Classification of nucleon decay experiments

CURRENT EXPERIMENTS

The current status (July, 1982) of nucleon decay experiments is that three experiments (Kolar [7], Soudan [8] and NUSEX in Mt. Blanc [9]) are all currently collecting data. These detectors all use ionization-dependent, tubular detectors and all have a density considerably more than one. To the experts, these detectors are all of the type known as "tracking calorimeters". Two detectors have completed their logistical preparations and are now in or about to enter the early stages of data collection. Both these experiments (IMB [10] and HPW [11]) use single tanks of very pure water to provide both the mass of decay candidates and the detection medium. The detection method in both cases is by Cerenkov effect. Future detectors which are in active construction include one of the water Cerenkov type at Kamioka in Japan [12], which may collect data early in 1983, and one of the tracking calorimeter type at the Frejus tunnel in France [13], which is still in the early stages of construction.

The multiplicity of detectors and even detector types is a direct function of the uncertainties surrounding the nucleon decay experiments. The nucleon lifetime is not known; the decay modes are not known; whether the process even occurs at all is not known. The only principles which are relatively certain are that the lifetime is at least 10^{30} years and that possible final states must include two or more of the 10 or so particles, which are less massive than the proton. What is known to experimentalists is how <u>not</u> to build a nucleon decay detector. Indeed, there are two possible directions for error in planning an experiment. The first is to have too little mass. The key point is that one metric ton contains 6×10^{29} nucleons, regardless of the material. At least 10 metric tons of useful mass are required for an experiment that will be sensitive at the 10^{30} year level, 100 metric tons for 10^{31} years and so on. The second possibility for error is more subtle. The mass of nucleons must be sufficiently well instrumented that nucleon decays can be differentiated from background events.

It is the consideration of the required degree of instrumentation that leads to the inevitable question, "How do you know a decay event when you see one?" All current experiments use three criteria. Conservation of energy--the rest mass of the nucleon is known and this energy should appear in the final state (except for escaping neutrinos and other complications). Conservation of momentum--the nucleon decay event should appear as a balanced explosion radiating out from a center (except for neutrals, fermi momentum and other complications). No external source--the decay event should not be time correlated with any energy entering the detector from the outside. The most serious sources of background are those which can in part evade some or all of these criteria. It is these background events which set a limit on how long a proton lifetime can possibly be measured.

At any deep underground location, background events are limited to three possible sources, local radioactivity, muon-induced events and neutrino-induced events. The first category poses no serious problem because the MeV energy regime typical of these events is far from the nucleon mass. The muon-induced events can be discussed in four categories. Three of these, through muons, stopping muons and charged secondaries from muon interactions in rock, can be eliminated arbitrarily well with a charged particle shield. The rejection factors required are orders of magnitude smaller than what is routinely obtained for rejecting charged particles in accelerator experiments. It is the neutral secondaries from muon interactions which pose the most serious problems. These problems, however, are no worse than those caused by neutrinos, so when the rate of the muon-induced background becomes small compared to the rate of neutrino-induced background, the muon problem can be safely considered as a small adjunct to the neutrino problem. A Monte-Carlo calculation by Grant [14] indicates that neutral, muon secondary-induced events become about 10 percent of neutrino-induced events at a depth of 1500 to 2000 m water equivalent.

Of the neutrino-induced reactions, it is clear that the process $\nu N \rightarrow (\mu/e)\Delta$ is the most troublesome. The pion from the N^* can go in any direction, even one appearing to conserve momentum with the lepton. Both the target nucleon and the nucleon from the N^* will be unobserved in a Cerenkov detector and may not be observed in an ionization detector. This background probably limits known techniques to lifetimes of less than 10^{34} years. Exactly how long a lifetime can be measured depends on the quality of the instrumentation in the detector. The key parameters for judging this quality are listed below.

a. Total energy resolution: Good energy resolution is required to enforce the criterion of total energy conservation. All proposed detectors have resolutions better than 20 percent for electromagnetic modes and 40 percent for hadronic modes, but resolutions as good as three or four percent would be very useful. Fermi momentum effects make resolutions better than this level not so important.

b. Spatial resolution: Although total energy resolution is important, the enforcement of the criterion of conservation of momentum necessitates localized energy resolution. This goal can only be attained if the detector has good spatial and angular resolution. Fine grain is especially important for sorting out the different decay modes, if nucleon decay were to be observed.

c. Directionality: The ability to resolve track direction is also essential to the requirement of conservation of momentum. This quality is particularly important for two-body decay modes, which may be difficult to resolve from straight tracks in some detectors. Directionality is an obvious property of the Cerenkov detectors, but it can also be achieved in ionization-type detectors with sufficiently fine-grained measurements of dE/dx.

d. Particle identification: This property appears to be more important for determining branching modes than for identifying nucleon decay events in the first instance. However, the application of energy and momenta criteria may require knowledge of the rest mass. For example, in water Cerenkov detectors only the kinetic energy above the Cerenkov threshold is visible, not the total energy. In these cases, good particle identification is essential to selection of decay events.

e. Acceptance: A good detector should be efficient, that is decays in any direction in most of its mass should produce recognizeable events. Three parameters contribute to the acceptance fraction. First, the outside mass in any detector is not useful, because the energy from a nucleon decay in this outer margin will not be contained within the sensitive volume and energy-momentum conservation cannot be enforced. The ratio of useful to total mass is often called the fiducial mass ratio and is highest for detectors which are either large or dense or both. Another acceptance factor is isotropic response. Detector designs which may work well in neutrino beams are not necessarily useful for nucleon decay, because the latter occur at random angles while the former are conveniently aligned. A final major acceptance factor is nuclear effects. Rescattering either within the decay nucleus or in other nuclei may prevent recognition of a large fraction of events. This effect is clearly worst in heavy nuclei, although there is considerable controversy over its exact magnitude.

THE SOUDAN 1 DETECTOR

Now that I have discussed detector characteristics in general, I shall report on a specific project which I know quite well. The Soudan 1 detector is the work of an Argonne-Minnesota collaboration. It is located in the Soudan Mine in northeastern Minnesota at a depth of 625 m, which implies the same overhead mass as 1800 m of water. The detector is of the ionization-sensitive, tracking calorimeter type, using an array of 3,456 proportional tubes to record the topologies of events. The tubes are each 2.8 cm in diameter and about 3 m in length. They are arranged in 48 layers of 72 tubes each, with alternate layers turned by 90°, as indicated schematically in Fig. 2. The mass of Soudan 1, 31 metric tons, is largely provided by a heavy concrete, made by mixing Portland cement with purified iron ore and water. The overall dimensions of the detector are 3 m in each horizontal direction and 2 m in height. The detector is covered on the top and four sides by a scintillation counter veto shield.

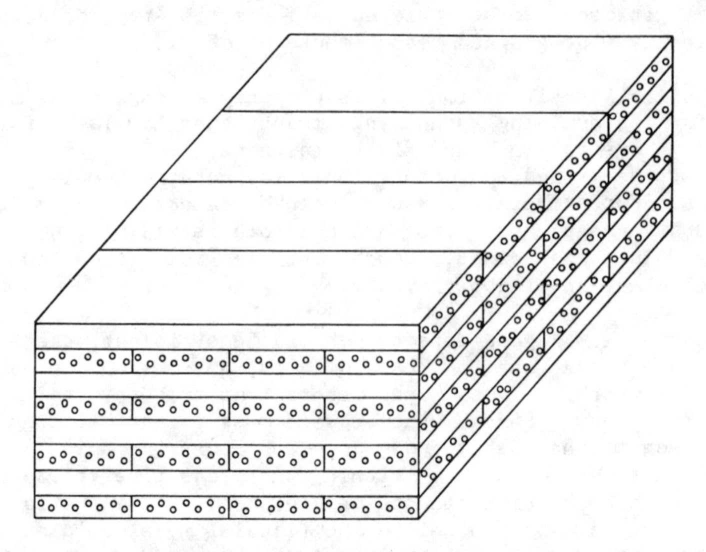

Figure 2: Schematic diagram of the Soudan 1 detector.

The characteristics of the Soudan 1 design were studied by exposing a sample portion of the detector in a test beam at the Argonne IPNS. The behavior of pions, muons and electrons in the test module were used to verify a Monte Carlo program, which was then used to generate sample nucleon decay events in the actual detector. These events were treated in the same manner as actual data (scanning first by computer and then the best events scanned manually) in order to determine the acceptance of the detector, which is expressed as a fiducial mass. These fiducial mass results, which are listed in Table 1, thus contain not only edge or containment effects, but also losses due to nuclear effects, limited resolution and sampling fluctuations.

The data reported here were collected between September, 1981, and July, 1982. During this period the detector was mainly operated remotely and an actual lifetime of 0.382 years was achieved. The data sample contained 462,438 straight-through single muon tracks, which have been used to calibrate and check the operation of the detector. The nucleon decay analysis used a computer to reconstruct those events which left no energy within 20 cm of any side or 30 cm of the top of the detector, which did not fire a scintillation counter in time and in line with the event and which contained more than eight proportional tube hits. Events which met these criteria, several hundred in all, were manually scanned for having a coherent pattern of hits and for not being a stopping track coming up from the bottom or a bottom corner. The result of this process was the isolation of the one event shown in Fig. 3.

Table 1: Fiducial Masses and Lifetime Limits (90% confidence) for the Soudan 1 Detector (July, 1982)

Branching Mode	Fiducial Mass (metric tons)	Lower Lifetime Bound (x 10^{30} yrs.)
$p \to e^+\pi^o$	10.8	1.08
$p \to e^+\rho^o$	9.7	0.97
$p \to e^+\omega^o$	12.0	1.19
$p \to \mu^+K_s$	12.0	1.19
$n \to e^+\pi^-$	10.0	1.00
$n \to e^+\rho^-$	11.1	1.11
$p \to \nu K$	2.2	0.22
$n \to \nu K^o$	5.6	0.56
$n \to \mu^+K^-$	8.4	0.84

```
48  .......:.......:.......:.......:.......:.......:.......:.......:........
46  .......:.......:.......:.......:.......:.......:.......:.......:........
44  .......:.......:.......:.......:.......:.......:.......:.......:........
42  .......:.......:.......:.......:.......:.......:.......:.......:........
40  .......:.......:.......:.......:.......:.......:.......:.......:........
38  .......:.......:.......:.......:.......:.......:.......:.......:........
36  .......:.......:.......:.......:.......:.......:.......:.......:........
34  .......:.......:.......:.......:.......:.......:.......:.......:........
32  .......:.......:.......:.......:.......:.......:.......:.......:........
30  .......:.......:.......:.......:.......:.......:.......:.......:........
28  .......:.......:.......:.......:.......:.......:.......:.......:........
26  .......:.......:.......:.......:.......:.......:.......:.......:........
24  .......:.......:.......:.......:.......:.......:.......:.......:........
22  .......:.......:.......:.......:.......:.......:.......:.......:........
20  .......:.......:.......:.......:.......:.......:.......:.......:........
18  .......:.......:.......:.......:.......:.......:.......:.......:........
16  .......:.......:.......:...86..:.......:.......:.......:.......:........
14  .......:.......:.....4.:.......:.......:.......:.......".......:........
12  .......:.......:.A.....:.......:.......:.......:.......:.......:........
10  .......:.......:.......:.......:.......:.......:.......:.......:........
 8  .......:.......:.......:.......:.......:.......:.......:.......:........
 6  .......:.......:.......:.......:.......:.......:.......:.......:........
 4  ...9...:.......:.......:.......:.......:.......:.......:.......:........
 2  .......:.......:.......:.......:.......:.......:.......:.......:........

47  .......:.......:.......:.......:.......:.......:.......:.......:........
45  .......:.......:.......:.......:.......:.......:.......:.......:........
43  .......:.......:.......:.......:.......:.......:.......:.......:........
41  .......:.......:.......:.......:.......:.......:.......:.......:........
39  .......:.......:.......:.......:.......:.......:.......:.......:........
37  .......:.......:.......:.......:.......:.......:.......:.......:........
35  .......:.......:.......:.......:.......:.......:.......:.......:........
33  .......:.......:.......:.......:.......:.......:.......:.......:........
31  .......:.......:.......:.......:.......:.......:.......:.......:........
29  .......:.......:.......:.......:.......:.......:.......:.......:........
27  .......:.......:.......:.......:.......:.......:.......:.......:........
25  .......:.......:.......:.......:.......:.......:.......:.......:........
23  .......:.......:.......:.......:.......:.......:.......:.......:........
21  .......:.......:.......:.......:.......:.......:.......:.......:........
19  .......:.......:.......:.......:.......:.......:.......:.......:........
17  .......:.......:.......:.......:.......:.......:.......:.......:........
15  .......4.4.....:6......:.......:.......:.......:.......:.......:........
13  .......:.......:....66.:.......:.......:.......:.......:.......:........
11  .......:.......:.......:.......:.......:.......:.......:.......:........
 9  .......:.......:.....4.:.......:.......:.......:.......:.......:........
 7  .......:.......:.......:.......:.......:.......:.......:.......:........
 5  .......:.......:.......:.......:.......:.......:.......:.......:........
 3  .......:.......:.......:.......:.......:.......:.......:.......:........
 1  .......:.......:.......:.......:.......:.......:.......:.......:........
```

Figure 3: Contained event in Soudan 1 detector. Symbols refer to pulse heights measured in proportional tubes.

The event shown in Fig. 3 meets the criterion of no outside source, since there is neither a track connection to the outside nor did any of the scintillation counters fire in coincidence with the event. The proportional tubes between the event and the outer wall are known to have been working at the time of the event from an examination of the straight-through muon tracks recorded contemporaneously. Although the energy of the event appears as about 700 MeV, the possibility of statistical fluctuation and possible rescattering with energy losses make this event possibly consistent with conservation of energy in a nucleon decay. On the other hand, the event fails on conservation of momentum, since the two prongs make an apparent angle of 120° with each other. While it is true that due to rescattering, a nucleon decay could look like Fig. 3, it is more likely that the event is the result of a neutrino interaction. About one such event should have been observed during the live time reported here. The conclusion that should be drawn from this event is that it is difficult for Soudan 1 and probably for any detector to separate decay events from neutrino-interaction backgrounds on an event-by-event basis. A large sample of such events will be required before any statistically valid inferences could be drawn. Since the world's supply of contained events is now 5 (3 for Kolar, 1 each for Soudan and NUSEX, as of August, 1982), some of which must be neutrino events, some time may be required before the question of whether nucleon decay has been observed can be resolved.

Taking the point of view that Soudan 1 has observed no events, we can deduce 90 percent confidence level (2.3σ) bounds on the nucleon lifetime. These limits are listed in Table 1. Although similar limits have been reported earlier, these results are independent of theoretical assumptions about branching ratios. Experimentalists will need be content with reporting such lower bounds until two criteria are met. 1) There must be a sufficient number of events, such that valid statistical tests can be made and 2) experiments cannot claim the detection of nucleon decay events until they observe and report a reasonable number of contained atmospheric neutrino events.

The Soudan 1 nucleon decay experiment represents the work of my colleagues at both Argonne and Minnesota, including D. Ayres, J. Bartelt, H. Courant, J. Dawson, T. Fields, K. Heller, E. May, E. Peterson, L. Price, K. Ruddick and M. Shupe. I appreciate their allowing me to report preliminary results prior to publication.

References

1. M. Goldhaber and L. Sulak, Comments on Nucl. and Part. Phys. 10, 215 (1981).
2. J.C. Evans, Jr. and R.I. Steinberg, Science 197, 989 (1977).
3. C.L. Bennett, Proceedings of the Second Workshop on Grand Unification, Ann Arbor, ed. by J. Leveille, L. Sulak, D. Unger, (Birkhauser, 1981).
4. E. Fireman, Neutrino 77, Elbus, USSR, Vol. 1, p. 53; R.I. Steinberg and J.C. Evans, Jr., ibid, Vol. 2, p. 321.
5. J. Learned, F. Reines and A. Soni, Phys. Rev. Lett. 43, 907 (1979).
6. M.L. Cherry, et al., Phys. Rev. Lett. 47, 1507 (1981).
7. M.R. Krishnaswamy, et al., "The Kolar Gold Field Baryon Stability Experiment," by the Tata Institute-Osaka-Tokyo Collaboration, 1980 (unpublished).
8. H. Courant, et al., "A Dense Detector for Baryon Decay," U. of Minnesota report, 1979 (unpublished).
9. G. Battistoni et al., "Proposal for an Experiment on Nucleon Stability with a Fine Grain Detector (NUSEX)," by the Frascati-Milano-Torino collaboration, 1979 (unpublished).
10. M. Goldhaber, et al., "Proposal for a Nucleon Decay Detector," by the Irvine-Michigan-Brookhaven collaboration, 1979 (unpublished).
11. J. Blandino, et al., "A Decay Mode Independent Search for Baryon Decay Using a Volume Cerenkov Detector," by the Harvard-Purdue-Wisconsin collaboration, 1979 (unpublished).
12. M. Koshiba, report to the International Conference on High Energy Physics, Paris, 1982.
13. P. Bareyre, et al., "Proposition d'une Experience pour l'Etude de le'Instabilite du Nucleon au Moyen d'un Detecteur Calorimetrique," by the Orsay-Pailaiseau-Paris-Saclay collaboration, 1980 (unpublished).
14. A. Grant, in proceedings of Argonne Summer Workshop on Proton Decay, 1982.

NUCLEON DECAY EXPERIMENTS IN EUROPE

M. Rollier
Istituto di Fisica dell'Università di Milano, Italia
Istituto Nazionale di Fisica Nucleare Sezione di Milano, Italia.

ABSTRACT.

The european experimental situation for what concern the nucleon decay is presented. The more recent results give a limit on the nucleon lifetime of the order of 10^{30} years. New data soon will come from the "Nusex" experiment which is already running in the Mont Blanc Tunnel.

The Frejus experiment with a mass of 2KTons will hopefully take the first data at the end of 1983.

The new large laboratory which will be escavated in the Gran Sasso Tunnel will provide the possibility to run a calorimetric experiment with a mass up to 10KTons.

INTRODUCTION.

Speculations on the possible decay of the nucleons were absent until 1967 when this hypothesis was firstly introduced trying to give an explanation of the barion excess in the Universe[1].

More recently the unified theories[2](GUT) of electroweak and strong interactions, including both quarks and leptons in the same multiplets, naturally introduced again the possibility of nucleon instability. Theoretical predictions for these new models are very uncertain either for the nucleon lifetime ($10^{31\pm1}$ years) or for the decay modes and branching ratios. The "standard model" (SU_5) predicts[3] the proton to decay mainly in the channel $p \rightarrow e^+\pi^o$ but recently new supersymmetric models (SU-SY)[4] also proposed other possible decay modes like $p \rightarrow K^+ \nu_\mu$

RECENT EXPERIMENTAL LIMITS.

After the first limits obtained by geochemical techniques[5] new direct measurements of the nucleon lifetime have reached values near to the theoretical predictions.

I will just review the last two results. The first has been obtained by a Pennsylvania-Brookhaven collaboration in the Homestake mine (South Dakota) at a depth of 4400 meter of water equivalent (m.w.e.). The detector consisted of Cerenkov modules with an upper layer of scintillators in anticoincidence. The detector was only sensitive to nucleon decay giving a muon as one of the final products. The limit obtained ranges from 1.5 to $3.\times 10^{30}$ years.

The second experiment has been carried out by a Tata-Osaka-Tokyo collaboration, with a calorimetric detector made with 34 horizontal iron plates, interleaved with planes of proportional counters. The detector was installed in the indian mine Kolar Gold Field at a depth equivalent to 7000 m.w.e. After a run of about one year the KGF collaboration has recorded ∿600 muons crossing the apparatus, 7 neutrino interactions and 3 nucleon decay candidates. Two of them (fig.1) are two prongs events with one of the two tracks leaving the detector. Also if it is difficult to evaluate the background, such events can also be interpreted as scattered tracks coming from a neutrino interaction in the surrounding rocks.

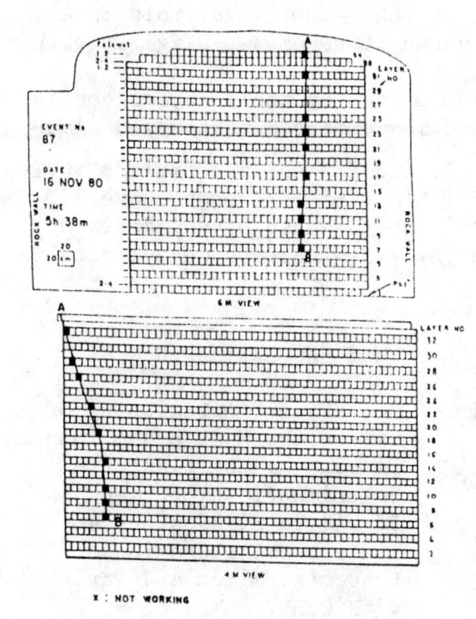

Fig.1. A nucleon decay candidate in the KGF experiment.

A third event has also been found which is completely confined in the detector and looks like a single electromagnetic shower. This event can also be interpreted as a shower produced by an atmospheric ν_e.

Also if no definitive conclusion can be drawn, the nucleon lifetime corrisponding to these 3 events in the KGF experiment is 8×10^{30} years in the hypothesis that the correction for branching ratios and efficiency are of the order of 50%.

BACKGROUND.

The problem of background is a very serious one in the nucleon lifetime measurements since very massive detectors are supposed to detect few events per years. The sources of background are:

a) natural radioactivity. The counting rate in a big detector can be considerably increased by the low energy γ's and even the neutrons from radioactivity. Also if these events cannot simulate nucleon decays, they can represent a problem for triggering the detectors.

b) muons. The flux of atmospheric muons in a unshielded detector is of the order of 10^5 muons per m^2 per day.

This can be drastically reduced in deep underground laboratories. In the Mont Blanc Tunnel at a depth of ∿5000 m.w.e. the measured muon rate is 0.1 muons per m^2 per day.

c) neutrals from rocks. Neutrons or neutral kaons produced by muons or neutrinos interacting in the surrounding rocks can simulate by their interaction a nucleon decay. This background has been calculated[5] to be three order of magnitude lower than the corresponding muon flux. At a depth less than 4000 m.w.e. this is a serious background, but at lower depth it becomes negligible with respect to the neutrino interactions.

d) atmospheric neutrinos. Neutrinos produced by the decay of cosmic pions, kaons and muons can interact inside the detector and simulate a nucleon decay. From the predicted fluxes[6] one expects ∿0.2 neutrino events per ton per year with a visible energy above 300 MeV. This figure set an upper limit to the mass of any detector, since it becomes very hard to recognize few nucleon decays out of $\sim 10^3$ background neutrino events.

THE "NUSEX" EXPERIMENT.

From the beginning of may in the Mont Blanc Tunnel is running the first big calorimetric experiment in Europe carried out by an italian collaboration of Frascati, Milan and Turin with CERN.

The laboratory at a depth of ∿5000 m.w.e. is very well protected by the high mountains and the total rate of muons crossing the detector is ∿20 per day. The total mass of the detector is 160 tons ($\sim 10^{32}$ nucleons). The apparatus (fig. 2) is a cube 3.5 m side done by iron layers of 1 cm interleaved with plastic tubes operating in the limited streamer mode[7].

Two sets of strips placed along the tubes and orthogonally to them allows to read the position of the streamer within about

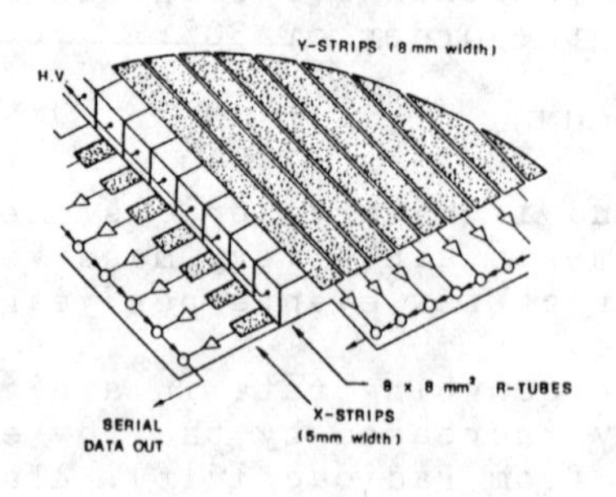

Fig. 2. The NUSEX apparatus

1 cm. The detector contains $4.3\ 10^3$ tubes with $8.6\ 10^3$ read-out channels.

A test module of reduced mass (15 tons) has been exposed at CERN to pions and electrons of momenta ranging from 150 to 2000 $\frac{MeV}{c}$ In fig.3 the **tracks** of a pion and an electron of 500 MeV/c are shown in both projected planes parallel and orthogonal to the pick up strips.

It is clearly visible that in the YZ plane frequently a track can give a signal in more than one strip due to the dimension of the

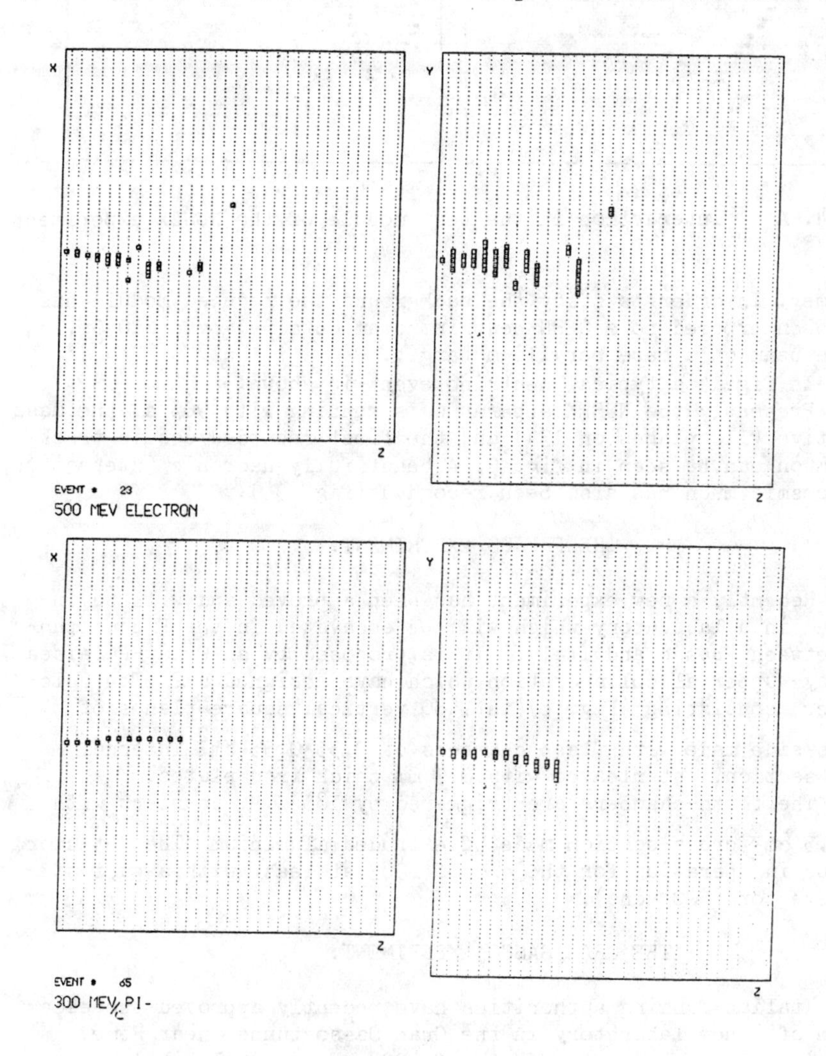

Fig. 3. An electron and a pion tracks in the test module of the NUSEX experiment.

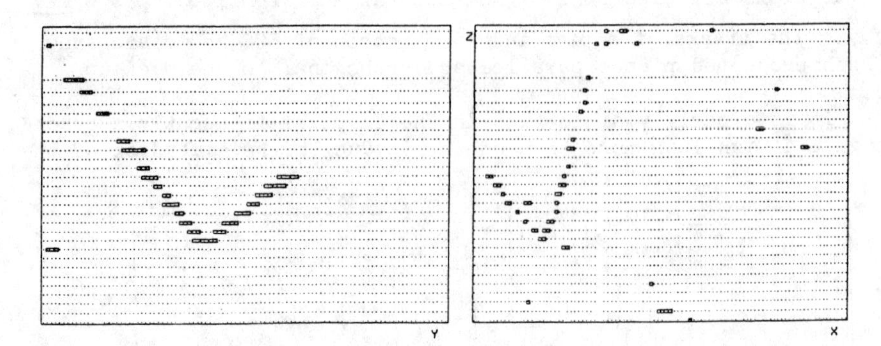

Fig. 4. A ν_μ interaction in the test module of the NUSEX experiment.

streamer. In order to study the background the reduced module has also been exposed to a CERN neutrino beam obtained by a ∿10 GeV proton beam on a bare beryllium target.

In fig. 4 a typical neutrino event is shown.

Presently the NUSEX apparatus is running with 90% of the mass sensitive (112 planes on 134) and the first recorded cosmic track of a muon can be seen in **fig.** 5. A beautifully hadronic interaction of a cosmic muon has also been recorded (fig. 5).

THE FREJUS EXPERIMENT.

Recently a new experiment has been approved and will be installed in a laboratory which will be escavated in the Frejus tunnel between France and Italy. It is proposed by an Orsay-Palaiseau -Saclay-Wuppertal collaboration which have designed a 2 kTon calorimeter consisting (Fig. 6) in 1480 vertical square planes of 6.2 meters side made with flash chambers of 0.55×0.55 cm^2 internal cross section, interleaved with 0.3 mm thick iron plates.

The flash-chambers are triggered by 185 G.M. counters with 1.5×1.5 cm^2 cross-section made of extruded aluminium. The detector will be in operation for the end of 1983 at least with a part of the detector (500 tons).

THE GRAN SASSO EXPERIMENT.

Italian funding authorities have recently approved the escavation of a new laboratory in the Gran Sasso tunnel near Rome.

This laboratory at a depth of 4000 m.w.e. will likely have a dimension of 25×20×100 m^3 and will house not only a proton decay

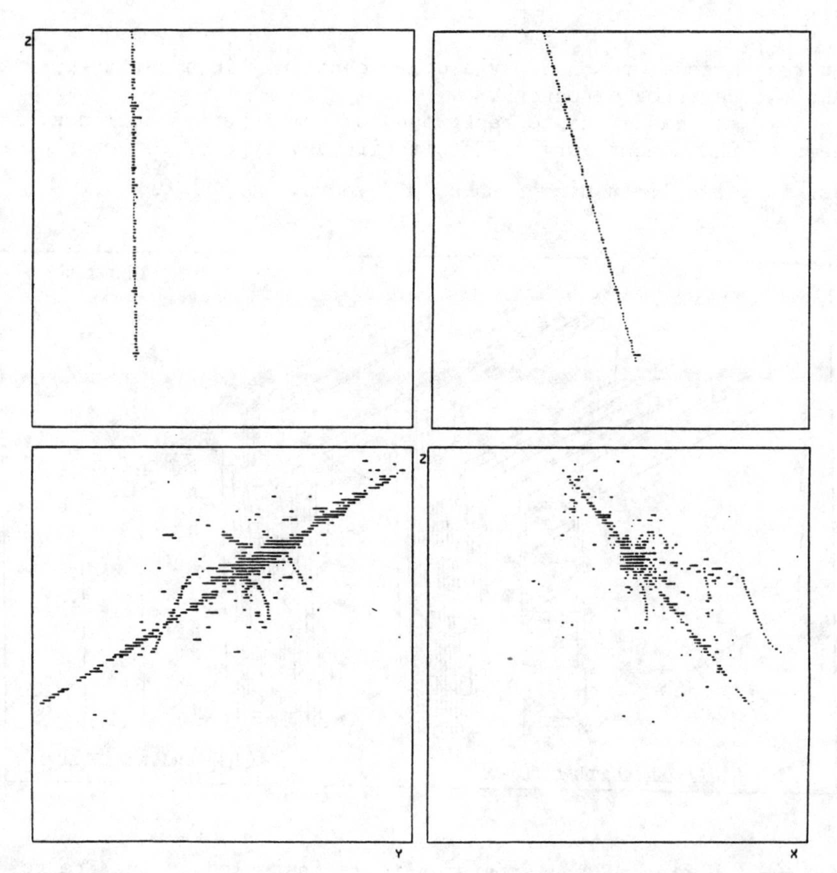

Fig. 5. One of the first recorded track of a muon in the NUSEX detector and an hadronic interaction of a cosmic muon.

set up, but also other experiments requiring a low background.

A giant calorimetric detector of a mass of up to 10 kTons is under study by a Frascati-Milan-Rome-Torino collaboration.

Three type of detectors are presently being studied and tested: flash chambers, resistive plate chambers and new type of G.M. counters proposed in Frascati, where the origin of the discharge can be reconstructed by measuring the time of its propagation in the tube.

CONCLUSIONS.

To conclude I want to enphasize that in a relatively short time it will be possible to have the first results from the european experiments bringing the limit on the nucleon lifetime up to $\sim 5\ 10^{31}$ years and the american Cerenkov experiments with higher mass possibly still will increase this limit.

If the new decay modes proposed by the supersymmetric models are dominant, I think that still calorimeters with good granularity can recognize a proton decay and are centainly in a better position than the Cerenkov detectors.

In any case, due to background of atmospheric neutrinos, it seems at the moment very difficult with any type of detector of any mass to get a limit higher than 10^{33} years.

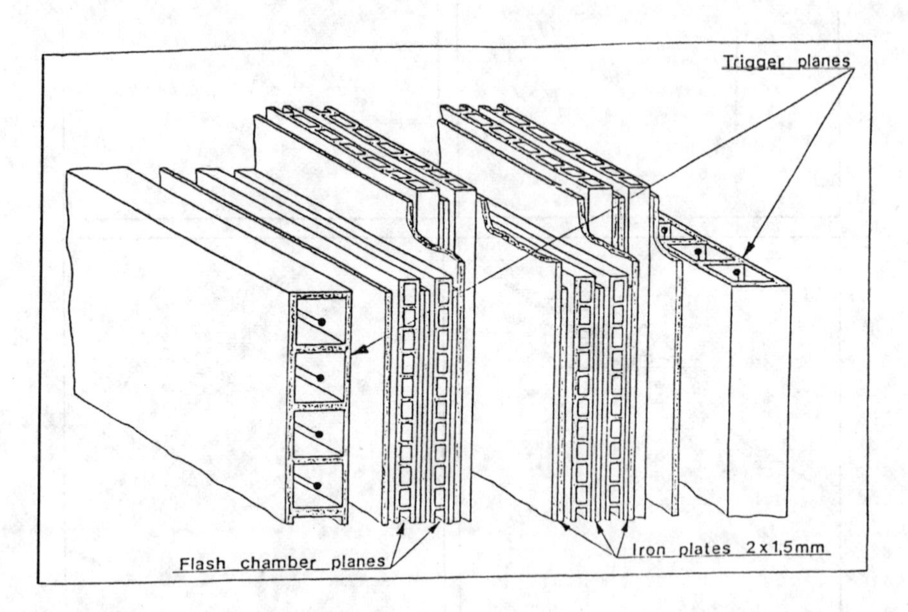

Fig. 6. A sketch of the apparatus to be installed in the Frejus tunnel.

REFERENCES.

1. A.Sacharov, Zh. Expt. Theor. Fiz. SSSR Pis'ma Redakt, 5, 32 (1967)
2. A review of theories is given in : "Proceedings of the Royal Society Meeting on Gauge Theories of the Foundamental Interactions" Phil. Trans. R. Soc. London A 304, 1 (1982)
3. Proceeding of the II Workshop on Grand Unification, Ann Arbor 24 - 26 April 1981
4. D.V.Nanopoulos, K.Tamrakis TH 3247 - CERN 18/2/82
5. D.Perkins: muon induced background in the Mont Blanc and Frejus Tunnels, CERN preprint, September 1979
6. Proceedings of the International Colloquium on Baryon-Non Conservation, Bombay, January 9-14 1982, to be published on Pramana
7. G.Battistoni et al., Nuclear hist. and Method, 164 (1979) 453 and 164 (1979) 57
8. Proceedings of the GUD Workshop for Giant Underground Detector Rome, November 1981, (M.Conversi ed.)

NEUTRINO OSCILLATIONS, NEUTRINOLESS DOUBLE BETA DECAY

F. Boehm
California Institute of Technology, Pasadena, CA 91125

ABSTRACT

We review the experimental evidence for neutrino mixing and neutrino mass. Searches for possible branches into heavy neutrinos do not reveal evidence for static mixing with branching ratios larger than 10^{-4} to 10^{-6}. Similarly, neutrino oscillation experiments show no evidence for dynamic mixing in various oscillation channels. Stringent limits for $\bar{\nu}_e$ disappearance from a recent reactor experiment are presented. Results from neutrinoless double beta decay provide sensitive test for Majorana mass and right-hand couplings, the present limits being 3-10 eV and 10^{-5}, respectively.

INTRODUCTION

In the frame work of grand unified theories the weak interaction states describing physical neutrinos are predicted to be mixed states[1]. If, in addition, finite rest mass is associated with these states, several new phenomena are predicted to occur. Neutrino mixing including light and heavy neutrinos, can lead to observable effects in weak decays, such as new monochromatic lepton lines. Another consequence is the well known neutrino oscillation process, i.e. the transition from one neutrino type to another. I would like to first review the static aspect of neutrino mixing, its experimental consequences and current results and then turn to the dynamics of a mixed neutrino system discussing our present knowledge of neutrino oscillations. Finally, I shall discuss the neutrinoless double beta decay, an old field, which has reappeared as a powerful and sensitive way of testing lepton non-conservation and neutrino mass.

1. STATIC NEUTRINO MIXING AND HEAVY NEUTRINOS

Following general practice[1] we write the weak-interaction neutrino state ν_ℓ ($\ell = e,\mu,\tau$) as a superposition of mass-eigenstate neutrinos $\nu_i (i = 1,2,3)$,

$$\nu_\ell = \sum_i U_{\ell i}\, \nu_i \, . \tag{1}$$

The expansion coefficients $U_{\ell i}$ which are the elements of a unitary matrix represent the mixing amplitudes.

It may be argued that the state ν_ℓ is predominantly composed of a light neutrino ν_1, with small admixtures of one or two heavy neutrinos ν_2 and ν_3, restricting $|U_{\ell\, 2,3}|^2 << |$. As a consequence, there will be a branch proceeding via the heavy neutrino, with branching ratio $B_{\ell i} \approx |U_{\ell i}|^2$. Shrock[2] has pointed out that the two-body decays,

K-μν, π-μν, π-eν offer sensitive tests in a search for these branchings. What would be expected is a monochromatic lepton line, due to the decay of the heavy neutrino, at some energy below the regular lepton peak. For example, in the decay $K^+ \rightarrow \mu^+\nu$ the muon momentum range between 100 MeV/c and 220 MeV/c was explored[3] with good momentum resolution, and stringent limits could be set for heavy neutrino masses. In a similar fashion the decays $\pi \rightarrow \mu\nu$[4] and $\pi \rightarrow e\nu$[5] were explored spectroscopically. No evidence for heavy neutrinos was seen and the resulting limits of the branching ratios $B_{\ell i}$ are depicted in Fig. 1.

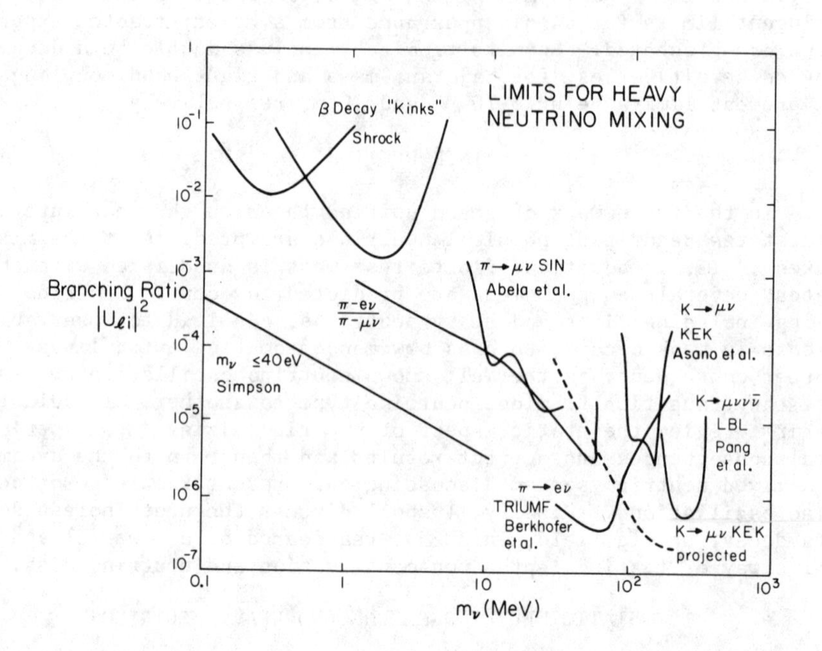

Fig. 1. Limits for branching ratios into heavy neutrino decay as a function of heavy neutrino mass as established from various experiments.

In nuclear beta decay a heavy neutrino may give rise to a discontinuity or "kink" in the beta spectrum. Shrock[2] has analyzed the existing spectroscopic data finding no evidence for kinks. These results are also shown in Fig. 1. In a similar fashion, by analyzing the ^{3}H decay spectrum, Simpson[6] finds branching ratios of less than 10^{-3} to 10^{-1} for $m_{\nu\ \mathrm{heavy}}$ of 0.1 to 10 keV.

Another relationship capable of supplying information on branching ratios into heavy neutrinos can be taken from double-beta decay. Simpson[7] argues that the mass limit from double-beta decay which is about 10-40 eV (see part 3 of this talk) can be interpreted as an average Majorana mass which for small mixing is given by

$$m_\nu = m_{\nu\,light}\cos^2\theta + m_{\nu\,heavy}\sin^2\theta \leq 40\ \text{eV}.$$

With $m_{\nu\,light} << m_{\nu\,heavy}$ it follows that

$$\sin^2\theta \leq 4 \times 10^{-5}/m_{\nu\,heavy}\ (\text{MeV}),$$

an argument which holds for $m_\nu < p_\nu$. The resulting limit is also shown in Fig. 1.

Other limitations shown in Fig. 1 were derived from branching ratios and are quoted in Refs. 2 and 8. We conclude that there is no evidence for heavy neutrinos that have masses between 10^{-1} and 10^3 MeV, with branching ratios larger than 10^{-4} to 10^{-6}.

2. DYNAMIC NEUTRINO MIXING - NEUTRINO OSCILLATIONS

The dynamics of a mixed neutrino state follows from Eq. 1 by considering the evolution in time of a mass eigenstate $\nu_i(t)$,

$$\nu_i(t) \sim e^{-iE_i t};\quad E_i = \sqrt{p^2 + m_i^2}.$$

Assuming that at least one of the masses m_i of the pure states ν_i is non-zero, neutrino oscillations will occur.

For the simplified case of 2 neutrinos, ν_{ℓ_1} and ν_{ℓ_2}, the matri $U_{\ell i}$ contains only one independent parameter, the mixing angle θ. The probability for appearance of a new state ℓ_2 in a detector a distance L (m) away from the source emitting neutrinos in the state ℓ_1 is given by

$$P(\nu_{\ell_1} \to \nu_{\ell_2}) \sim \frac{\sin^2 2\theta}{2}\left[1 - \cos(2.53\ \Delta m^2 (L/E_\nu))\right]. \quad (2)$$

The quantities $\sin^2 2\theta$ and $\Delta m^2 \equiv |m_1^2 - m_2^2|$ (eV^2) are the two parameters characterizing neutrino oscillation in this 2 neutrino model; E_ν (MeV) is the neutrino energy. Similarly, the probability of disappearance of a state ℓ_1 is given by

$$P(\nu_{\ell_1} \to X) \sim 1 - P(\nu_{\ell_1} \to \nu_{\ell_2}). \quad (3)$$

The oscillations are characterized by an oscillation length Λ (m) = 2.48E (MeV)/Δm^2 (eV^2).

For illustration, the values of L/E_ν to be explored by various experiments are shown in Fig. 2.

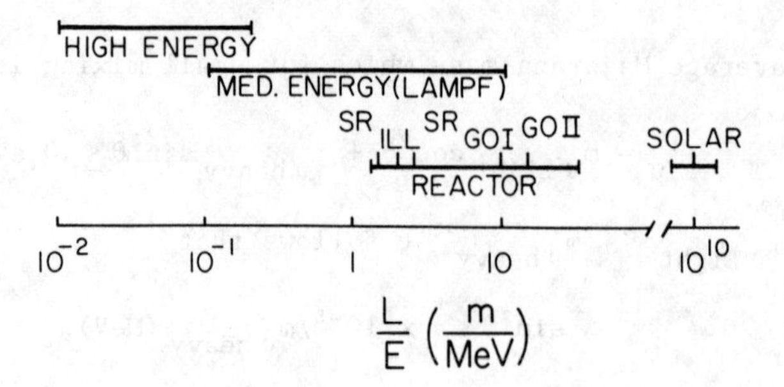

Fig. 2. Illustration of the L/E_ν value for various oscillation experiments.

We now review the experimental results on oscillations in the appearance channels $\nu_\mu \to \nu_e$ and $\nu_\mu \to \nu_\tau$. No evidence for oscillations has been reported and the current best upper limits of the parameters Δm^2 and $\sin^2 2\theta$ are also shown by the solid lines in Fig. 3.

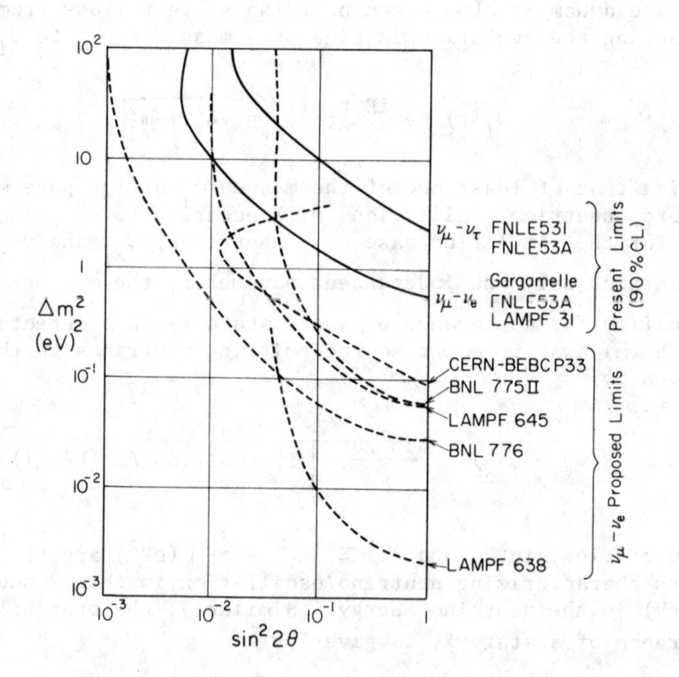

Fig. 3. Limits for neutrino oscillations $\nu_\mu \to \nu_e$ and $\nu_\mu \to \nu_\tau$. The solid curves represent current experimental limits at 90% c.ℓ. The dashed curves illustrate forthcoming experimental limits.

The curves which are 90% confidence limit contour lines have been composed from the data referred to by the label. A detailed list of references is contained in the recent review by Baltay[8] in Neutrino 81. The regions to the right of the curves are excluded.

As for forthcoming results, several experiments have been proposed with the aim of exploring the region of smaller Δm^2 and $\sin^2 2\theta$. The sensitivities expected to be reached in these experiments are shown in Fig. 3 by dashed lines (The experiments are identified by Laboratory and proposal number[9]). It appears that an improvement in sensitivity of more than an order of magnitude should be attainable in the next 2-3 years.

Disappearance experiments have been conducted at high energy accelerators as well as with the help of fission reactors. Their results, in terms of the two parameters Δm^2 and $\sin^2 2\theta$ are summarized in Fig. 4. By far the most stringent limits come from a recent $\bar{\nu}_e \to X$ experiment[10] at the Gösgen power reactor. Since this work has not yet appeared in publication, I shall briefly describe it here.

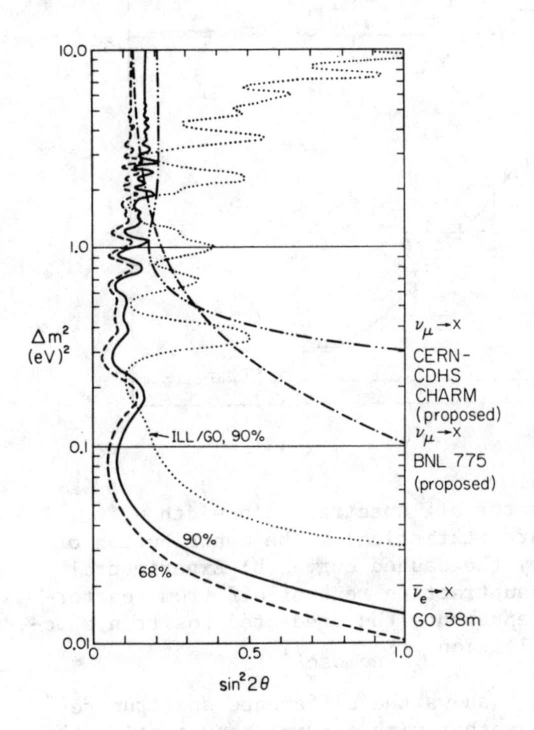

Fig. 4 Limits for neutrino oscillations $\nu_\mu \to X$ and $\nu_e \to X$. The solid curve represents current experimental limits at 90% c.ℓ. from the Gösgen reactor experiment. Limits obtained from the ratio of the ILL (8.7 m) and Gösgen (L = 38 m) data are shown by the dotted curve labelled ILL/GO. Forthcoming experimental limits are shown by broken lines.

The Caltech-Munich-SIN group[10] has just completed a measurement of the neutrino spectrum at a distance of 38 m from the core of the 2800 MW reactor at Gösgen, Switzerland. The neutrinos were detected by the reaction $\bar{\nu}_e p - e^+ n$ using a composite liquid scintillation detector and ^{3}He-multiwire proportional chambers[11]. A time correlated e^+,n event constituted a valid signature. Compared to our earlier experiment at Grenoble[11] the signal-to-noise was markedly improved by enforcing position-correlation cuts for the time-correlated events in the scintillation cells and the ^{3}He chamber. Cosmic ray related fast neutron background was effectively reduced by means of pulse-shape discrimination. Events from charged cosmic-ray particles were eliminated with a tight liquid scintillator-veto.

About 11,000 neutrino induced events were recorded in a six-months reactor-on period. Background was recorded during a one-month reactor-off period. Figure 5 (a) depicts the time correlated

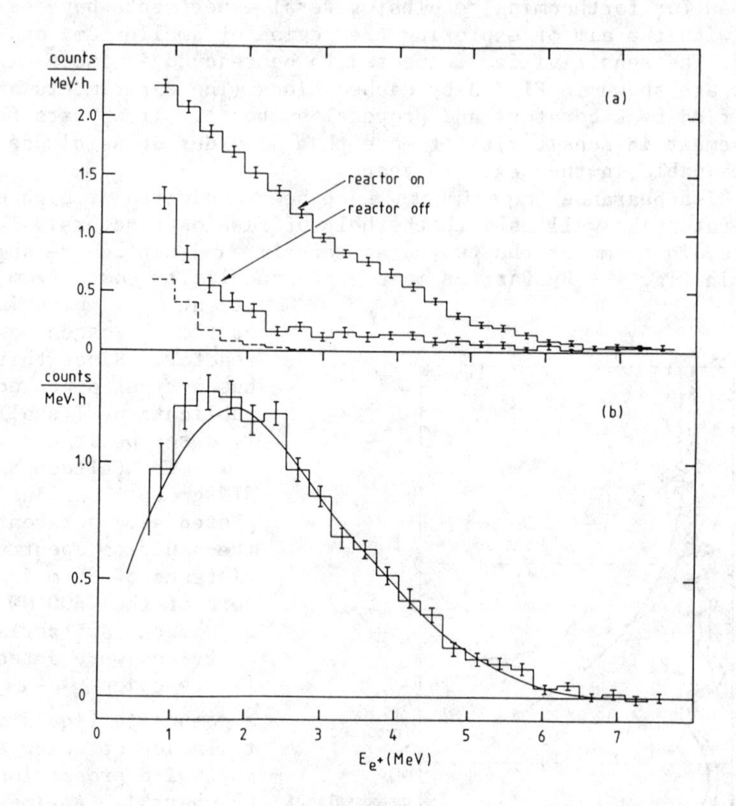

Fig. 5. (a) Reactor-on and reactor-off spectra. Bin width = 0.305 MeV. The errors shown are statistical. The contribution of the accidentals is indicated by the dashed curve. (b) Experimental positron spectrum obtained by subtracting reactor-off from reactor-on spectra. The solid curve represents the predicted positron spectrum assuming no neutrino oscillations ($\tilde{Y}_{no\ osc}$).

positron spectrum. Figure 5 (b) shows the difference spectrum reactor-on minus reactor-off together with a curve representing the expected spectrum for no oscillations. The latter was obtained from the on-line beta spectroscopic measurements by Schreckenbach *et al*[12] studying ^{235}U and ^{239}Pu fission targets. Small contributions from fission of ^{238}U and ^{241}Pu were taken into account with the help of the calculations of ref. 13. The relative contributions to the neutrino spectrum of each of the fissioning isotopes were well known at any time.

Figure 6 shows the ratio of the observed yield to the yield expected for the no-oscillation case, as a function of $L/E_{\bar{\nu}}$, together with the best fit for $\Delta m^2 = 0$ and for two possible $\Delta m^2 \neq 0$ solutions. A systematic χ^2 test to all possible values of Δm^2 and $\sin^2 2\theta$, including those shown in Fig. 6, followed by a maximum likelihood ratio test resulted in the 68% and 90% confidence limit contour lines displayed in Fig. 4. The same procedure was also applied to the ratio of the data of ref. 11 taken at L = 8.7 m to the present data (L = 38 m) and the resulting contour lines are also shown in Fig. 4. This ratio test is somewhat less restrictive because of the moderate statistical accuracy of the data of ref. 11.

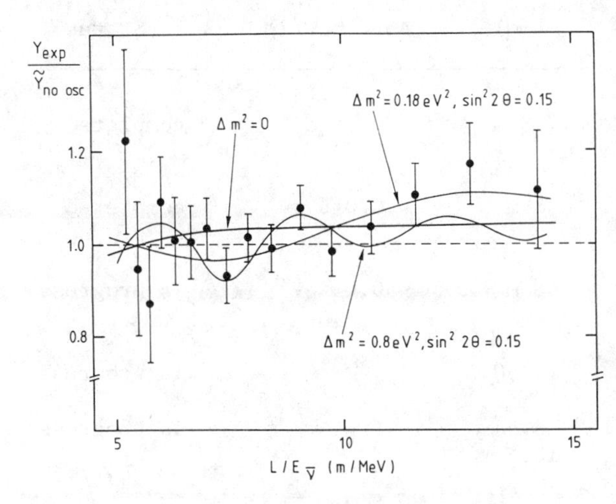

Fig. 6. Ratio of experimental to predicted (for no oscillations) positron spectra. The errors of the data points shown are statistical. Oscillation functions consistent with the data are shown for various sets of oscillation parameters.

It must be concluded from this experiment that there are no neutrino oscillations with parameters larger than those given by the curves GO in Fig. 4. For large mass parameters it is found that $\sin^2 2\theta \leq 0.17$ (90% c.ℓ.), and for maximum mixing the limit is $\Delta m^2 < 0.016\ \text{eV}^2$ (90% c.ℓ.). The results are in clear disagreement with those by Reines *et al.*[14] For the integral yield it is found that the experimental value divided by the predicted no-oscillation value is 1.05 ± 0.05 (68% c.ℓ.).

A brief summary of present and future reactor experiments is contained in Table I.

The limits for ν_μ disappearance are not very restrictive at present, the constraint on Δm^2 for full mixing being about 7 eV^2 [8], but several experiments have now been proposed to search for oscillations in the $\nu_\mu \rightarrow X$ channel. As an example the curves in Fig. 4 illustrate the expected sensitivity limit (90% c.ℓ.) for the CDHS and CHARM experiments[9] at CERN and for a Brookhaven experiment[9]. These experiments will be able to provide an almost 2 order of magnitude improvement in sensitivity to Δm^2. Another proposed experiment at Los Alamos[9] may further improve that limit.

Table I. Status of Reactor $\bar{\nu}_e \to X$ Oscillation Experiments

Reactor, Collaboration	L(m)	$\Delta m(\theta = \pi/2)$	Status
ILL (Grenoble), Caltech-Munich-ISN[11]	7.8 m	$< 0.15\ eV^2$	completed
Gösgen I, Caltech-Munich-SIN[10]	38 m	$< 0.016\ eV^2$	completed
Gösgen II, Caltech-Munich-SIN[10]	48 m	$< 0.01\ eV^2$	in progress
Savannah River, UC Irvine[15]	15-50 m		in progress
Bugey, Annecy-ISN[16]	13.5 m, 18.5 m		in progress

Recent searches[17] for particle-antiparticle transitions $\nu_\mu \to \bar{\nu}_e$ and $\nu_e \to \bar{\nu}_e$ revealed no evidence having oscillation parameters Δm^2 (full mixing) larger than 0.7 eV^2 and 7 eV^2, respectively.

3. NEUTRINOLESS DOUBLE-BETA DECAY

Double-beta decay, a second-order semileptonic process accompanying a transition from a nucleus Z to Z + 2, may be observable if the transition from Z to Z + 1 is energetically forbidden. The process is known to proceed in two ways, (1) via standard second-order beta decay $Z \to (Z+2) + e_1 + \nu_1 + e_2 + \nu_2$, known as the 2ν-double-beta decay, and (2) by the no-neutrino process $Z \to (Z+2) + e_1 + e_2$. This second process violates lepton number and, if observed, would signal that neutrinos have a finite Majorana mass, or that there are right handed currents, or both[18]. It should be noted that other mechanisms not involving neutrino exchange have also been discussed[19]. It is this second process which is receiving much attention today but I shall have to include the first process in the discussion as well, since it is of help to elucidate the nuclear parameters.

Inasmuch as double-beta decay is somewhat outside the general interest of particle physics, I would like to be more explicit in the discussion that follows.

For the 2ν $\beta\beta$-process the situation is illustrated in Fig. 7. The transitions from Z to Z + 2 proceeds via a set of giant Gamow-Teller states in Z + 1. Its strength, as far as the nuclear physics is concerned, is given by a second-order perturbation matrix-element between initial state i, intermediate state m, and final state f,

$$\sum_m \frac{\langle f|H_w|m\rangle\langle m|H_w|i\rangle}{E_m - E_i}.$$

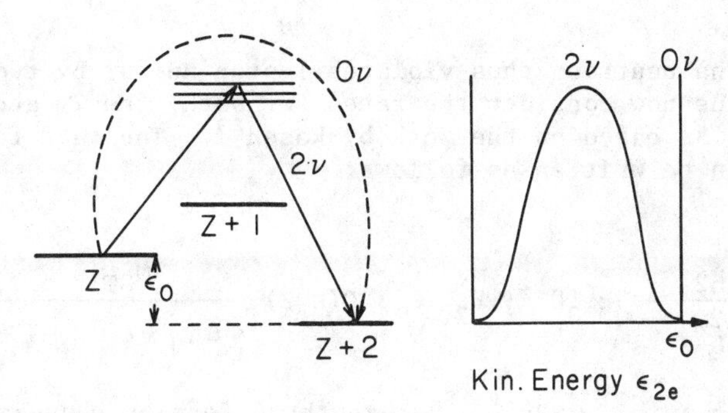

Fig. 7. Energy levels in double-beta decay.

The decay energy is shared by four leptons which results in a distribution of the energy of both electrons sketches in the figure.

The 0ν decay, on the other hand, may proceed by virtual neutrino exchange, as illustrated in Fig. 8. The neutron n_1 emits an electron e_1 and a neutrino. The latter is absorbed by the neutron n_2 which then emits an electron e_2. If the neutrino is a Dirac particle emitted as a right-handed $\bar{\nu}_R$ it cannot be absorbed without breaking helicity and charge conjugation. In order for the process to go, it will be required that (a) the neutrino be a Majorana particle ($\nu^M \equiv \bar{\nu}^M$), (b) there will be a helicity flip ($\nu^M_R \rightarrow \nu^M_L$). Two mechanisms have been identified[20,21] which can give rise to a helicity flip.

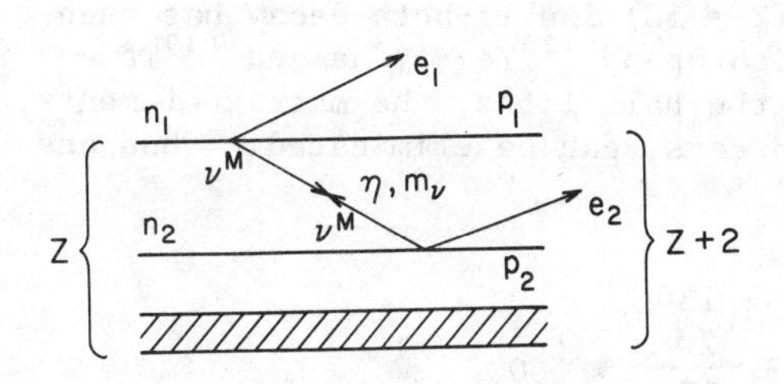

Fig. 8. Illustration of neutrinoless double-beta decay.

1. The lepton wave function has an explicit right helicity admixture with the lepton current given by

$$\ell_\lambda \approx e\gamma_\lambda[(1+\gamma_5) + \eta(1-\gamma_5)]\,\nu^M.$$

We assume here that the hadronic current is purely left handed.
2. There is a neutrino mass term in the standard left current of

the form

$$\ell_\lambda \approx \frac{m_\nu p_\nu}{m_\nu^2 + p_\nu^2} e\gamma_\lambda (1+\gamma_5) \nu^M.$$

The Majorana neutrino thus violates lepton number by two units.

Let us now consider the rates for both, the 2ν and the $o\nu$ processes, as based on the work by Rosen[20]. The rate for the 2ν process can be written as follows:

$$\Gamma^{2\nu} = \frac{\ell n2}{T_{1/2}^{2\nu}} = 3\times10^{-21} F_{2\nu}(\varepsilon_o)\ C(\varepsilon_o Z) \frac{|M_{GT}|^2}{<\Delta E_N + \varepsilon_o/2+1>^2} \text{ (year}^{-1}\text{)}$$

where $F_{2\nu}(\varepsilon_o) \sim \varepsilon_o^{11} + 22\varepsilon_o^{10} + \ldots$ is the 4-fermion phase-space factor, $C(\varepsilon_o Z)$ is the Coulomb function, M_{GT} is the second-order Gamow-Teller matrix-element describing the transitions via states in $Z+1$ (see Fig. 7) and ΔE_N is the average nuclear energy difference between these states and the initial state. With a lepton energy ε_o of about 5 m_e, the energy denominator is roughly 25 m_e. The main uncertainty in this decay rate stems from the nuclear matrix element $|M_{GT}|^2$ and may be as large as a factor of $10^{\pm2}$.

A rough estimate for the 2ν half-life yields $T_{1/2}^{2\nu} \simeq 10^{22\pm2}$ years.

In the element tellurium (Z = 52) double-beta decay has been experimentally studied for two isotopes, $^{128}Te \rightarrow {}^{128}Xe$ and $^{130}Te \rightarrow {}^{130}Xe$. By forming the ratio of the half lifes, the matrix elements, assumed to be the same in both decays, can be eliminated. Thus one has

$$\frac{T_{1/2}^{2\nu\ 128}}{T_{1/2}^{2\nu\ 130}} \simeq \frac{F_{2\nu}^{130}}{F_{2\nu}^{128}} = 5600.$$

A similar estimate for the $o\nu$ process can be made. Both right-hand current (RHC) and mass mechanisms have been included[20,21]. For a transition between groundstates $(0^+ - 0^+)$ this rate is given by[20]

$$\Gamma^{o\nu} = \frac{\ell n2}{T_{1/2}^{o\nu}} = 5\times10^{-16} F_{o\nu}(\varepsilon_o)\ C(\varepsilon_o Z) \left|\frac{M_{GT}}{<r_{ij}>m_p}\right|^2 \text{ (year}^{-1}\text{)}.$$

The quantity $F_{o\nu}(\varepsilon_o)$ is the 2-fermion phase-space factor. It has terms in mass (m in units m_e) and RHC (η) in the following form,

$$F_{o\nu}(\varepsilon_o) = m^2 f_m(\varepsilon_o) + m\eta f_{m\eta}(\varepsilon_o) + \eta^2 f_\eta(\varepsilon_o).$$

$f_m \sim \varepsilon_o^5 + 10\,\varepsilon_o^4 + \ldots$ has the same energy dependence as in allowed beta decay (s-wave) and $f_\eta \sim \varepsilon_o^7 + \ldots$ has higher powers in energy reflecting the content of orbital angular momentum (p-wave) in the RHC process. The quantity $|\ \ |^2$ represents the Gamow-Teller matrix-element divided by an average nucleon separation distance (~ 0.5 nuclear radius), m_p being the proton mass.

Again, a rough estimate gives $T_{\frac{1}{2}}^{o\nu} \simeq 10^{15\pm 2}\,\eta^{-2}\,(\text{or } m^{-2})$ (years).

We note that the $o\nu$ decay is 10^7 times faster than the 2ν decay, if $\eta = 1$ or $m = 1$. The reason for this is twofold: the matrix-elements are larger and the phase space factor is larger, the latter having to do with the fact that the virtual neutrino is constrained to the nucleus and thus can have a momentum given by $p_\nu = \hbar c/R \simeq 35$ MeV/c.

As before we can form the half-life ratio for the tellurium isotopes and find[20],

$$\frac{T_{\frac{1}{2}}^{o\nu\,128}}{T_{\frac{1}{2}}^{o\nu\,130}} = \begin{cases} 320 & (m = 0) \\ 31 & (\eta = 0) \\ 80 & (m = \eta) \end{cases} .$$

Let us have a closer look at the tellurium results. Geochemical lifetime measurements have been conducted over a long period of time. The results are shown in Fig. 9. The data points arranged in chronological order are labelled by author and provenance of the ore. Clearly the lifetime values scatter considerably in excess of the error limits given by the authors. This seems to point to the fact that there are unresolved difficulties in our understanding of the xenon abundance in tellurium ore.

Until recently the only ^{128}Te measurement was that by Hennecke[22] giving a half-life ratio $R = T_{\frac{1}{2}}^{128}/T_{\frac{1}{2}}^{130} = 1600 \pm 50$. This ratio can not be explained by 2ν decay alone and has therefore inspired several researchers[20,21,23] to invoke the presence of $o\nu$ decay. A recent measurement by Kirsten *et al.*[24] however gives a ratio of $R = 10{,}000^{+\infty}_{-5000}$ invalidating that claim.

What is measured are the numbers of ^{128}Xe and ^{130}Xe atoms assumed to be formed by double-beta decay in Te ore. From these numbers the half-life ratio:

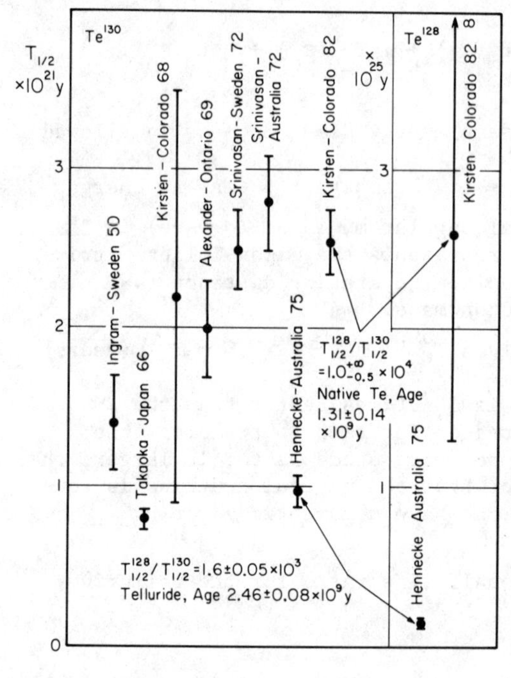

Fig. 9. Overview of the geochemical life-time data for ^{128}Te and ^{130}Te arranged in chronological order. The references are M.G. Inghram and J.H. Reynolds, Phys. Rev. 78, 822 (1950); N. Takaoka and K. Ogata, Z. Naturf. 219, 84 (1966); T. Kirsten *et al.*, Phys. Rev. Lett. 20, 1300 (1968); E. Alexander *et al.*, Earth Planet Sci. Lett. 5, 478 (1969); B. Srinivasan *et al.*, J. Inorg. Nucl. Chem. 34, 2381 (1972); E. W. Hennecke, Ref. 22; T. Kirsten Ref. 24.

$$R = \frac{T_{1/2}^{128}}{T_{1/2}^{130}} = \frac{\Gamma_{2\nu}^{130} + \Gamma_{0\nu}^{130}}{\Gamma_{2\nu}^{128} + \Gamma_{0\nu}^{128}} = \frac{\Gamma_{2\nu}^{130}}{\Gamma_{2\nu}^{128}} \left(\frac{1 + (\Gamma_{0\nu}^{130} / \Gamma_{2\nu}^{130})}{1 + (\Gamma_{0\nu}^{128} / \Gamma_{2\nu}^{128})} \right)$$

is obtained. This ratio involves (besides the 2ν ratio) the ratio of the matrix-elements

$$\rho \equiv \frac{|M_{GT}|}{\left|\frac{M_{GT}}{\langle r \rangle}\right|} \quad \frac{m_p}{\left\langle \Delta E_N + \frac{\varepsilon_o}{2} \quad 1 \right\rangle} = \begin{cases} 1.2 \text{ Rosen}^{20} \\ 0.5 \text{ Haxton}^{23} \end{cases} .$$

The two quoted calculations show that ρ is not very sensitive to details of the evaluation. The neutrino mass then is approximately given by[20]

$$m_\nu \simeq \frac{\rho}{10^4} \sqrt{\frac{6{,}600 - R}{21.5\,R}} ; \quad \eta = 2.3\, m_\nu .$$

With the R value of Kirsten[24] and $\rho = 0.5$ one obtains the following limits

$$m_\nu < 3.1 \text{ eV}; \ \eta < 1.4 \times 10^{-5}.$$

In contrast, the best limit for η from beta-decay experiments is $\eta < 0.1$. Figure 10 shows the allowed values for m and η from the experiments by Kirsten[24] and Hennecke[22]. The scales correspond to the evaluations of Haxton[23], Rosen[20], and Doi[21].

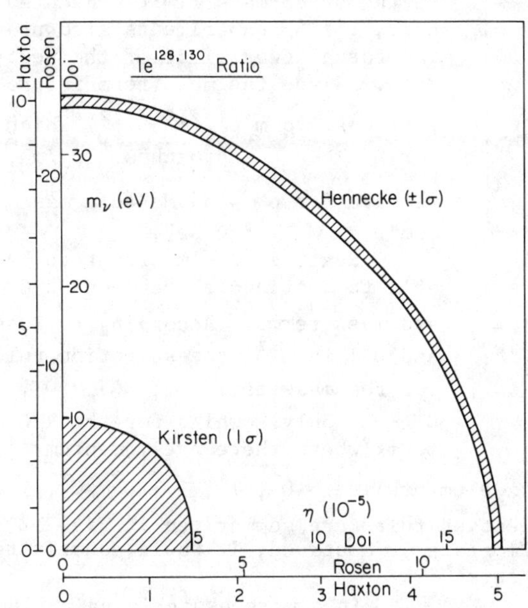

Fig. 10. Values of m_ν and η from double-beta decay life-times of ^{128}Te and ^{130}Te. The three axes correspond to the calculations of Ref. 20,21, and 23.

I shall briefly mention several other $\beta\beta$ results notably those in ^{82}Se and ^{76}Ge. No ratio argument is possible in these cases and the calculations have a larger uncertainty stemming from the uncertainty in the matrix elements.

In ^{82}Se geochemical lifetimes[26] $T^{0\nu+2\nu}$ of $1.45 \pm 0.15 \times 10^{20}$ y[25] and $2.8 \pm 0.9 \times 10^{20}$ y[26] have been reported. Laboratory experiments[27] give a much shorter lifetime of $T^{2\nu} = 1.0 \pm 0.4 \times 10^{19}$ y, while $T^{0\nu} > 3.1 \times 10^{21}$ y has been reported[28]. The calculation[29] gives $T^{2\nu} = 1.7 \times 10^{19}$ in agreement with the laboratory experiment, but not with the geochemical results. Based on the calculated $T^{2\nu}$ it follows that $T^{0\nu} = 1.6 \times 10^{12}/m_\nu^2$ or $m_\nu < 12$ eV. If instead the matrix element is taken from the geochemical lifetime, then $m_\nu < 52$ eV.

In the case of ^{76}Ge there is an excellent laboratory limit by Fiorini *et al.*[30] of $T^{0\nu} > 5 \times 10^{21}$. Based on Haxton's calculation[31] the mass limit is $m_\nu < 15$ eV. The results for Se and Ge are summarized in Fig. 11.

The results[31] in ^{48}Ca of $T^{0\nu} > 1.6 \times 10^{21}$ y and $T^{2\nu} > 3.6 \times 10^{19}$ y are more difficult to interpret in terms of neutrino mass limit. Recent work by Bellotti *et al.*[32] gives limits for $T^{0\nu}$ of $10^{18} - 10^{19}$

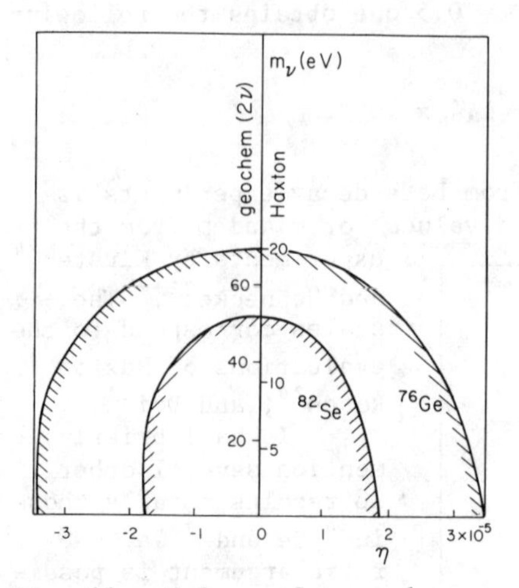

Fig. 11. Values of m_ν and η from double-beta decay lifetimes of ^{82}Se and ^{76}Ge. The axes correspond to the calculations of Ref. 29.

years for the cases of ^{58}Ni, ^{92}Mo, ^{100}Mo, ^{148}Nd, and ^{150}Nd.

Before concluding this section, I would like to mention several further considerations. First, It should be noted[20] that in addition to the upper mass limit presented here, the ββ experiments also provide for a lower limit of the neutrino mass. Since the mass term is proportional to $m_\nu p_\nu/(m_\nu^2 + p_\nu^2)$ which for $m_\nu >> p_\nu$ approaches p_ν/m_ν, we have for $p_\nu \simeq 35$ MeV/c a mass limit of $m_\nu >$ few GeV.

Next, it is important to be able to distinguish between RHC and mass terms. According to Rosen[20] and Doi *et al.*[21] the selection rules for the mass term are: $\Delta J = 0^+$, $0^+ \to 0^+$ only. While for the RHC term (where there are momentum dependent terms in the matrix elements) $\Delta J = 0^+, 1^+, 2^+; 0^+ \to 0^+, 1^+, 2^+$. The effects of these selection rules are summarized in Fig. 12. Thus if a ββ branch to the 2+ state is observed, it would constitute evidence for right-hand currents.

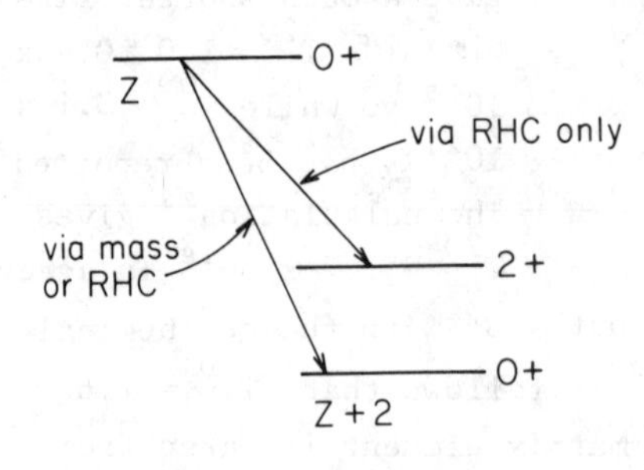

Fig. 12. Selection rules for double-beta decay to 0^+ and 2^+ states.

In a more esoteric vain, it was pointed out by Rosen[20] that the angular correlation between the two electrons is different for RHC and mass mechanisms. Preferential opposite emission is predicted in the latter case since $\Delta J = 0$ is required, while a deviation from this will occur for right-handed particles.

As to the outlook, there are several experiments with ^{76}Ge in progress now, including one at Milano[33], at Caltech[34], and Battelle[35], using large high resolution Ge detectors and improved background suppression. It can safely be predicted that an order of magnitude improvement of the present sensitivity will be achieved soon, leading to a more stringent limit of the neutrino mass, as illustrated in Fig. 13. Work is in progress on a new ^{82}Se measurement by the Irvine group[27]. An experiment on ^{136}Xe is in preparation[36] using a liquid

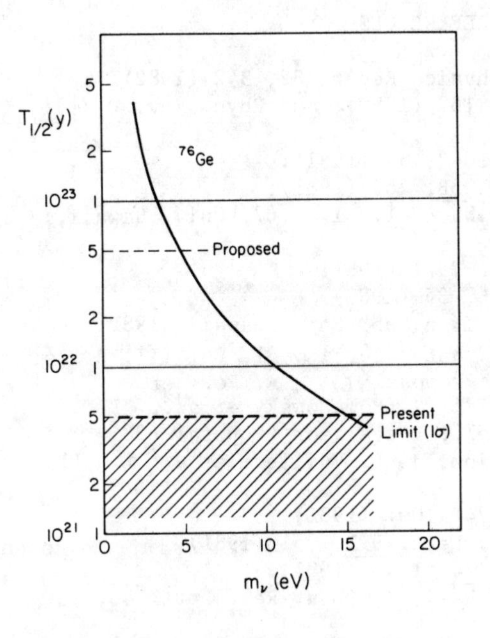

Fig. 13. Illustration of neutrino mass limits which can be achieved from life-time measurements in ^{76}Ge.

Xe TPC. Compared to the geochemical experiment, clearly these laboratory experiments are more decisive if evidence for the important neutrinoless ββ process should be established, and should be pursued vigorously. If neutrinos are of the Majorana type, present evidence points to the fact that their masses cannot exceed 3 or 10eV.

In conclusion, I am summarizing in Table II the development of some of the evidence in support of neutrino mass. Most of the claims for finite mass of 1980 have proved to be invalid, except for the ^{3}H spectrum, an experiment which is waiting to be challenged.

Table II Evidence for Neutrino Mass

Experiment	Claim, 1980	Present Status, 1982
$\bar{\nu}_e \rightarrow X$ (Reines)	Oscillations $\Delta m^2 \sim 1eV^2$	No Evidence, $\Delta m^2 < 0.016eV^2$
^{3}H Spectrum (ITEP)	$14 < m_\nu < 46eV$	Unchallenged
Solar ν Puzzle (Davis)	Supporting Oscillations	Possible Support
$\nu_\mu \rightarrow X$ Beam Dump (CERN)	Osc. large Δm^2	No Evidence
0_ν -ββ $Te^{128,130}$ (Hennecke)	$m_\nu^M \sim 10eV$	No Evidence, $m_\nu^M < 3eV$

REFERENCES

[1] P.H. Frampton and P. Vogel, Physics Report 82, 342 (1982).
[2] R.E. Shrock, Phys. Lett. 96B, 159 (1980), and Phys. Rev. D 24. 1232 and 1275 (1981).
[3] Y. Asano *et al.*, Phys. Lett. 104B, 84 (1981).
[4] R. Abela *et al.*, Phys. Lett. 105B, 263 (1981).
[5] D. Berkhofer *et al.*, Neutrino 81, Vol. II, p.67, Univ. Hawaii, 1982.
[6] J.J. Simpson, Phys. Rev. D 24, 2971 (1981).
[7] J.J. Simpson, Phys. Lett. 102B, 35 (1981).
[8] C. Baltay, Neutrino 81, Vol. II, p.295, Univ. Hawaii, 1982.
[9] For CERN Proposals, see A.L. Grant, Neutrino 81, Vol. II, p.214; M. Murtagh *et al.*, Brookhaven Proposal E775. A. Pevsner *et al.*, Brookhaven Proposal E776; T. Romanowsky *et al.*, LAMPF Proposal 645; T. Dombeck *et al.*, Los Alamos Proposal 638.
[10] Caltech-Munich-SIN Collaboration, J.-L. Vuilleumier *et al.*, Phys. Lett. B, to be published.
[11] H. Kwon *et al.*, Phys. Rev. D 24, 1097 (1981).
[12] K. Schreckenbach *et al.*, Phys. Lett. 99B, 251 (1981).
[13] P. Vogel *et al.*, Phys. Rev. C 24, 1543 (1981).
[14] F. Reines, H.W. Sobel, and E. Pasierb, Phys. Rev. Lett. 45, 1307 (1980).
[15] M. Mandelkern, Neutrino'81, Vol. II. p.203 (1981).
[16] J.F. Cavaignac *et al.*, Raport Annuel 1980, ISN 81.01, p.71.
[17] A.M. Cooper *et al.*, Phys, Lett. B 112, 97 (1982).
[18] Detailed references are contained in the recent reviews of ref. 20, and in Yu. G. Zdesenko, Sov. J. Part. Nucl. 11 (6), 582 (1981); D. Bryman and C. Picciotto, Revs. Mod. Phys. 50, 11 (1978).
[19] R.N. Mohapatra and J.D. Vergados, Phys. Rev. Lett. 47, 1713 (1981).
[20] S.P. Rosen, Neutrino 81, Vol. II, p.76, Univ. Hawaii, 1982, and Workshop on Low Energy Tests on High Energy Physics, I.T.P., Santa Barbara, Jan. 1982. See also, H. Primakoff and S.P. Rosen, Phys. Rev. 184, 1925 (1969); Halprin *et al.*, Phys. Rev. D 13, 2567 (1976).
[21] M. Doi, T. Kotani, H. Hishiura, K. Okuda, and E. Takasugi, Phys. Rev. 103B, 219 (1981).
[22] E. Hennecke, O. Manuel, and D. Sabu, Phys. Rev. C11, 1378 (1975).
[23] W.C. Haxton, G.J. Stephenson and D. Strottman, Los Alamos preprint LA-UR-81-3454, Phys. Rev. D., to be published.
[24] T. Kirsten, E. Jessberger, E. Pernicka, and H. Richter, to be published (1982), presented at Workshop on Low Energy Tests of High Energy Physics, I.T.P., Santa Barbara, Jan. 1982.
[25] T. Kirsten and H. Muller, Earth and Plan. Sci. Lett. 6, 271 (1969).
[26] B. Srinivasan *et al.*, Econ. Geol. 68, 252 (1973).
[27] M. Moe and D. Lowenthal, Phys. Rev. C22, 2186 (1980).
[28] B. Cleveland *et al.*, Phys. Rev. Lett. 35, 737 (1975).
[29] W.C. Haxton, G.J. Stephenson, and D. Strottman, Phys. Rev. Lett. 47, 153 (1981).
[30] E. Fiorini *et al.*, Nuovo Cim. A13, 747 (1973).
[31] R. Bardin *et al.*, Phys. Lett. 26B (1967), Nucl. Phys. A158, 337 (1970).

[32] E. Bellotti, E. Fiorini, C. Liguori, A. Pullia, A. Sarracino, and L. Zanotti, preprint, Milano 1981.
[33] E. Bellotti, Workshop, I.T.P., Santa Barbara, Jan. 1982.
[34] F. Boehm, Workshop, I.T.P., Santa Barbara, Jan. 1982.
[35] F. Avignone, Workshop, I.T.P., Santa Barbara, Jan. 1982.
[36] H. Chen, private communication.

FIRST RESULTS FROM THE CERN-SPS PROTON-ANTIPROTON COLLIDER

M. Yvert
LAPP, Annecy, France

ABSTRACT

After a short description of the experimental apparatus involved, a review is given of the results obtained so far by three experiments at CERN (UA1, UA2, and UA4) on elastic scattering, quark search, multiplicities and KNO scaling, total transverse energy spectra, and inclusive charged and neutral single-particle spectra.

INTRODUCTION

The first collisions of antiprotons on protons in the CERN SPS used as a collider were observed in July 1981. The energy of 540 GeV obtained in the centre of mass opens up a new path in experimental particle physics, since $\sqrt{s}$ is thus increased by one order of magnitude as compared to the energy available at the ISR. In this talk I will review the physics results obtained so far by the UA1, UA2, and UA4 experiments; the UA5 experiment will be discussed by H. Mulkens in the next talk. I will divide my presentation into three sections; in Section 1 I will give a brief description of the machine, in Section 2 I will describe the experimental arrangements, and in Section 3 I will discuss the physics results obtained up to the time of this conference.

1. THE MACHINE

1.1 History

In 1976 the idea was launched by Rubbia, Cline and McIntyre[1] of using the SPS as a $p\bar{p}$ collider, the main goal being to reach a high enough energy to discover the intermediate vector bosons predicted by the Weinberg-Salam theory. In the same year several working groups started to study both the machine and experimental aspects of this project. In 1977 the working groups came to their final conclusions, and in 1978 the project received official authorization. In the same year five Underground Area experiments were approved.

On 10 July 1981 the first $p\bar{p}$ collisions at 270 + 270 GeV were reported by the UA1 Collaboration. During machine development sessions in October and November the UA1, UA3, UA4, and UA5 Collaborations took data, the UA5 Collaboration completing its data taking during this period. In the December period about 100 h of $p\bar{p}$ collisions were available for physics and data were collected by the UA1, UA2, UA3, and UA4 Collaborations. The total available integrated luminosity over the 1981 periods was about 1.8×10^{32} cm^{-2}, which corresponds to the predicted production of 0.1 W boson decaying into leptons. The same integrated luminosity could be reached in 10 s at the ISR for pp collisions; nevertheless, as I will show later, a lot of precious data have been successfully collected by all the experiments. The next $p\bar{p}$ period will start in October 1982 with a much higher expected luminosity.

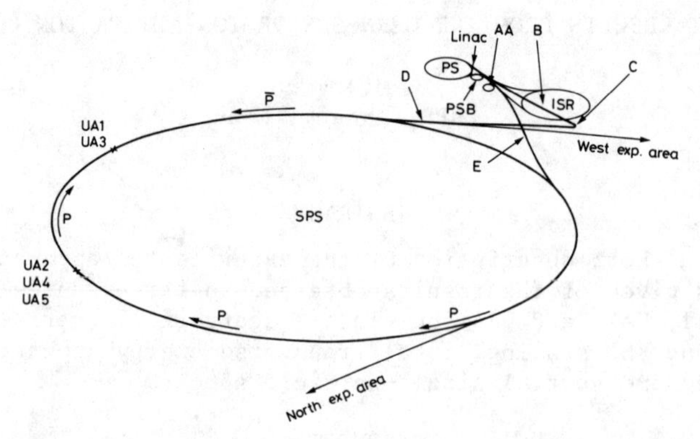

Fig. 1 Schematic layout of the SPS $p\bar{p}$ collider.

Table I: Machine performance during 1981

	Achieved (the best)		Design	
	p	$\bar{p}$	p	$\bar{p}$
Injection momentum	26 GeV/c		26 GeV/c	
Particles/bunch	7×10^{10}	5×10^{9}	10^{11}	10^{11} $(2 \times 5 \times 10^{10})$
Number of bunches	2	1	6	6 (12 → 6)
Colliding momentum	270 GeV		270 GeV	
β_H	1.5 m		2 m	
β_V	0.75 m		1 m	
Luminosity (cm^{-2} s^{-1})	5×10^{27} cm^{-2} s^{-1}		1×10^{30} cm^{-2} s^{-1}	

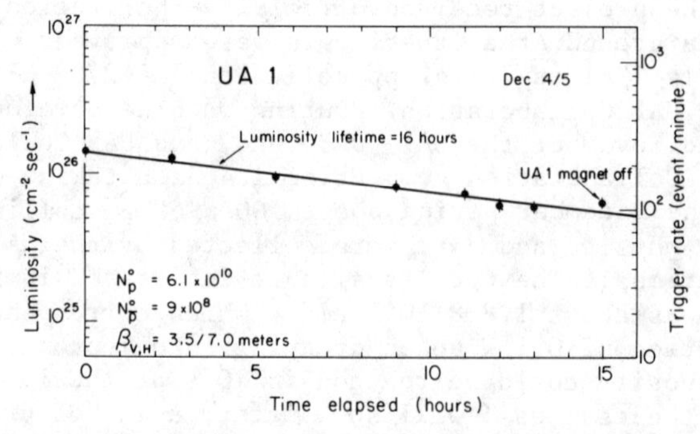

Fig. 2 The luminosity lifetime in $p\bar{p}$ collisions as measured in the UA1 experiment

1.2 Principle of operation

In Fig. 1 is shown the layout of the SPS $p\bar{p}$ complex. The PS proton beam, accelerated at 26 GeV (10^{13} protons every 2.4 s), hits a copper target, from where antiprotons produced at 3.5 GeV/c are directed towards the Antiproton Accumulator (AA) in which they are stored, cooled, and accumulated during about 1 day. When the AA is full, the antiprotons (at 3.5 GeV/c) are sent to the PS, where they are accelerated to 26 GeV/c before being transferred to the SPS (via the lines B, C, D), where the necessary proton bunches have been injected already via line E (at 26 GeV/c). Once protons and antiprotons are circulating clockwise and anticlockwise into the SPS, they are both accelerated to 270 GeV, which is the maximum energy that can be obtained from the power supplies in d.c. mode.

1.3 Performances

In Table 1 the best performances of the machine during 1981 are briefly summarized. For the October 1982 run improvements are being carried out concerning mainly the antiproton transfer efficiency, the increase in the number of antiproton bunches to three instead of one, and in the number of antiprotons per bunch. These improvements, combined with a better reliability, allow us to believe, with good confidence, that a peak luminosity of 5×10^{28} cm^{-2} s^{-1} will be obtained. A very good lifetime of 16 h was achieved for the luminosity already at the beginning of December 1981 (Fig. 2).

2. THE EXPERIMENTS

The Institutes involved in the five approved experiments are listed in Table II. The main physics objectives covered by the various

Table II: Institutes involved in the CERN SPS $p\bar{p}$ experiments

Experiment (Spokesman)	Institutes involved
UA1 4π detector (C. Rubbia)	Aachen, Annecy, Birmingham, CERN, London (QMC), Paris (CdF), Riverside, Rome, Rutherford, Saclay, Vienna (127 physicists)
UA2 Large solid angle detector (P. Darriulat)	Bern, CERN, Copenhagen, Orsay (LAL), Pavia, Saclay (46 physicists)
UA3 Search for monopoles (P. Musset)	Annecy, CERN (4 physicists)
UA4 Elastic scattering and σ_{tot} (G. Matthiae)	Amsterdam, CERN, Genoa, Naples, Pisa (27 physicists)
UA5 Large streamer chamber (J.G. Rushbrooke)	Bonn, Brussels, Cambridge, CERN, Stockholm (39 physicists)

Table III: Physics objectives covered by the UA experiments

Physics objectives	Experiments
Observation of Weinberg-Salam intermediate bosons $W^{\pm} \to e^{\pm}\nu$, $Z^0 \to e^+e^-$; $\mu^{\pm}\nu$, $\mu^+\mu^-$	UA1 UA2
Interactions of the hadron constituents: jet physics, large p_T	UA1 UA2
Conventional hadronic physics (large s → large Δy)	UA1 UA2 UA5
Elastic scattering, σ_{tot}	UA1 UA4
Search for free quarks	UA1 UA2
Search for monopoles	UA3
New types of events, e.g. Centauros seen in cosmic-ray experiments	UA1 UA2 UA5

experiments are summarized in Table III. I will give a brief description of the parts of the experimental set-ups that have been used to obtain the physics results I will discuss later.

2.1 UA1 experiment

The UA1 set-up[2] covers the whole solid angle around the interaction point. Figure 3 shows the experimental arrangement; inside a large magnet giving a horizontal magnetic field of 0.7 T for a maximum current of 10 kA, is installed an electromagnetic calorimeter which surrounds a large track detector. The yoke of this magnet used to filter muons is equipped as a hadron calorimeter.

2.1.1 *Central detector*

The tracks emitted from the interaction point are detected in an image chamber[3] of 6000 resistive wires covering a useful cylinder 2.2 m in diameter and 6 m long (Fig. 4). The pulse height is sampled every 30 ns at both ends of each wire; these wires are aligned with the magnetic field and arranged in planes such that the curvature of the tracks is measured using the drift-time measurement, thus giving the best accuracy in momentum reconstruction. An argon + ethane mixture (40% + 60%) was used with a drift field of 1.5 kV/cm. The coordinate along the wire is measured by charge division. Typical accuracies obtained in the December run were 300 μm in the drift direction and 2.8% of the wire length along the wire.

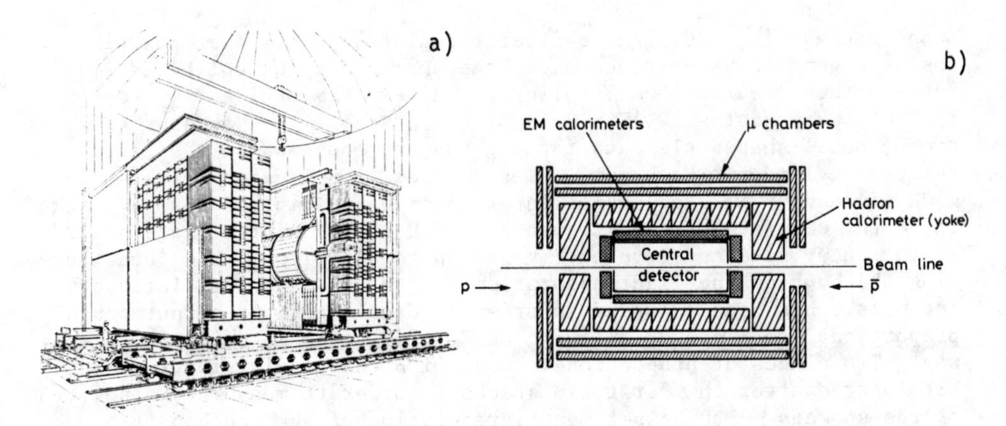

Fig. 3 a) the UA1 apparatus as installed in the LSS5 interaction region; only shown here are the parts covering an angle from 5° to 175° with respect to the beam line. b) Schematic cut of the UA1 apparatus in a vertical plane containing the beam line.

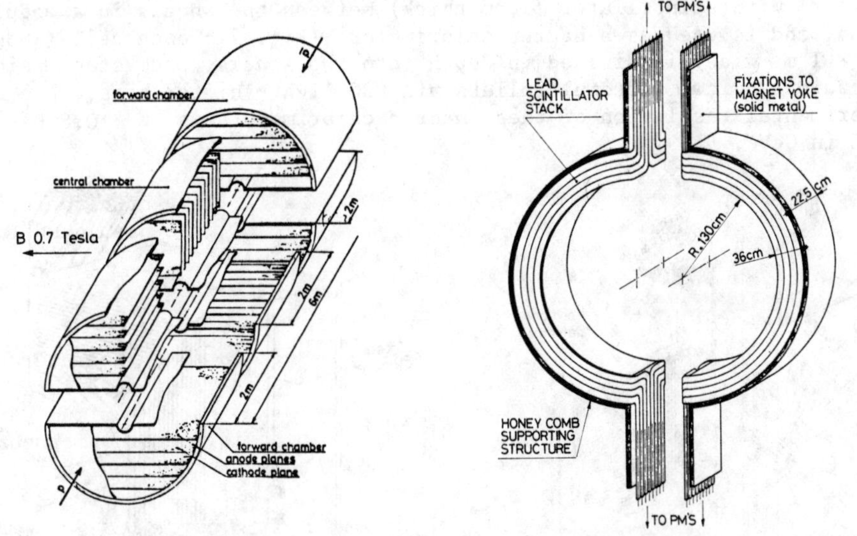

Fig. 4 The UA1 central detector.

Fig. 5 The gondolas: schematic view of a pair of elements.

2.1.2 *Central electromagnetic calorimetry*

There are two kinds of electromagnetic calorimeters: the "gondolas" covering an angle about the beam direction from 25-155° and the "bouchons" covering the 5-25° and 155-175° end-cap regions.

The gondolas (Fig. 5) are 48 half-cylinder elements, 25 cm wide, covering a barrel around the central detector; each gondola is a 26 radiation lengths thick sandwich of lead (1.2 mm) and scintillator (1.5 mm) divided into 4 stacks in depth; each stack is read out using

4 phototubes with BBQ light-shifter technique. The shower position inside a gondola is obtained using the sharing of the collected light between these four photomultipliers. The experimental energy resolution of the gondolas is $\Delta E = 0.15\ E^{\frac{1}{2}}$ (E in GeV). The bouchons (Fig. 6) are 64 petal-shaped elements forming the two end-caps of the central-detector cylinder. Each petal is a 27 radiation lengths thick sandwich of lead (4 mm) and scintillator (6 mm) also divided into 4 stacks in depth, each stack being viewed by a BBQ light-shifter plate placed at the outer edge and read out by one photomultiplier. When constructing the bouchon the light-absorption length in the scintillator of the petals has been adjusted in order to obtain a light output roughly proportional to E sin θ, allowing an E_T trigger to be built in an easy way. Two planes of proportional tubes with charge division read-out[4] are located after the first two stacks in order to measure the position of the showers -- the experimental resolution of the bouchon is $\Delta E_T = 0.12\ E_T^{\frac{1}{2}}$ (E_T in GeV).

2.1.3 *Hadron calorimetry*

The return yoke of the magnet is built of sheets of iron (5 cm thick) with scintillator (1 cm thick) between the sheets in a modular way, and is used as a hadron calorimeter[5] (Fig. 7); each cell (about $1 \times 1\ m^2$ wide) is divided in depth into two stacks, each stack being read out by two photomultipliers via BBQ light-shifter bars. The experimental resolution of these hadron calorimeters is $\Delta E = 0.8\ E^{\frac{1}{2}}$ (E in GeV).

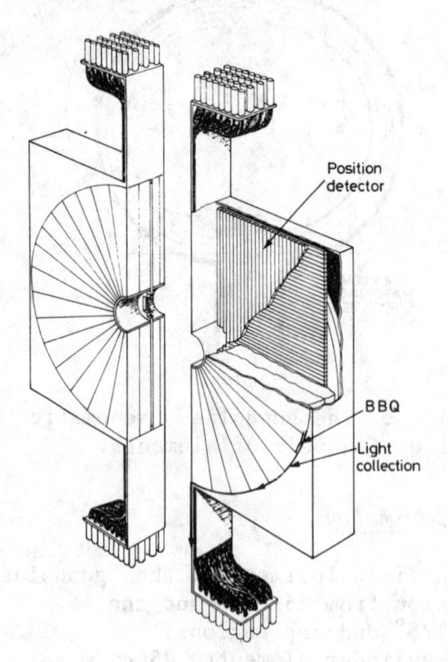

Fig. 6 The "bouchons": UA1 end-cap electromagnetic calorimeters

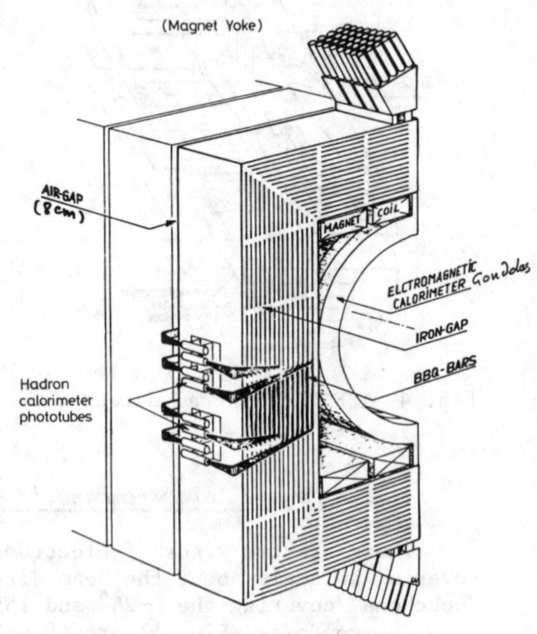

Fig. 7 A module of the UA1 hadron calorimeters.

2.2 UA2 experiment

In Fig. 8 is shown the general layout of the UA2 detector[6,7]. The interaction region is surrounded by a vertex detector, of which the outer layer is used as an electromagnetic shower-position detector; then comes an assembly of 240 cells of electron-hadron calorimetry (17 radiation lengths of lead/scintillator sandwich followed by iron/scintillator sandwich). The light of each cell is collected via BBQ light-shifter technique. This calorimeter covers a rapidity domain of ±1 unit around zero. In the angular regions 20-37.5° and 142.5-160° about the beam line are installed 24 sectors of toroid magnet spectrometers followed by electromagnetic calorimeters.

During the December 1981 period, a wedge of 30° angle in azimuth was open in the central calorimeter and instrumented as a 1.1 T•m magnetic spectrometer as shown in Fig. 9. Covering the open wedge twelve

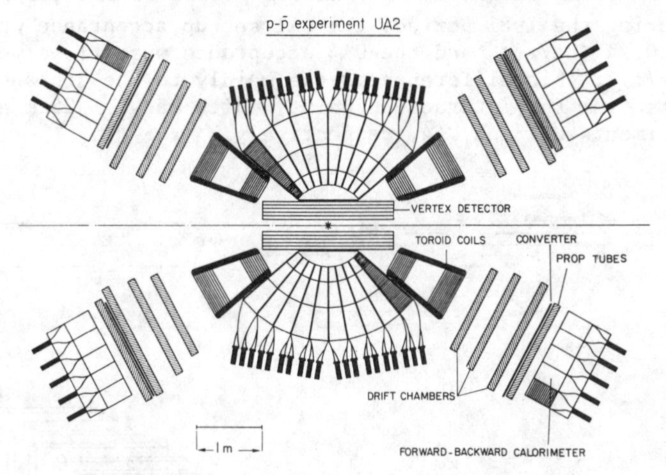

Fig. 8 Schematic cut of the UA2 experiment.

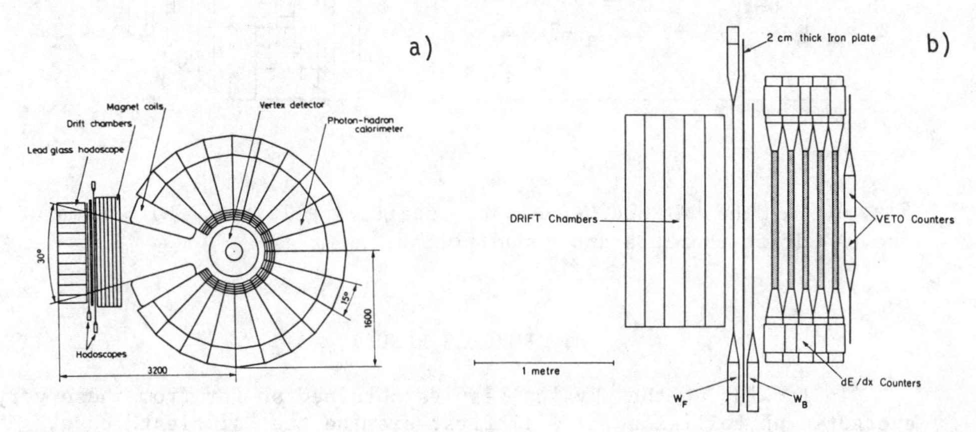

Fig. 9 a) The UA2 large-angle magnetic spectrometer. b) The UA2 dE/dx counters.

planes of drift chambers were followed by a wall of lead-glass blocks (15×15 cm^2) 14 radiation lengths in depth. Two columns of this lead-glass array were replaced by a dE/dx set-up of 5 scintillators, each of them being 4 cm thick and read out by two photomultipliers.

2.3 Elastic scattering in UA1 and UA4

The elastic scattering set-ups of UA4 [8,9] and UA1 are based on the same principle (Fig. 10); the outgoing elastic-scattered proton and antiproton in the forward diffraction cone are emitted at very small angle with respect to the beam line; the experimental idea is to measure them using a set of drift chambers installed in "Roman pot" devices which allow the active volume of the drift chambers to be placed at distances as small as 16 mm from the beam axis. The trigger signal is built up using scintillator counters placed behind the drift chambers. During the 1981 periods the UA1 set-up acceptance was $|t|$ between 0.16 and 0.24 (GeV/c)^2 and the UA4 acceptance was $|t|$ between 0.06 and 0.17 (GeV/c)^2; this difference is due mainly to the distance of the Roman pots from the interaction point, which was $\pm \sim 25$ m and $\pm \sim 40$ m for experiments UA1 and UA4, respectively.

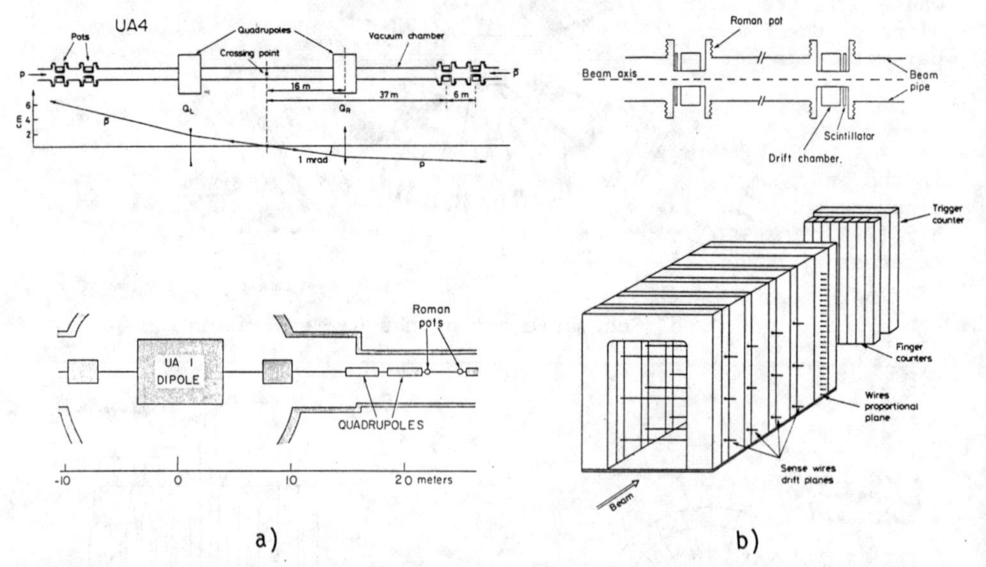

Fig. 10 a) The UA1 and UA4 elastic scattering layout. b) Schema of the UA4 drift chambers and Roman pot devices.

3. PHYSICS RESULTS

In looking at the physics results obtained so far from these very energetic $p\bar{p}$ collisions, I will first examine the "simplest" case, i.e. that of elastic scattering. From there, I will move to what we learn when the colliding particles are broken, going from the search for free quarks to the multiplicity distribution of the produced particles. Going deeper into the dynamics of the interactions, I will then discuss the total transverse energy and single-particle transverse momentum spectra and say a few words about some attempts that have been made to hunt for jets and strangely behaving events.

3.1 Elastic scattering

From the ISR data, we know since 1972 that the proton-proton elastic scattering cross-section[10], in the $|t|$ range 0.05-0.25 $(\mathrm{GeV}/c)^2$ presents an exponential shape $d\sigma/dt \propto e^{bt}$ with a change in the slope parameter b around $|t| = 0.1\ (\mathrm{GeV}/c)^2$, the slope being greater at low-t values. In 1981, with the ISR working as a $p\bar{p}$ collider, Carboni et al.[11] reached the same conclusion at a centre-of-mass energy of 53 GeV for the $p\bar{p}$ elastic cross-section. A very interesting analysis has been made recently by Burq et al.[12], which shows that the slope b is, at large momentum, consistent with the hypothesis of a universal logarithmic shrinkage of the hadron-diffraction cone

$$b(p) \simeq b_0 + 2\alpha'_p \log(p) ,$$

where p is the momentum of the incident particle in GeV and α'_p the slope of the Pomeron trajectory. This group found that the shrinkage parameter has the same value for all particles ($p^\pm$, $K^\pm$, $\pi^\pm$),

$$2\alpha'_p \simeq 0.28\ (\mathrm{GeV}/c)^{-2},$$

in the $|t|$ region of 0.2 $(\mathrm{GeV}/c)^2$. The results of the UA4 Collaboration are shown in Fig. 11a. Their value of the slope parameter is $b = 17.2 \pm 0.8\ (\mathrm{GeV}/c)^{-2}$ for $0.06 < -t < 0.18$; this value is shown together with the results of other experiments in the same t range in Fig. 11b.

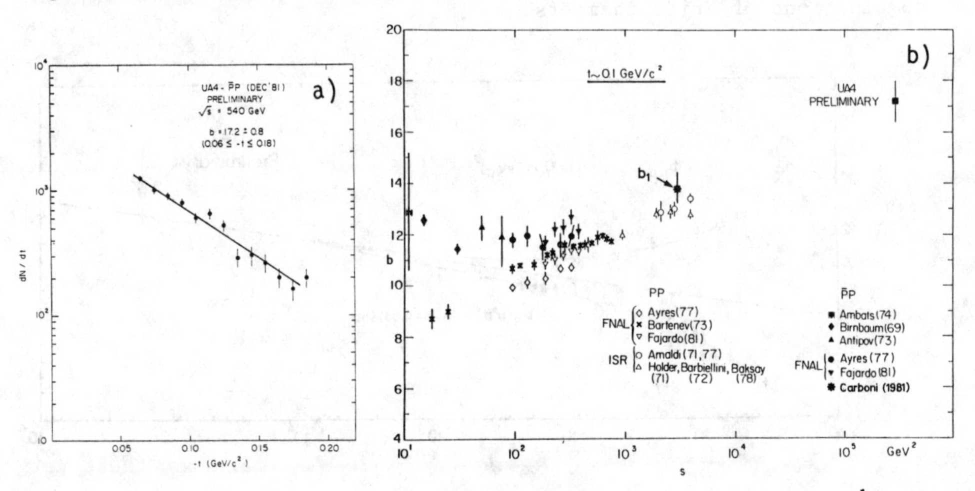

Fig. 11 a) UA4 elastic scattering results fitted with $d\sigma/dt \propto e^{bt}$. b) UA4 value of the slope b compared with other experiments in pp and $p\bar{p}$.

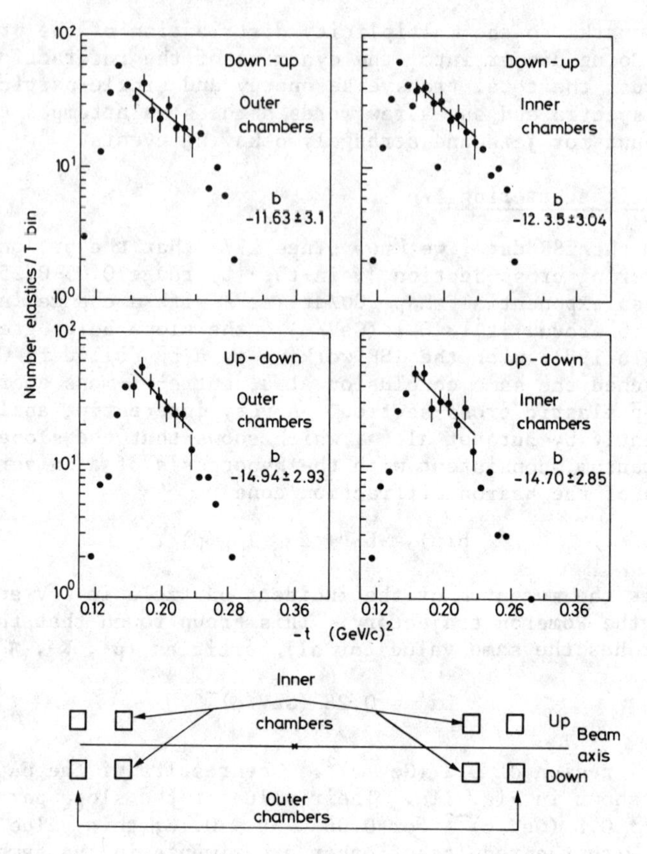

Fig. 12 UA1 elastic scattering results fitted with $d\sigma/dt \propto e^{bt}$ for four combinations of drift chambers.

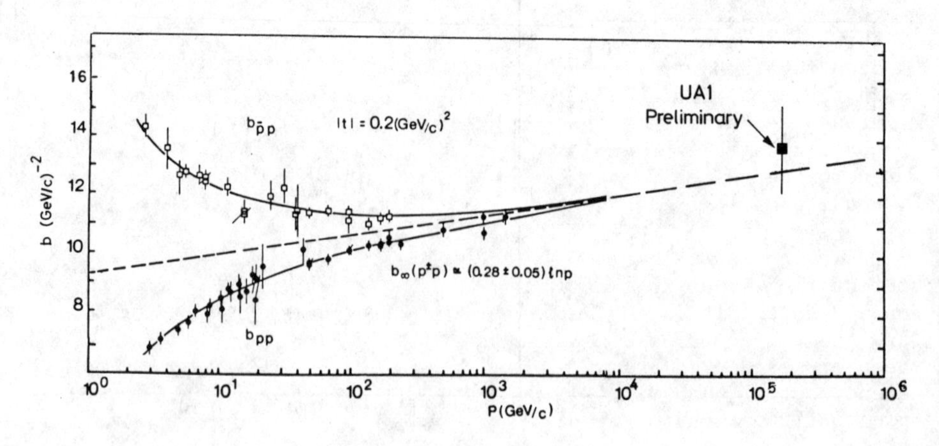

Fig. 13 UA1 value of the slope b compared with an extrapolation of the Burq et al. analysis.

The results of the UA1 Collaboration are shown in Fig. 12 for the 4 combinations of the chambers. The combined results give a slope

$$b = 13.6 \pm 1.5\ (\mathrm{GeV}/c)^{-2}$$

for a t range $0.16 < -t < 0.24\ (\mathrm{GeV}/c)^2$; this result falls in the t range of the Burq et al. analysis and is shown in Fig. 13 to be consistent with an extrapolation of this fit.

The results of both the UA4 and UA1 Collaborations are still preliminary and do not yet give an absolute cross-section. They both extrapolate well from already existing data; the statistical errors are still large but will improve with the 1982 runs; however we can already conclude that there is no dramatic surprise coming from these elastic-scattering measurements.

3.2 Inelastic trigger

To study inelastic collisions the best way is to record each interaction and then to look in detail at all the possible phenomena; this implies the use of an unbiased trigger. During the December 1981 run the interaction rate per bunch crossing was about 10^{-3}. Therefore it was impossible to use the obviously unbiased trigger given by the beam-crossing signal; it was chosen so as to require in the trigger the coincidence of two planes of hodoscopes installed on each side of the interaction region covering a substantial part of the rapidity domain. In the UA1 experiment these trigger hodoscopes were at ±6 m from the interaction point covering $3.6 < y < 5.1$. In the UA2 experiment the distance was ±10 m, with a $4.2 < y < 5.3$ coverage. These requirements exclude the elastic and single-diffractive events. Some more conditions were imposed during part of the data taking:

- In the UA1 experiment, data were taken with different magnetic field values (0, 0.28, and 0.56 T) and with either a threshold on the total energy deposited in the calorimeters ($E_T > 0$, 20, and 40 GeV) or a threshold on e.m. calorimeters. A total of 0.5×10^6 triggers have been recorded with a non-zero magnetic field.
- In the data reported by the UA2 Collaboration some specific conditions have also been made on the requirements depending on the physics goal: a charged particle detected in the open wedge; electromagnetic energy deposited in the lead-glass wall facing the wedge; pulse-height conditions in the dE/dx counters.

The characteristics of these inelastic reactions are very different from event to event. In Fig. 14, are shown some events as seen on a Megatek display of data from the UA1 central detector; the lines are the trajectories reconstructed by the analysis program which has been cross-checked by a scanning done by physicists to have an efficiency better than 91%. As we see, these events range from very simple configuration to very complex ones with a very high number of tracks.

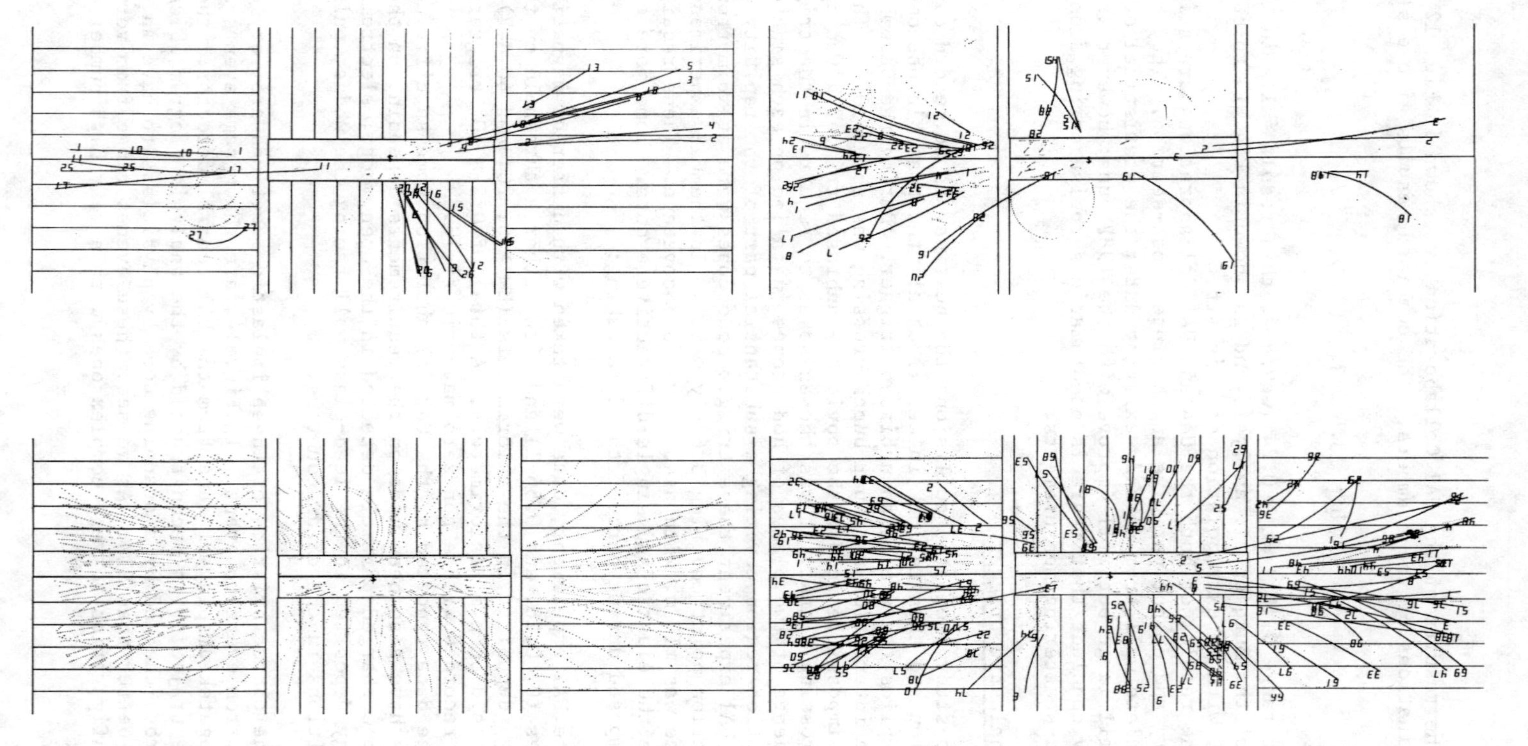

Fig. 14 Typical UA1 events as seen using a Megatek display of the central detector information

3.3 Quark search

The UA2 Collaboration, using the pulse height of the five dE/dx counters, computes the most probable energy loss I_0 from the three smallest measurements:

$$I_0 = \sqrt[3]{I_1 I_2 I_3} \ .$$

After rejection of inconsistent events they are left with the distribution of this most probable energy loss normalized to minimum ionizing particles shown in Fig. 15. Seven events are possible quark candidates which are rejected because they give a pulse height greater than that of a minimum ionizing particle in the counter placed in front (6 events), or because they give an inconsistent impact position (top/bottom ratio) in the different counters (1 event).

No candidates are left out of 6374 charged particles. Assuming a p_T distribution for the quark production the same as for pions and a number of charged particles produced per rapidity unit dn/dy = 3, they reach the 90% C.L. upper limits for quark/charge 1:

Mass / $\lvert Q \rvert$	0	m_π	$2m_\pi$
1/3	1.5×10^{-4}	1.7×10^{-4}	2.0×10^{-4}
2/3	1.7×10^{-4}	2.5×10^{-4}	3.3×10^{-4}

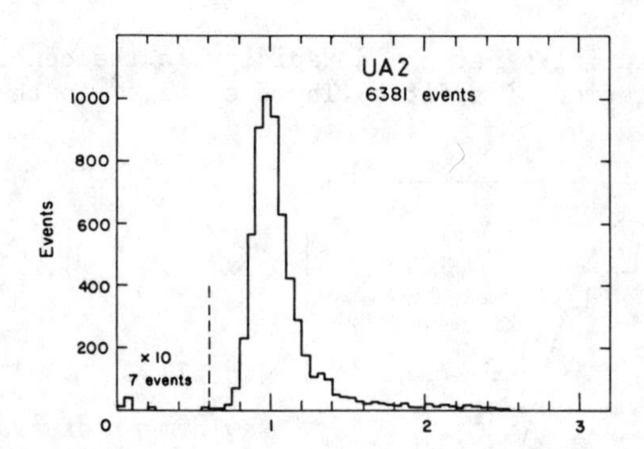

Fig. 15 Most probable energy loss I_0/I_{mip} obtained by the UA2 Collaboration in their set of dE/dx counters.

3.4 Multiplicities

Among the simplest variables to measure in inelastic collisions are the number of charged tracks and their direction; this already gives a lot of information about the collision process. From the pp collisions at the ISR the main observations about multiplicities were[13]:

- The distribution of the number of charged particles versus the rapidity y^* is almost flat.
- The height of this plateau grows linearly with log (s).
- The distribution of the number of tracks produced follows the scaling law proposed by Koba, Nielsen and Olesen[14] (KNO).

In November 1981 the UA1 and UA5 Collaborations[15,16] already reported a measurement of $dn/d\eta$ in the central region. For $|\eta| < 1.3$, the UA1 Collaboration found $dn/d\eta = 3.9 \pm 0.3$ when excluding zero track events from the fiducial region and $dn/d\eta = 3.6 \pm 0.3$ when including zero tracks; these numbers are corrected for γ-ray conversion but not for K_S^0 or Λ^0 decays. In the UA5 experiment this value is found to be $dn/d\eta = 3.0 \pm 0.1$. In Fig. 16 are shown these results together with ISR pp data and cosmic-ray data[17]; we see that they extrapolate nicely, still growing like log (s).

The next question one can ask is, Are these particles produced randomly? In Fig. 17 is shown the distribution of the number of tracks observed in the fiducial volume of experiment UA1 for $|\eta| < 1.3$;

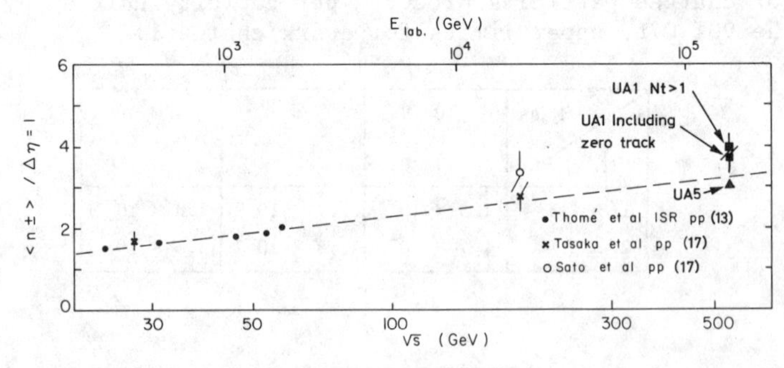

Fig. 16 Multiplicity per pseudorapidity in the central region; the line is an extrapolation of a fit by Thomé et al.[13] to their data.

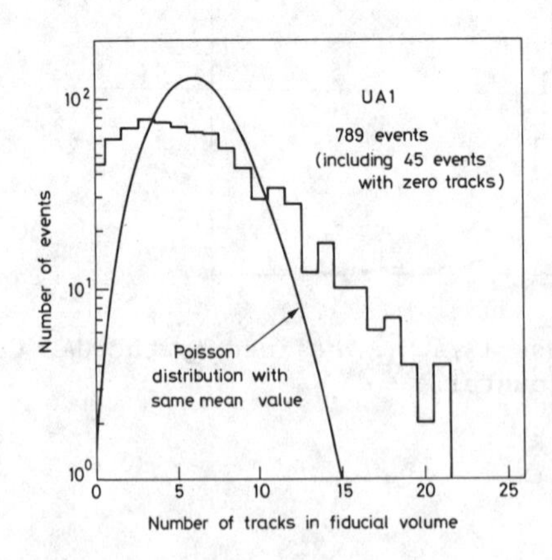

Fig. 17

Raw number of tracks observed in the UA1 fiducial volume[15]. The curve is a Poisson distribution with the same mean value and normalized to the same number of events.

*The rapidity y is given by $y = 1/2 \log[(E + p_L)/(E - p_L)]$. In the approximation $E \simeq p$ we have $y \simeq \eta = -\log \mathrm{tg}\, \theta/2$; where η is the pseudorapidity and θ the angle of the track with respect to the beam axis.

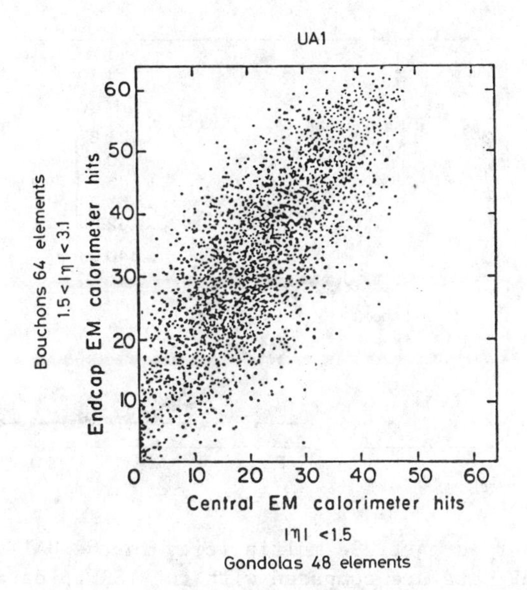

Fig. 18 Number of cells of electromagnetic calorimeter hit; forward + backward region versus central region.

on the same plot the Poisson distribution with the same mean clearly does not reproduce the data. Another piece of evidence is given in Fig. 18, where the number of electromagnetic calorimeter cells above threshold in the central region (gondolas covering $|\eta| < 1.5$) are plotted versus the number of cells in the forward-backward region (bouchons covering $1.5 < |\eta| < 3.1$). This plot shows a clear correlation between the central and forward-backward region, indicating again a non-random process.

The scaling hypothesis proposed by Feynman in 1969[18] led KNO[14] to show that the multiplicity distribution of the particles produced in high-energy hadron collisions is asymptotically a function of only one variable $z = n/\langle n \rangle$. This property of the multiplicity distribution has been checked by Slattery[19] against experimental data and is surprisingly well satisfied even at relatively low centre-of-mass energies ($\sqrt{s}$ from 9.8 to 24 GeV). Later the work of Thomé et al.[13] gave confirmation of this property in the ISR energy range ($\sqrt{s}$ from 23 to 63 GeV) for the total multiplicity and also for a range in η limited to the central region (they used $|\eta| < 1.5$).

The UA1 results are shown in Fig. 19 together with the Thomé et al. data[13]. It is striking to see that this KNO scaling is still satisfied and therefore holds from 50 GeV to 155 TeV lab. energy.

In the work of KNO, the multiplicity distribution $\psi(z)$ has no predicted shape and is only constrained by the normalization condition

$$\int_0^\infty \psi(z)\,dz = \int_0^\infty z\psi(z)\,dz = 1 \;.$$

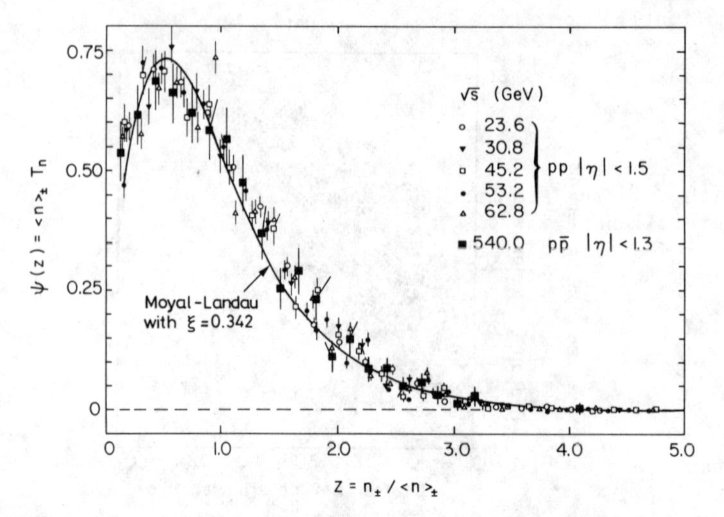

Fig. 19 The charged-particle multiplicity in the UA1 fiducial region[15]. The $\sqrt{s}$ = 540 GeV data are compared with the ISR pp data of Thomé et al.[13]. The curve is an overall fit of a Moyal-Landau shape.

The shape of $\psi(z)$ reflects the mechanism of the multiparticle production. It is interesting to note that this multiplicity distribution can be quite well described using the Moyal distribution of the number of ion pairs produced by a charged particle traversing an absorber[20]. Moyal starts from the basic idea that along its trajectory a charged particle ejects some electrons which, if they are energetic enough, eject some other electrons which in turn produce a cascade effect. He found that the distribution of the number n of particles in the cascade is given by

$$P_n = \frac{1}{A}\,\frac{1}{\sqrt{2\pi}}\,e^{-\frac{1}{2}(\omega+e^{-\omega})} \quad \text{with} \quad \omega = \frac{n - n_m}{A}\,,$$

n_m and A being two constants describing the ionization process. The main input in Moyal's calculation is the use of the classical Thomson cross-section

$$\sigma(E) \propto \frac{1}{E^2}$$

for the emission of a recoil electron with energy E during ionization. If one assumes a $1/E^2$ dependence for the quark-quark basic interaction, it is natural to expect a multiplicity distribution

$$\psi(z) = \frac{1}{I}\,e^{-\frac{1}{2}(\omega+e^{-\omega})}$$

$$\text{with} \quad \omega = \frac{z-z_0}{\xi} \quad \text{and} \quad I = \int_0^\infty e^{-\frac{1}{2}(\omega+e^{-\omega})}dz\,.$$

The normalization condition

$$\int_0^\infty z\psi(z)\ dz = 1\ .$$

gives a relation between z_0 and ξ which leaves a multiplicity distribution with only one free parameter.

An overall best fit to the data of Thomé et al. ($\sqrt{s}$ from 23 to 63 GeV, $|\eta| < 1.5$) and of the UA1 Collaboration ($\sqrt{s}$ = 540 GeV, $|\eta| <$ < 1.3) gives $\xi = 0.342 \pm 0.002$ with a χ^2/DF = 3.8 for 132 data points; the result is shown in Fig. 19.

I find it very surprising that such a simple argument gives a one-parameter formula which describes the data in the central region; this could be due to some analogy between the ionization and the high-energy hadronic collision processes.

3.5 Transverse energy

Using information from the central calorimeters, the UA1 Collaboration obtains a flat transverse energy density as a function of the pseudorapidity, see Fig. 20. This distribution, combined with the multiplicity measurement, leads to a mean E_T per particle of 0.5 GeV ±20% (the large error is mainly due to the lack of the momentum measurement at this early stage of the UA1 experiment). In Fig. 21 is shown the smooth dependence of the deposited transverse energy as a function of the track multiplicity observed in the central detector. In Fig. 22 is shown the transverse total energy distribution obtained from the UA1 central calorimeters for different η and ϕ cuts, since the experiment is symmetric in ϕ (around the beam line) and the dn/dη density is flat as a function of η; the product $\Delta\eta \times \Delta\phi$ is proportional to the number of particles. We see that the mean E_T is roughly proportional to the number of particles involved in the distribution. A detailed analysis comparing UA1 data with the ISR and NA5 experiments[21] would

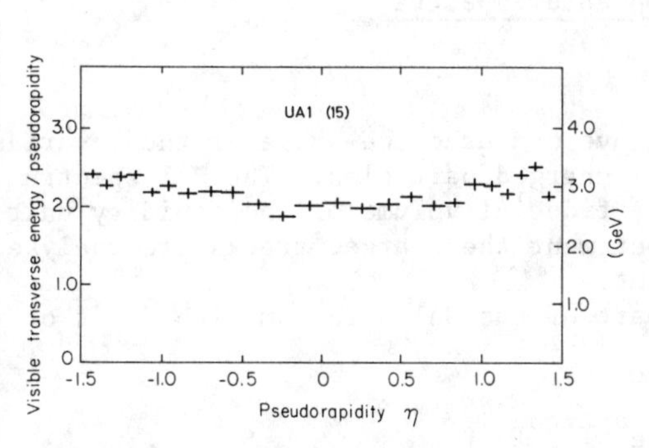

Fig. 20 Transverse energy as a function of pseudorapidity (UA1 data)

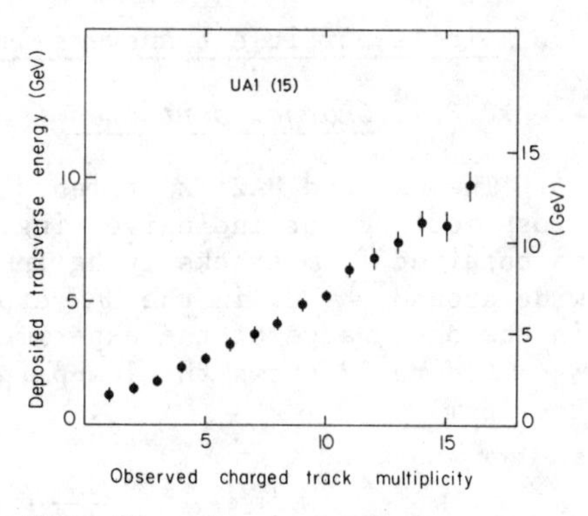

Fig. 21 Mean transverse energy as a function of the observed charged-track multiplicity in UA1[15]

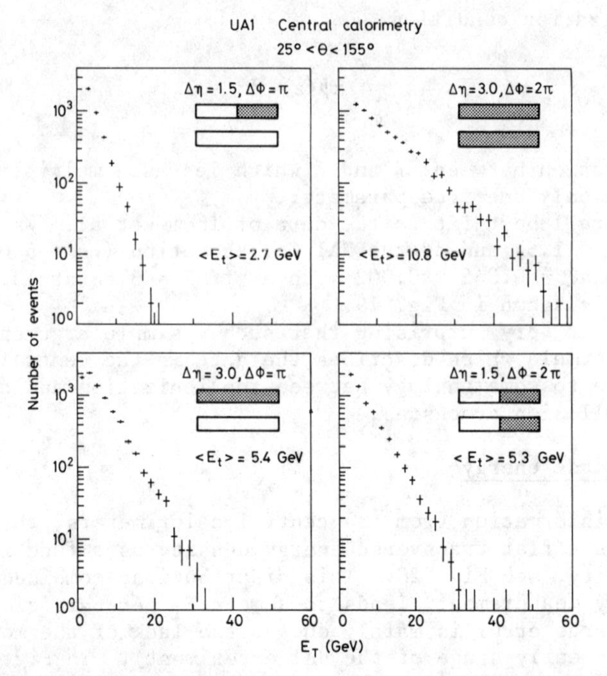

Fig. 22 Total transverse energy from the UA1 calorimeters for different η and ϕ cuts for an angular region $25^{\circ} < \theta < 155^{\circ}$ about the beam direction

be welcome in order to obtain a better understanding of this behaviour, which might not be linear at all energies.

From these observations, the main conclusion is that events with a large total transverse energy are mainly built from many particles, each with a rather small transverse energy.

3.6 Single-particle transverse momentum spectra

3.6.1 Charged particles

The UA1 and UA2 experiments have measured the shape of the invariant cross-section for inclusive single charged particles. The UA1 spectrum is obtained from tracks lying in a fiducial volume of one rapidity unit wide around y = 0; in the UA2 experiment the charged tracks are analysed in the open wedge of the experiment.

Figure 23 shows the low-p_T part of the UA1 spectrum with a fit of the form

$$\frac{dN}{dp_T^2} \propto e^{bp_T+cp_T^2}$$

up to p_T = 1 GeV/c. The fit gives b = −7.18 GeV/c^{-1} and c = = 1.82 GeV/c^{-2}, values which are in quite good agreement with the

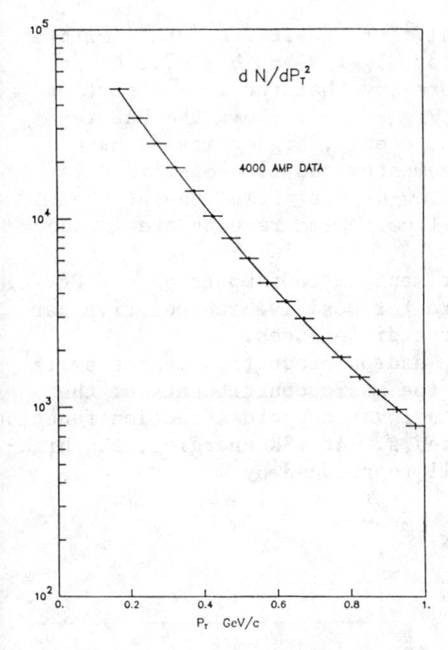

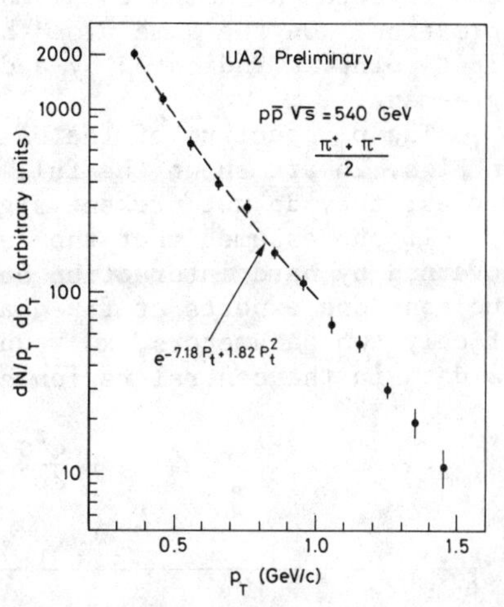

Fig. 23 Low-p_T part of the UA1 single charged-particle spectrum at 4000 A. The curve is a fit to the data $dN/dp_T^2 \propto e^{-7.18p_T+1.82p_T^2}$.

Fig. 24 UA2 single π and K spectrum. The dashed line shows the result of the UA1 fit normalized to the first point.

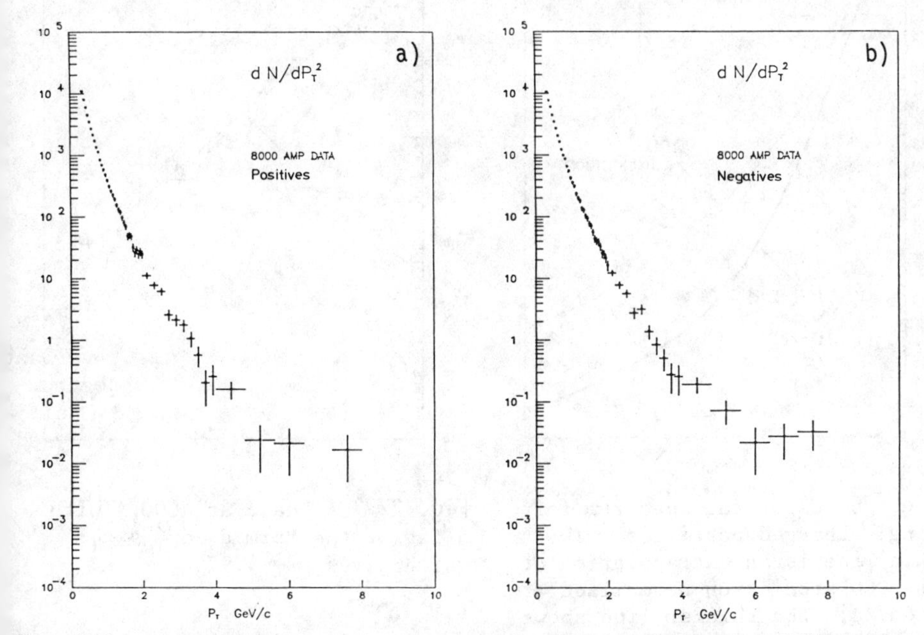

Fig. 25 UA1 single-particle spectra at 8000 A: (a) for positive tracks, (b) for negative tracks.

extensive fit performed by Alper et al.[10] to their ISR data[22] with a result, for example, at $\sqrt{s}$ = 53 GeV for π^+ or π^- of b = -7.3 GeV/c^{-1} and c = 1.1 GeV/c^{-2}. I would therefore say that the low-p_T part of the spectrum is almost ISR-like. In Fig. 24 are shown the UA2 results; here the spectrum is for π and K only, higher masses have been rejected by using the combined momentum and time-of-flight information. On the same plot the UA1 low-p_T fit normalized at the first point is indicated by a dotted line. Both results are in good agreement.

The p_T spectrum of the UA1 experiment extends up to p_T = 8 GeV/c. In Figs. 25 are shown the full spectra for positive and negative particles; they do not present significant differences.

If one assumes that the large-p_T hadron production mechanism is governed by hard interaction between the basic constituents of the nucleon, one expects at fixed angle an invariant cross-section function of only two parameters, $x_T \simeq 2p_T s^{-\frac{1}{2}}$ and s. At ISR energies, the high-p_T data in the central region are well reproduced by

$$E \frac{d^3\sigma}{dp^3} \propto (1-x_T)^m p_T^{-n} .$$

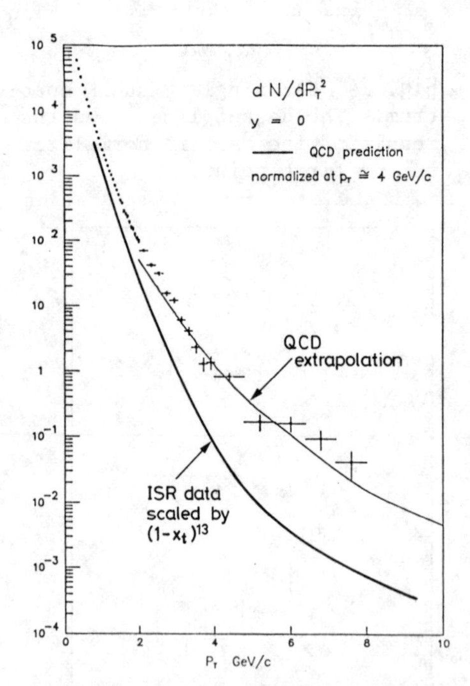

Fig. 26 UA1 total spectrum for single charged particles. The thin line is an extrapolation of the QCD prediction normalized at 4 GeV/c; the thicker line shows the ISR data scaled to 540 GeV by $(1-x_T)^{13}$.

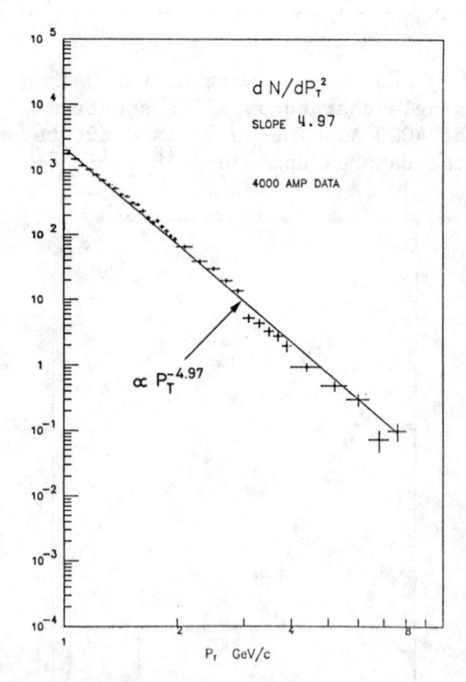

Fig. 27 UA1 data at 4000 A with a fit of the form $dN/dp_T^2 \propto p_T^{-n}$ which gives n = 4.97

In Fig. 26 is shown the full UA1 spectrum with positive and negative particles combined compared with the ISR data scaled to $\sqrt{s}$ = = 540 GeV and normalized at the point with lowest p_T; there is a significant difference at large p_T showing a non-ISR-like behaviour. On the same plot is shown a QCD extrapolation normalized at 4 GeV/c. At $\sqrt{s}$ = 540 GeV the term $(1-x_T)^m$ is nearly constant over the p_T range covered and one should have $dN/dp_T^2 \propto p_T^{-n}$, where n is expected to be four if we deal with a pure parton-parton collision. In Fig. 27 is shown the fit of the slope n, giving a value of n = 4.97 which is preliminary but indicates that we are reaching p_T values where QCD hard scattering of partons might start to show up.

3.6.2 Neutral particles

The UA2 Collaboration measured the single π^0 spectrum in the lead-glass wall facing the open wedge of the experiment. In Fig. 28 is shown the invariant $\gamma\gamma$ mass spectrum they obtained where the π^0's clearly show up, and in Fig. 29 the corresponding invariant single π^0 cross-section, which also shows clearly a "non-ISR" behaviour. Note that the lack of η's in the $\gamma\gamma$ mass spectrum is due to the acceptance and does not allow one to conclude that there is a change in the large η^0/π^0 ratio found at the ISR.

At much larger p_T, the UA1 Collaboration obtained also a neutral spectrum which is shown on the same graph. This spectrum includes π^0's, η^0's and single γ's, which are not resolved, detected in the end-cap electromagnetic calorimeters, covering the rapidity range $1.6 < y < 2.5$. This spectrum is preliminary because of the calorimeter calibration which has not yet been completed at this time, thus giving an uncertainty in the E_T scale of about 30%.

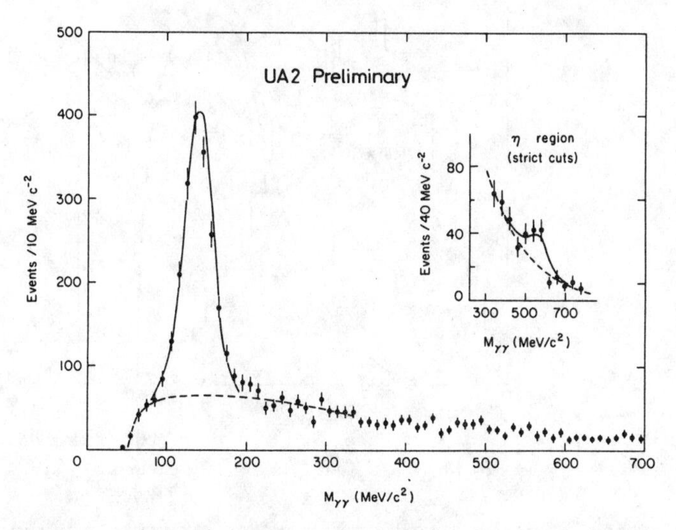

Fig. 28 Two-photon invariant $\gamma\gamma$ mass obtained by UA2 in the lead-glass calorimeter.

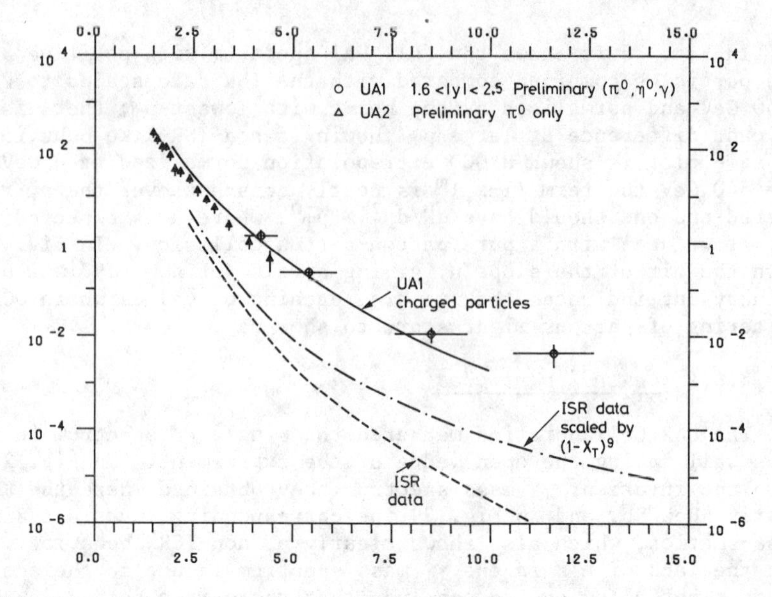

Fig. 29 UA1 and UA2 single neutral-particle spectrum, the full line represents the UA1 charged-particle spectra; the dotted line, the π^0 ISR data; and the dashed-dotted line, the ISR data scaled to 540 GeV by $(1-x_T)^9$.

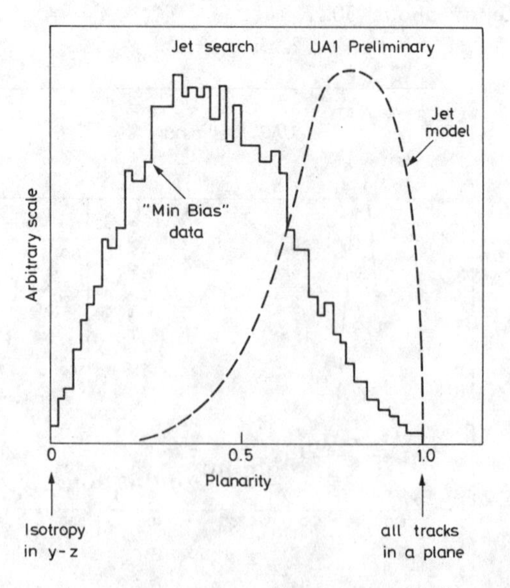

Fig. 30 Planarity of events in UA1 "minimum bias" data; a preliminary search for jets.

3.7 Jet search

I will briefly mention a preliminary search performed in a small subsample of the UA1 data to investigate if jets are produced in a copious way. The planarity P is defined as the maximum value obtained by rotation in a plane (y,z) perpendicular to the beam axis:

$$P = \max\left(\frac{a - b}{a + b}\right) \quad \text{with} \quad a = \sum P_y^2,\ b = \sum P_z^2 \ .$$

A planarity of 0 means an event fully isotropic, and P = 1 is obtained if all the tracks are in a place containing the beam axis. Figure 30 shows the data compared with the prediction of a 4-jet model; clearly jets are not produced in large numbers; a much more refined search is under way.

3.8 Centauros?

Since 1972, in cosmic-ray experiments[23], events have been found at very high energy (200 TeV) in which a large number (60-70) of charged particles are produced, but no γ-rays or electrons. These reactions are interpreted as the production of baryons (or antibaryons) without any mesons.

In the UA1 data a search is under way for such strange events using both the central track detector and the information from the surrounding calorimeters; no events have yet been reported.

3.9 High-p_T electrons

A search has been performed among all the UA1 data to extract electrons or positrons with a transverse energy greater than 10 GeV; among 4×10^5 triggers (including the high total E_T triggers) no candidates survive a reasonable set of cuts. This result, compatible with the expected production of 0.1 W, gives good confidence that an unexpectedly large background level of large-p_T electrons or positrons will not put the forthcoming W search into danger.

4. CONCLUSIONS

A lot of new information emerges already from these highly energetic $p\bar{p}$ collisions, studied by UA1, UA2, and UA4, even though they are based on a rather low integrated luminosity and many of them are of a preliminary nature.

i) The elastic scattering slope extrapolates from lower energy data and is consistent with a logarithmic shrinkage of the hadron diffraction cone.

ii) The multiplicity density $dn/d\eta$ also extrapolates well from ISR data and grows like log (s).

iii) The multiplicity distribution shows a non-random behaviour indicating correlations between the produced particles and, as in the ISR data, obey the KNO scaling in the central region.

iv) The total transverse energy density $dE_T/d\eta$ is almost flat as a function of η and indicates a mean E_T per particle of 0.5 GeV (±20%).

v) The deposited transverse energy grows with the multiplicity of charged tracks. Large total E_T events are mainly built of a large number of particles, each of them with a low-p_T value.

vi) The single particle invariant cross-section for charged particles below 1 GeV shows the same behaviour as ISR data. At large p_T this cross-section does not extrapolate from the ISR data with a simple $(1-x_T)^m$ scaling and it behaves above 1 GeV according to a power law $dN/dp_T^2 \propto p_T^{-5}$, indicating that if the parton-parton basic interaction is not yet reached we are probably approaching this regime.

vii) The single neutral particle spectrum also does not extrapolate from ISR energies with a simple $(1-x_T)^m$ scaling.

viii) Free quark production compared to charge-1 particles is less than 2×10^{-4}.

I would like to conclude by emphasizing that in my opinion the most striking result is that the great majority of events with a large total transverse energy are built of many low-p_T particles and not of jets. This probably indicates a large contribution of gluon radiation. More will be known soon about the $p\bar{p}$ reactions because the analysis of the 1981 run is still in progress and giving results more and more accurate and statistically significant.

ACKNOWLEDGEMENTS

I would like to thank my colleagues from the UA1 Collaboration for asking me to represent them at this conference and also for many fruitful discussions.

REFERENCES

1. C. Rubbia, P. McIntyre and D. Cline, Proc. Int. Neutrino Conf., Aachen, 1976 (eds. H. Faissner, H. Reithler and P. Zerwas) (Vieweg, Braunschweig, 1977), p. 683.
2. A. Astbury et al., A 4π solid angle detector for the SPS used as a proton-antiproton collider at a centre-of-mass energy of 540 GeV, CERN/SPSC/78-06 (1978).
3. M. Calvetti et al., The UA1 central detector, preprint CERN EP/82-44 (1982), presented at the Int. Conf. on Instrumentation for Colliding Beam Physics, SLAC, 1982.
4. B. Aubert et al., Nucl. Instrum. Methods 176, 195 (1980).
5. M.J. Gorden et al., Physica Scripta 25, 11 (1982).
6. M. Banner et al., Proposal to study antiproton-proton interactions at 540 GeV c.m. energy, CERN/SPSC/78-8 and CERN/SPSC/78-54 (1978).
7. P.A. Dorsaz, Preliminary results from the UA2 experiment at the SPS $p\bar{p}$ collider, to appear in Proc. 17th Rencontre de Moriond, Les Arcs, March 1982.
8. M. Battiston et al., The measurement of elastic scattering and of the total cross-section at the CERN $p\bar{p}$ collider, CERN/SPSC/78-105 (1978) and CERN/SPSC/79-10 (1979).
9. M. Bozzo, First results from the UA4 experiment at the CERN collider, to appear in Proc. 17th Rencontre de Moriond, Les Arcs, March 1982.
10. G. Barbiellini et al., Phys. Lett. 39B, 663 (1972).
11. G. Carboni et al., Phys. Lett. 108B, 145 (1982).
12. J.P. Burq et al., Phys. Lett. 109B, 124 (1982.
13. W. Thomé et al., Nucl. Phys. B129, 365 (1977).
14. Z. Koba, H. B. Nielsen and P. Olesen, Nucl. Phys. B40, 317 (1972).
15. G. Arnison et al., UA1 Collaboration, Phys. Lett. 107B, 320 (1981).
16. K. Alpgård et al., UA5 Collaboration, Phys. Lett. 107B, 310 and 315 (1981).
17. S. Tasaka et al., Proc. 17th Int. Cosmic-Ray Conf., Paris, 1981 (CEA, Saclay, 1981), Vol. 5, p. 126.
Y. Sato et al., J. Phys. Soc. Japan 41, 1821 (1976).
18. R.P. Feynman, Phys. Rev. Lett. 23 1414 (1969).
19. P. Slattery, Phys. Rev. Lett. 29, 1624 (1972).
20. J.E. Moyal, Philos. Mag. 46, 263 (1955).
21. K. Pretzl, paper presented at the Topical Conference on Forward Collider Physics, Madison, December 1981.
22. B. Alper et al, Nucl. Phys. B100, 237 (1975).
23. C.G.M. Lattes, Y. Fujimoto and S. Hasegawa, Phys. Rep. 65, 151 (1980).

FURTHER RESULTS FROM THE UA5 COLLABORATION AT THE SPS COLLIDER

Presented by H. L. Mulkens
Bonn-Brussels-Cambridge-CERN-Stockholm
Free University of Brussels, Brussels, Belgium

ABSTRACT

A summary of physics results obtained so far by the UA5 Collaboration from the analysis of the first $\bar{p}$-p events recorded at $\sqrt{s}$ = 540 GeV at the CERN SPS Collider is presented.

This includes pseudo-rapidity distribution, multiplicity distribution and pseudo-rapidity correlations for charged particles; pseudo-rapidity distribution and multiplicity distribution for photons; rapidity distribution and multiplicity distributions for strange particles in the Central Region and search for Centauro-like phenomena.

1. THE EXPERIMENTAL APPARATUS

The UA5 detector (Ref. 1) is installed in the LSS4 underground area of the CERN SPS accelerator which was modified during the year 1980 to be able to work in Collider mode.

The UA5 experiment is a visual technique experiment (Fig. 1) based on the use of two large Streamer Chambers, each of 6 x 1.25 x .50 cubic meters and situated 4.5 cm above and below an elliptical stainless steel beam pipe of 0.4 mm. thickness. The lower chamber was

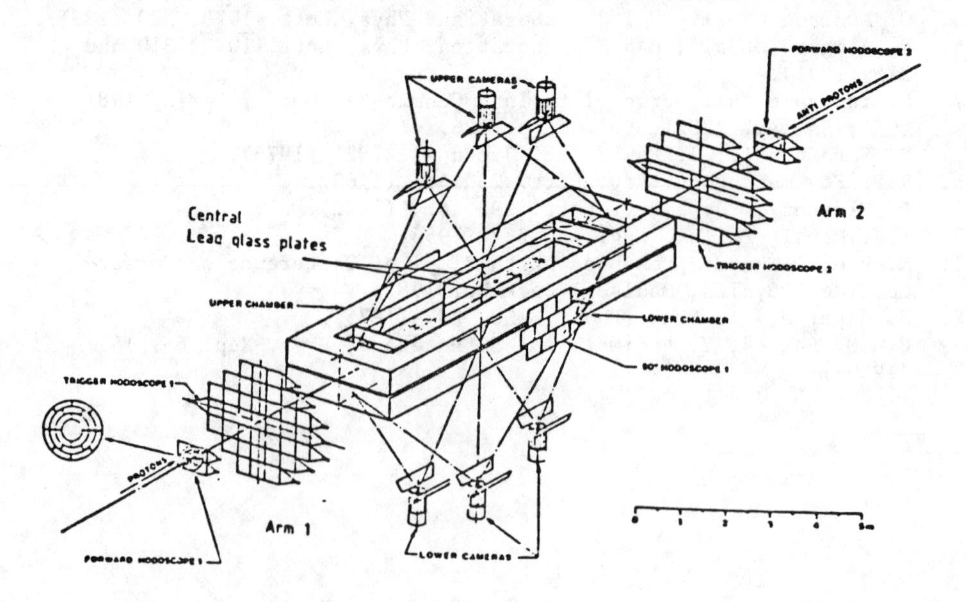

equipped with forward and lateral lead glass conversion plates of one radiation length thickness for photon identification. The upper chamber had only one such forward plate. Two sets of three stereo camera were used to photograph the entire volume of the chambers. Pictures recorded showed very good track quality over the entire length of the chambers and very good two-tracks resolution down to 2 mm of separation.

There is no magnetic field, so all charged particles are seen as straight tracks spreading out radially from the unseen interaction vertex, and the only quantities we could measure were the track angles from which we compute the pseudo-rapidity $\eta = -\ln\tan\theta/2$. Tracks were recorded in the acceptance region of the chambers of $|\eta| \leq 5$.

The trigger consisted of six planes of scintillation counters constituting four trigger "arms", two forward arms covering $2.1 < \eta < 5.7$ and two lateral arms covering $|\eta| < 0.5$.

2. DATA TAKING

A normal minimum bias trigger would require one hit in any one arm and would be sensitive to about 95% of the total inelastic cross section. However during the first Collider runs in October-November 1981, to which the data reported here refers, the achieved luminosity was so poor ($L \cong 10^{25}$ cm^{-2} sec^{-1}) that we had to restrict the trigger to a two forward arms coincidence, thus excluding all single diffraction events; we had even to increase to two the hit multiplicity in the photon arm to further reduce background events.

A summary of data taking during that period is given in Table 1.

TABLE 1

Run	Number of Pictures	Trigger (required multiplicity)	Fraction of good $\bar{p}$-p events	Number of $\bar{p}$-p interactions	Remarks
7-9 Oct	1500	Varia	--	--	Single p beam
21-22 Oct	20000	arm1(1)*arm2(1)	16%	3200	
6-11 Nov	60000	arm1(2)*arm1(1)	33%	13000	

3. DATA ANALYSIS

To date 2800 events and a special sample of 480 high multiplicity events (with more than 40 charged primaries) have been completely measured and analysed. Data processing is done in the following way:

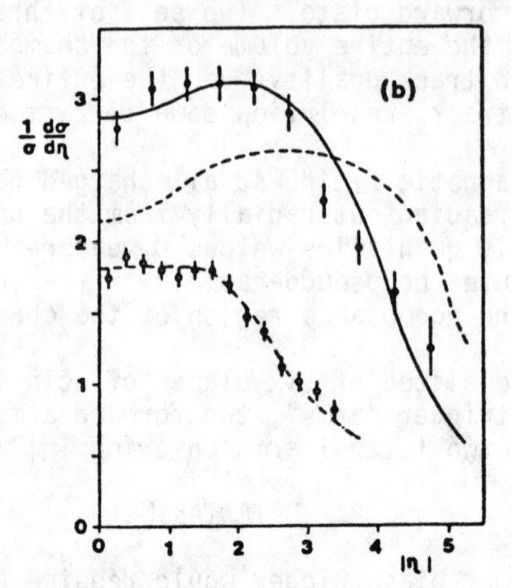

Figure 2. Data from 340 events at $\sqrt{s}$ = 540 GeV compared with earlier UA5 data obtained at the ISR at $\sqrt{s}$ = 53 GeV (open points). The dashed and dot-dashed curves are predicted from p_t limited phase-space for $\langle p_t \rangle$ = 350 MeV/c. The solid curve for $\sqrt{s}$ = 540 GeV is from the same model but for $\langle p_t \rangle$ = 500 MeV/c.

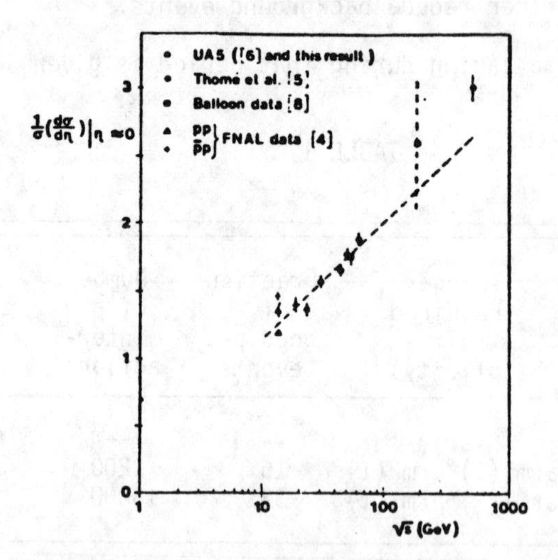

Figure 3. Charged particle density in pseudorapidity near η = 0 from various experiments up to Collider energy. Our Collider value is for single diffractive events excluded.

films are scanned for "good" events (on the basis of vertex position, topology, asymmetry); those are then measured and processed through a vertex finding program to determine the primary interaction vertex coordinates and associate all relevant primary tracks to it. Events are eventually selected if they have at least two primary tracks and if their vertex is well inside the beam pipe. All data were then subsequently corrected with the help of a Monte-Carlo program to take account of the various losses and inefficiencies.

4. PHYSICS RESULTS

The different samples corresponding to the topics covered in this report are described in Table 2.

TABLE 2

Topics	Event Sample
Charged Hadrons Production	340 Events
Charged Hadrons Correlations	1520 Events
Photon Production	1860 Events
Strange Particles Production	2100 Events

4a. CHARGED HADRONS PRODUCTION

Pseudo-rapidity distribution (Ref. 2): From Fig. 2 we conclude that η is flat for $|\eta| < 3$ and that the height of the plateau is

$$1/\sigma(d\sigma/d\eta)\Big|_{|\eta|<3} = 3.0 \pm 0.1$$

which confirms (Fig. 3) the logarithmic rise already observed at the ISR (provided one corrects for untriggered diffractive events).

Multiplicity distribution (Ref. 3): $\langle n_{ch} \rangle$ is found to be 27.4 ± 2.0 ruling out the $s^{1/4}$ law and in good agreement with an extrapolation of the ISR fit (Ref. 4). The Koba-Nielssen-Olesen scaling and the Wroblewski relation are still found to hold at those energies.

Left-Right Correlations (Ref. 5): Looking at the mean multiplicity of particles emitted in one hemisphere (which could be either the p or the $\bar{p}$ hemisphere - we checked that this made no difference) as a function of the number of particles emitted in the other, we observe a linear dependence whose slope b is the correlation strength between both hemispheres. Correlations are usually separated in short range ($|\eta| < 1$) whose dynamics is dominated by well known resonance production and long range ($4 > |\eta| > 1$) whose dynamics is still poorly understood. At $\sqrt{s}$ = 540 GeV, we observe

$$b_{|\eta|<1} = 0.52 \pm 0.03 \qquad \text{and} \qquad b_{|\eta|>1} = 0.42 \pm 0.03$$

(Fig. 4) in good agreement with the extrapolation of the ISR fit (Ref. 6). Increasing the pseudo-rapidity gap between both hemispheres leads to a reduction of this long range correlation and ends up with the expected perfect independence of the true fragmentation regions (Fig. 5).

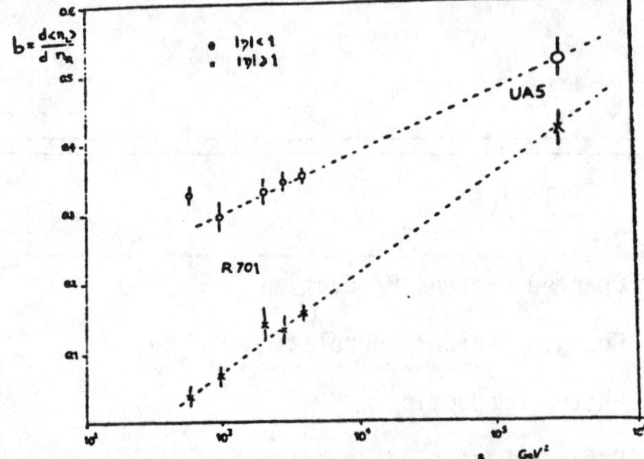

Fig. 4. Center-of-mass energy dependence of Short and Long Range correlation parameter, from ISR to SPS Collider.

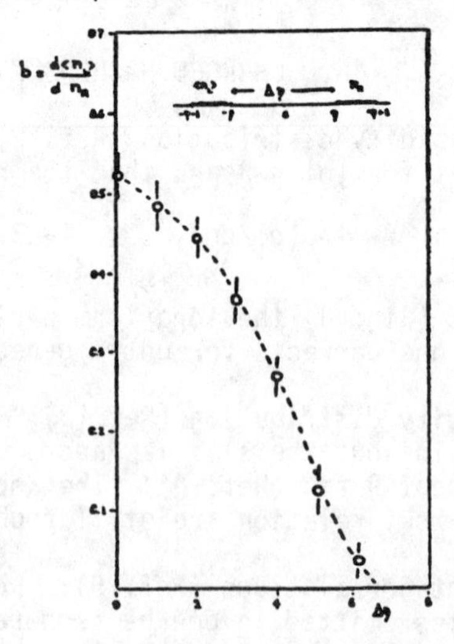

Fig. 5. Decrease of long range correlation strength as the pseudo-rapidity gap between both hemispheres is increased.

4b. PHOTON PRODUCTION

Photons converted in two different parts of the apparatus were used for this analysis: visible conversions in the lateral lead glass plates of the lower Streamer Chamber allowed the study of photons of $|\eta| < 1$, whereas invisible conversions in the beam pipe walls leading to showers of close-in tracks observed in both Streamer Chambers allowed the study of photons up to $|\eta| < 5$. Those two samples had different systematics and needed different analysis procedures.

Pseudo-rapidity distribution: The resulting corrected pseudo-rapidity distribution $1/\sigma(d\sigma/d\eta)$ for photons is shown in Figure 6,

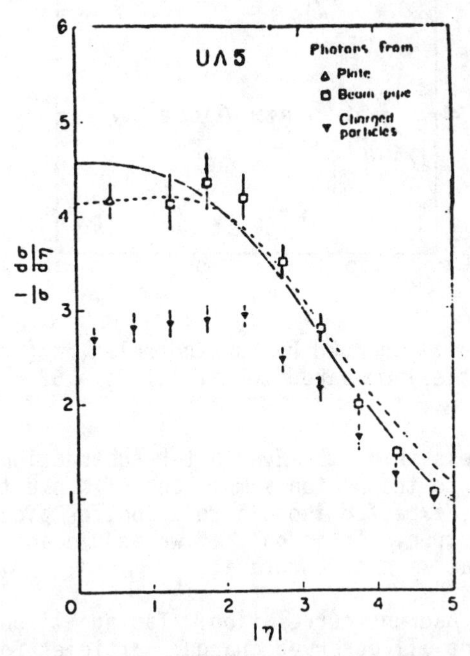

Fig. 6. Pseudo-rapidity distribution of photons (with K_S^0 contribution removed) and charged particles (with $K,p,\bar{p}$, contributions removed). Dotted line is Monte-Carlo prediction if all photons come from η decay. Solid curve is Monte-Carlo prediction assuming isospin relation for pions and remaining photons due to η decay.

where we check that we have at first good agreement between both photon samples. Integrating this distribution for $|\eta| < 5$ we derive an average photon multiplicity $\langle n_\gamma \rangle = 34 \pm 2$, whereas for the same sample the average charged particle multiplicity in that pseudo-rapidity range is $\langle n_{ch} \rangle = 26.5 \pm 1$, the excess of photons being concentrated at

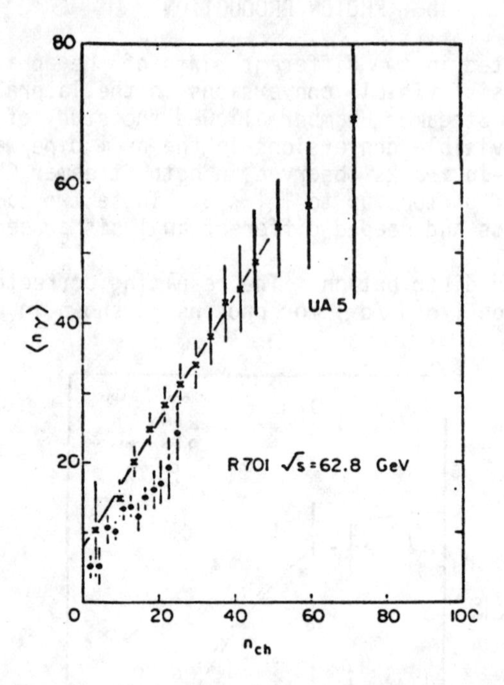

Fig. 7. Photon to Charged hadrons correlation (true multiplicities) compared to data at $\sqrt{s}$ = 62.8 GeV.

pseudo-rapidities less than 2. Even after subtraction of the contribution of K°s to the photon sample and of K,p,$\bar{p}$ to the charged particle sample, the expected isospin relation for pions $<\eta_{\eta^0}> = 1/2$ $<\eta_{\pi^0}>$ does not hold true. It is only if we assume an η/π^0 ratio of the order of 30% that we can restore it.

Photon-Charged hadrons correlation: Taking all observed photons for $1.5 < |\eta| < 4$ and all observed charged particles for $|\eta| < 4$, we deduce after detection and acceptance corrections the following linear dependence between photons and charged particles multiplicities (Fig. 7) $<n_\gamma> = (8.0 \pm 3.0) + (0.90 \pm 0.08)\, n_{ch}$. The energy dependence of the slope from $\sqrt{s}$ = 3 to 540 GeV is shown in Fig. 8 (Ref. 7).

Centauro search: from the characteristics of the five reported Centauro events found by the Brazil-Japan collaboration (Ref. 8) we estimate that in order to properly take account of the different acceptance and detection efficiencies, we should look for events in our data sample having more than 43 charged particles with $\eta \geq 2$ in one of the two hemispheres and look for the associated number of photons with $\eta \geq 2$ in that hemisphere. In 3600 half events of

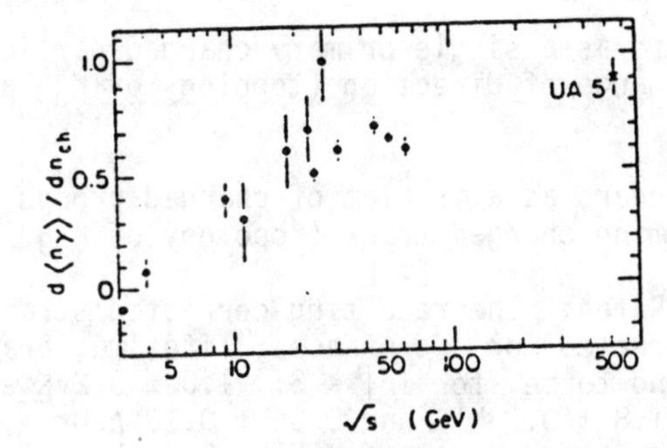

Fig. 8. Strength of Photon to Charged hadrons correlation as a function of centre-of-mass energy.

equivalent $E_{lab} \cong 150$ TeV we see no events with that charged multiplicity (Fig. 9).

4c. STRANGE PARTICLE PRODUCTION

We searched in the visible volume of both Streamer Chambers for topologies compatible with decays in flight of primary strange particles produced in the range $|\eta| < 3$.

- V° appears as two charged particles forming a vee without charged incoming (topology for K°s, K°_L, Λ, $\bar{\Lambda}$ and background from photons).

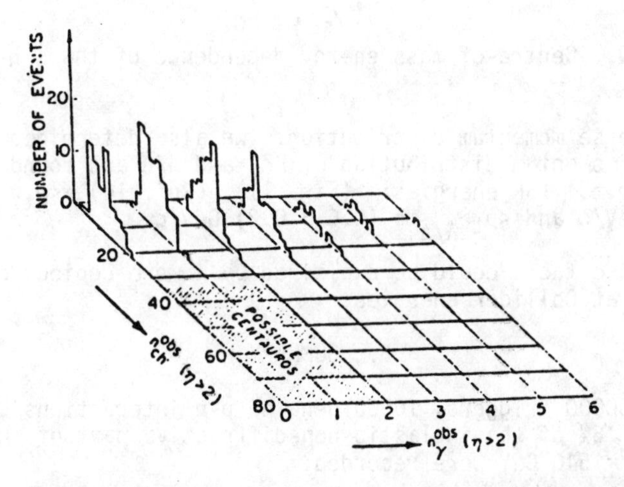

Fig. 9. Number of events against charged particle multiplicities at pseudo-rapidity > 2, in slices of increasing photon multiplicity in the same pseudo-rapidity range, showing that we have no evidence for a Centauro phenomenon at $\sqrt{s}$ = 540 GeV.

• Kink appears as a single primary charged particle suffering a sudden change of direction (topology of $K_{\mu 2}$ and background from $\pi^{\pm}$).

• Trident appears as a triplet of charged prongs connected to an incoming charged track (topology of $K_{\mu 3}$.

Production Rates: The resulting corrected strange particle production rates (corrected for acceptances, lifetime, branching ratios, etc..) were found to be, for $|\eta| < 3$: 1.0 ± 0.2 K°_{s} per event, 1.0 ± 0.3 K°_{L}, 1.8 ± 0.2 K^{+-} and 0.35 ± 0.10 Λ or $\bar{\Lambda}$, leading to a ratio $(K^{\circ} + \bar{K}^{\circ})/\pi^{ch}|_{|\eta|<3} = (11 \pm 2)\%$ higher than at ISR energies (Fig. 10)(Ref. 9). The pseudo-rapidity (actually true rapidity for K°_{s} and $\Lambda/\bar{\Lambda}$) distribution for the strange particles were found essentially flat for $|\eta| < 3$.

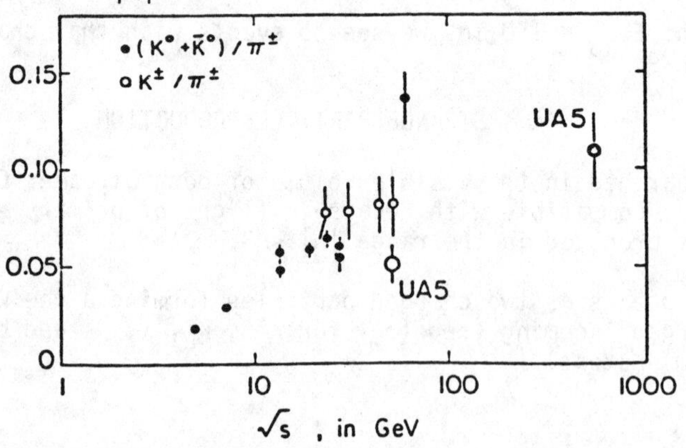

Fig. 10. Centre-of-mass energy dependence of the K/η ratio.

Transverse momentum distribution: we also determined the transverse momentum distribution of K°_{s} and $\Lambda/\bar{\Lambda}$ and found an increase of $<p_t>$ w.r.t. ISR energies: (Fig. 11), (Ref. 10) $<p_t>_{K^{\circ}_{s}} = (0.70 \pm 0.12)$ GeV/c and $<p_t>_{\Lambda/\bar{\Lambda}} = (0.6 \pm 0.2)$ GeV/c.

All those facts could be explained by a more copious charm production at Collider energies.

5. SUMMARY

Among 80000 triggers, 16000 genuine $\bar{p}$-p interactions corresponding to 93 ± 5% of the inelastic non-diffractive part of the cross section $\sqrt{s}$ = 540 GeV were recorded.

From the analysis of more than 3000 events, we confirm the logarithmic rise of the central charged particle density with energy, the validity of KNO scaling and the increase in charged particle multiplicity as extrapolated from ISR energies. Long range correlations between charged particles are at Collider energies three times more

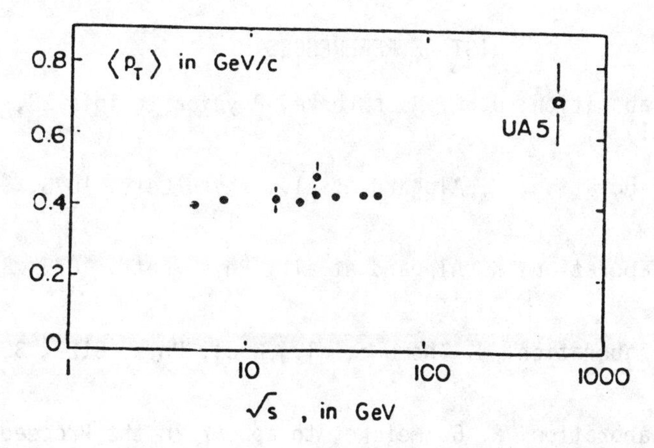

Fig. 11. Centre-of-mass energy dependence of K mean transverse momentum.

important than at ISR energies and become of the same order as the short range ones. The average photon multiplicity is higher than expected and could be explained by a sensible η production. The strength of the correlation between the production of photons and the producion of charged hadrons has increased from ISR energies. We have no evidence for a Centauro phenomenon at Collider energies (which are still an order of magnitude below the mean estimated energy of the observed Centauro's in Cosmic Rays). The mean transverse momentum of K_S^o and the K/π ratio are both higher than at ISR energies and could be due to a more copious charm particle production.

6. CONCLUSIONS AND FUTURE OUTLOOK

The UA5 experiment successfully took data during the early CERN SPS Collider Machine Development periods. First results were soon reported and revealed the essential features of $\bar{p}$-p inelastic interactions at the new record energy of $\sqrt{s}$ = 540 GeV.

A new data taking period is scheduled for the end of September 1982. The UA5 apparatus described here will include two new detectors, the HMT (High Multiplicity Trigger) consisting of a set of proportional tubes fixed around the beam pipe in the interaction region that will enable us to trigger on selected high multiplicity events and the NHD (Neutral Hadron Detector) being a Lead-Scintillator electromagnetic and Iron-Scintillator hadronic calorimeter optimised for the detection of low energy neutrons and antineutrons with good photon rejection capability.

LIST OF REFERENCES

1. UA5 Collaboration, J. G. Rushbrooke, Physica Scripta 23, 642 (1981).

2. UA5 Collaboration, K. Alpgard et al., Phys. Lett. 107B, 310 (1981).

3. UA5 Collaboration, K. Alpgard et al., Phys. Lett. 107B, 315 (1981).

4. R701 Collaboration, W. Thome et al., Nucl. Phys. B129, 365 (1977).

5. UA5 Collaboration, R. B. Meinke, to appear in the Proceedings of the 2nd Topical Conf. on Forward Collider Physics, Madison, USA (1981).

6. R701 Collaboration, S. Uhlig et al., Nucl. Phys. B132, 15 (1978).

7. J. Whitmore, Phys. Rep. 10C, 273 (1974) and Phys. Rep. 27C, 187 (1976).

8. C.M.G. Lattes, Y. Fujimoto, S. Hasegawa, Phys. Rep. 65C, 151 (1980).

9. H. Kikuchi et al., Phys. Rev. D20, 37 (1979), R. K. Kass et al., Phys. Rev. D20, 605 (1979), D. Drijard et al., Z. fur Phys. C12, 217 (1972), and W. Thome et al., Nucl. Phys. B129, 365 (1977).

10. V. Blobel et al., Nucl. Phys. B69, 455 (1973), M. Alston-Garnjost, et al., Phys. Rev. Lett. 35, 142 (1975), J. W. Chapman et al., Phys. Lett. 47B, 465 (1973), K. Jaeger et al., Phys. Rev. D11, 2405 (1975), A. Sheng et al., Phys. Rev. D11, 1733 (1975), and A. M. Rossi et al., Nucl. Phys. B84, 269 (1975).

PROBLEMS, PUZZLES AND PROSPECTS: A PERSONAL PERSPECTIVE ON PRESENT PARTICLE PHYSICS

H. David Politzer *
California Institute of Technology
Pasadena, California 91125

The organizers of this meeting on novel results asked that I make some closing remarks which would put in perspective some of the issues we discussed. Perspective depends, of course, on where you stand, so what follows will be of a highly idiosyncratic nature.

A few years ago, we solved everything. The "standard model" ($SU(3)_{color} \otimes SU(2)_L \otimes U(L)_Y$ accounted for all known and virtually all doable particle physics. It could be "unified" elegantly into a single compact theory,[1] and this unification predicts the Weinberg angle to uncanny accuracy [2] as well as accounting for the cosmological baryon excess.[3] (I register in passing my own deep suspicion of all cosmological, as opposed to astronomical, "facts." Historically, many have been based on extremely tenuous arguments rather than on observed data.) Ideas of unification also stimulated a new generation of proton decay and neutrino mass searches.

The confidence of the theory community was echoed in the April 1980 inaugural lecture of the new Lucasian Professor of Mathematics at Cambridge, Steven Hawking. (Mind you that this chair comes with no mean legacy of theoretical tradition.) Hawking's lecture was entitled "Is the end in sight for theoretical physics?" and his answer was yes.[4] Admittedly, there was some work yet to be done on gravity, but supersymmetrars were hot on the trail.

This is very sad. The end of theoretical physics means the end of all basic physics.

In the interim, this point of view might well have been refuted by experiment. But to date, there exists no outstanding conflict between the standard model and what we know. Rather, more evidence has accumulated: jets, scaling, μ-pairs, hadrons at large $p_\perp$, τ's, and bottom. (Even surprises such as D-decay rates in the end strengthened rather than weakened our faith.[5])

But give theorists enough rope, and they'll hang themselves.

As people contemplated the prospect that the standard model was actually part of a genuinely fundamental theory of the universe, they realized that it is absolutely unsatisfactory on many scores and necessarily incomplete. Depending on personal preference as to which of the model's problems seems most pressing, people go off on different tracks. But not one of these fundamental questions has been resolved to general satisfaction.

* Work supported in part by the U.S. Department of Energy under Contract No. DE-AC03-81ER40050.

Confusion reigns. Ask individual theorists "What are the most important theoretical or experimental issues?" And you will get a variety of contradictory answers. Here are some examples:

1) Are the gauge interactions that we believe we have identified necessarily fundamental[6] or are they obviously just an effective, low energy phenomenology, whose underlying dynamics could be just about anything?[7] (Such divergent views are held by the most respected of theorists.)

2) Is the supergap (the splitting, assuming supersymmetry, between the particles we know and their superpartners), necessarily a few GeV, TeV, 10^{19}GeV? Each of these possibilities is espoused by various superpeople as being necessary for supersymmetry to make any sense at all.[2]

3) Is the smallness of the observed astrophysical cosmological constant relative to the Planck or hadronic scales _the_ most important problem facing basic physics or is it simply a non-problem? The latter could be the case if the quantum cosmological constant were truly enormous,[8] but the parameter we observe in the Robertson-Walker metric is actually something else, e.g., the result of an averaging that we don't know how to perform because we don't understand quantum gravity.

I wish to share my confusion on a few topics to make it clear the the situation is indeed confused. But rather than attempt to straighten it all out, I will later try to address the question of what we might practically hope to do in the near future.

I. Gravity:

In the past few years, many theorists have seriously expressed the opinion that gravity is "the only thing left to work on." There are two essential problems posed by gravity: 1) It has no apparent relation to the other fundamental forces. Even if the standard model gets unified into a single, simple group, gravity remains an additional force of rather different structure. And 2) quantum gravity makes no sense --- or rather we have not understood how to make sense out of it. The symptom of this problem is non-renormalizability.

Supergravity answers problem 1).

In fact, supergravity doesn't solve problem 1), but instead relates gravity to yet other forces which, while definitely not the forces we know from particle physics, might be responsible for our familiar forces. Maybe. Regarding 2), even the most optimistic superenthusiasts expect that all known supergravities are non-renormalizable in perturbation theory[9]. Non-renormalizability means that our understanding is incomplete. Either non-perturbative effects or some new interactions must serve to give an ultraviolet cutoff. In either case, our present theory can only be a crude phenomenology, which leaves open the question of which of its properties we should take seriously. Needless to say, no one has suggested a way in which experiments (which occasionally have guided even our wisest sages), might shed light on some of these issues.

II. Unification:

The standard ideas on unification, e.g., the SU(5) model, present us with the hierarchy problem: there are two vastly different fundamental scales present in a single system. Some attempted solutions try to make the existence of the two variant scales natural or at least stable to small variations of parameters[10] (although neither of these criteria can be particularly well-defined). Another method of solution is to populate the enormous scale gap with many intermediate scales.[11] Of course, the problem exists only to the extent that one believes in both scales. What are these two scales? The lower one includes all the particles of the standard model. Few people are embarrassed that the relevant masses range over 0 for photons, a few eV for neutrinos (perhaps?), 5-10 MeV for u and d, 140 MeV for pions, 1 GeV for protons, 5 GeV for b quarks, $\sim$100 GeV for W's. The higher scale of $\sim 10^{16}$ GeV exists mostly on paper.

III. Substructure:

Noting that matter is made of molecules are made of atoms are made of nuclei are made of protons are made of quarks, we need no great imagination to ask what's next. But there exists as yet no compelling evidence for nor compelling theoretical success using substructure. There are actually several obstacles confronting any model of quark substructure.[12] Furthermore, virtually no substructure models seriously address the property of the standard model that cries out for substructure, i.e., the fermion spectrum.

To get onto a more positive track, I wish to ask:

When is the soonest something dramatic must turn up in experiment?

I believe the answer to this lies within the weak interactions. Are there W's and a Z^o exactly as predicted by the standard model? Both minor and major variants are definitely possible, e.g., more Z's or a confining SU(2) which gives strongly interacting W's and Z's.[13] If the standard model comes out successful, Z decay will reveal virtually all particles of mass less than $M_Z/2$. And as long as the weak standard model survives, precision measurement of its parameters is essential; from radiative corrections we can learn indirectly about yet much heavier particles.

An outstanding example is $M_W/M_Z \cos\theta_W$ and its deviation from 1, currently known to be accurate to a few percent. The value 1 reflects that before Z-γ mixing, the unmixed W^+, W^- and W^o form an isotriplet of what is often referred to as "custodial" SU(2). I would like to emphasize that custodial SU(2) is the weak interactions:[13] the physical fermions are doublets of the global SU(2). They are all singlets under the gauge $SU(2)_L$.[14] Transitions between them are affected by W emission or absorption, which is allowed because the W's are isospin 1. This global SU(2) is in some ways one of the deepest mysteries of known particle physics because as a symmetry it is only approximate. It is broken by electromagnetism and fermion doublet mass splittings and the non-existence of right-handed neutrinos. While exact symmetries are symmetries, approximate symmetries usually arise from underlying physics.

The standard model predicts Higgs bosons. If they are light, we will see them. If they are heavy, as M_H^2 increases, so must the Higgs self-coupling increase to keep $\langle\phi_H\rangle$ fixed at ∿300 GeV, which we know from M_W/e. Higgs bosons are required for the renormalizability and unitarity of the standard model. The Higgs particles may be much heavier than the W's, but then unitarity requires that W-W scattering be modified near 300 GeV, if not by Higgs exchange, then by the W's themselves becoming strongly interacting.[15] While this energy is post LEP, it is not astronomical and represents a scale by which something dramatic must happen. The search for Higgs particles may turn up other scalars in addition or instead, e.g., technipions or inos.

Let us now consider a different question:

When is the soonest that something dramatic <u>might</u> happen?

The answer here is clearly tomorrow. The answer might even be yesterday, with the serious possibilities of magnetic monopoles,[16] fractional charge,[17] or both.(!)

Existing or future experiments could at any time see evidence of new particles and/or new interactions, substructure, technipions, or inos. (The latter reminds me that one achievement of QCD is that we could fairly certainly identify gluinos[18] or, for that matter, any new colored object.) Rare decays (i.e., things that shouldn't happen but do or should happen but don't, typically involving neutral weak currents), should certainly be pursued further because they provide some of our sharpest constraints on speculations of new physics.

In fact, I think it is safe to say that anything that can be measured better or over a wider range must be useful in delineating alternatives.

Of course, there exist classes of experiments that are hard to interpret (usually those involving genuinely strong interactions). In such cases, unless the results show a dramatic break from earlier behavior, their significance is unclear. Yet some of these strong interaction phenomena are so conceptually simple that we must someday understand them. Examples are hadron-hadron total cross sections and hadron polarization effects. A few years ago I would have added hadron elastic scattering to this list, but today elastic processes are of clear and immediate theoretical interest because we now have some fairly explicit QCD theoretical expectations.[19]

Another question I would like to address, particularly because it occupies so much of so many people's efforts, is whether there is any value to the endeavor commonly known as "testing QCD."

The answer, I think, is definitely yes, but we must pay attention to precisely what we're doing. At present it seems unlikely but it is always possible that we will see a violation of our expectations that is so gross as to shake our confidence in the assumed underlying principles. More likely, by pursuing this question we can sharpen our theoretical and experimental understanding. This in principle would allow us to identify the first deviations that are a signal of new physics. We also need more guidance from experiment to understand the dynamics of strong interactions. An essential element in almost all speculative thinking, this subject remains largely a mystery.

A propos "testing QCD", I would like to give a plug for lattice Monte Carlo calculations.[20] There are many serious criticisms that can be leveled against present efforts.[21] However, I am optimistic that we will have in the near future a vast variety of predictions from this method that should be good to the 10% level. Some of the quantities that could thus be computed include magnetic moments, g_A's, glueball-quark-meson mixing, and inelastic structure functions (including higher twist).

There are, to date, no "clean tests" of QCD. I am as guilty as many others for having written a paper (or many), which implied as much in its title.[22] At the time we thought we had identified a particularly characteristic interference effect that manifested itself in angular distributions in leptoproduction. But, as with every other "clean test", there are competing effects of comparable magnitude in present experiments[23] (but which asymptotically go like $1/Q^2$). What we really need is an honest assessment of the uncertainties, theoretical and experimental --- which both tend to be quite large for the quantities of interest. At present, many people get needlessly worked up over dramatic successes or failures of QCD which are in fact neither.

If QCD perturbation theory were "exact," i.e., if there were no further corrections to worry about, the as yet uncomputed $O(\alpha_S{}^3)$ effects in the theory lead to an uncertainty in the measured Λ which even optimists would have to admit to $\pm$ 40%.

Of course, perturbation theory (especially as we use it), isn't the whole story. There are several possible areas of possible problems:

1) Even for $Q^2 \to \infty$, it is still debated[24] whether soft exchanges between spectators render the structure functions in hadron-hadron scattering different from those measured in lepton-hadron scattering.

2) Do there exist strong corrections to the perturbative picture of hard scatterings that don't go away like some power of $1/Q^2$? Gupta and Quinn argue yes,[25] but I do not find their logic compelling, even for the hypothetical, simplified model they discuss. But the challenge will exist until we really understand strong interactions.

3) There is an orthodox theory of $1/Q^2$ corrections to the scaling limit. But does it make any sense to talk about power corrections without first summing all logs? The latter are each asymptotically more important than any inverse power of Q^2. Furthermore, strictly speaking, our present calculations are only valid for the asymptotic behavior of amplitudes expanded about $Q=\infty$. I personally believe that power corrections can be defined operationally, and it is worth studying their systematics. However, the whole area is subject to this fundamental criticism.

a) This orthodox theory is fairly well developed for the problem of inclusive leptoproduction. The experimental situation is encouraging in that we are beginning to be in a situation where we can attempt a statistically significant separation of powers from logs.[26] This is not an easy task because the logs come with a great number of adjustable parameters and functions, while virtually nothing is known *a priori* about the structure of the powers. The theory of power corrections to leptoproduction is given by the operator product expansion. While it is understood in principle, no serious phenomenological analysis has been attempted. A popular and thoroughly misleading model of these power corrections attributes them all to an initial parton $k_\perp$ distribution. This may well give the correct order of magnitude, but it ignores such comparable effects as off-shellness, final state interactions, and quantum interference. This naive model also gives the wrong sign for scaling violations for large x.

b) We give considerable attention to final state hadrons, e.g., in e^+e^- annihilation. Power corrections to QCD perturbation theory are invariably implemented using a "hadronization" Monte Carlo, typically of the Feynman-Field or the Lund-string varieties.[27] These models are crucial to present attempts to extract underlying QCD parameters from the data, and most people seem confident of their reliability. After all, if one varies the parameters of the model by $\pm$ 100% or even goes to a different Monte Carlo and one still gets the same range of values for the underlying parameters, what more could one ask? A lot. These schemes are woefully naive and classical. There are many phenomena of comparable importance that are simply not included. So scanning the limited possibilities as described above in no way explores the whole possible range of $1/Q^2$ effects.

Here is just one example of such an effect: Consider "three jet" events. We normally think of this as arising to lowest order via a scaling amplitude in which one of the two initial quarks radiates a hard gluon. The standard Monte Carlo hadronization routines give $1/Q^2$ corrections to the energy flow. However, there are also several independent three-jet amplitudes, with their own energy and angular dependence, that go like $1/Q^2$ relative to the scaling process. Though we do not yet know how to compute them explicitly, pictorially they involve (among other things), the radiation of "diquarks" and "digluons" (neither of which are necessarily color singlets). Since these amplitudes have their own angular dependence, they are in no way simply a dressing of the individual outgoing partons of the scaling amplitude. That is to say that these corrections are *process-dependent* in a way that can't be mocked up by the existing hadronization algorithms.

What to do? I suggest a very phenomenological approach: collect data over a broad enough range to attempt a statistically significant separation of the scaling piece from $1/Q^2$ corrections. If that is not yet possible, I would simply use existing hadronization routines as an estimate of the relevant uncertainty, i.e., compare with no hadronization to see the magnitude of the effect in the quantity under study.

I'd like to point out that even in quantities where direct comparison with perturbation theory is possible, (i.e.,"infrared safe,") hadronization or $1/Q^2$ effects may be important. They are certainly important in present-day analyses of energy flow in e^+e^- annihilation.

The total e^+e^- annihilation cross-section is reputed to be one of the best understood QCD predictions. There, the operator product expansion can be used to estimate the first power correction to straightforward perturbation theory. Aside from calculable heavy quark mass effects, the first such correction to R is of order $(400\text{MeV}/Q)^4$. In fact, using ITEP operator matrix elements,[28] that is probably the value (with a plus sign), to within a factor of two, making it, in practice, rather small. However, one is always near thresholds--- light quark thresholds if not heavy ones. This necessitates smearing in energy-squared by an amount Δ.[29] A safe Δ is presumably some small fraction of Q^2, but no one really knows. This uncertainty, which must also be present in shape predictions, undermines the predictive power of perturbative QCD.

QCD phenomenological analyses are sufficiently complex that it is tempting to lie and/or be lazy about the inclusion of some effects. Things that we already understand in principle are often ignored in practice. Common examples are quark mass effects and the uncertainty of the glue distribution. And sometimes silliness prevails, as in the attempt to add up the electric charge of a jet. If we are to be serious about testing QCD, we must raise our standards.

Old references are not always models of careful analyses. Consider exhibit A.[30] The solid lines show QCD predictions for scaling violations in electroproduction, with the uncertainty reflecting the uncertainty of the glue distribution. No one today would take those 1974 fits seriously. Too many relevant phenomena were ignored. But have we really learned our lessons? Consider exhibit B.[31] (Let me emphasize that I think the experimental work and the method of presentation of ref. 31 are excellent.) The observation of the running or Q^2-dependence of α_s is subject to the same criticisms (and then some), as my 1974 scaling fits. This analysis of $\alpha_s(Q^2)$ came from studying energy moments of the cross section into a cone of variable opening angle. While the data fits the "theory" nicely, it is noted in the top figure that α_s must be varied from 0.16 to 0.18 to fit different moments. Hence one might conclude there is a 10%, i.e., $O(\alpha_s^2)$, uncertainty in α_s. However, the running of α_s, supposedly determined from the same analysis, is itself $O(\alpha_s^2)$. So I would say this is yet another situation where a variety of low Q^2 phenomena came to be described over a limited range by simply readjusting the perturbative QCD parameters.

To summarize briefly all that I've been trying to say: I think the experimental prospects are wide open. All we have to do is try.

References:

1. H. Georgi and S. Glashow, Phys. Rev. Lett. 32, 438 (1974).
2. See e.g., M.K. Gaillard, this conference.
3. A. Sakharov, JETP Lett. 5, 27 (1967); M. Yoshimura, Phys. Rev. Lett. 41, 281 (1978).
4. S. Hawking, "Is the end in sight for theoretical physics?" Cambridge University Press, Cambridge (1980).
5. See e.g., R. D. Field, this conference.
6. E.g., M. Veltman, unpublished.
7. D. Foerster, H.B. Nielsen, and M. Minymiya, Phys. Lett. 94B, 135 (1980).
8. S. Hawking, Nucl. Phys. B144, 349 (1978).
9. M. Grisaru and W. Siegel, Caltech preprint CALT-68-892 (Jan. 1982).
10. E.g., B. Sathiapalan and T. Tomaras, Caltech preprint CALT-68-922 (May 1982).
11. S. Dimopoulos, S. Raby, and L. Susskind, Nucl. Phys. B169, 373 (1980).
12. See J. Preskill, this conference.
13. See L. Abbott, this conference.
14. S. Elitzur, Phys. Rev. D12, 3978 (1975).
15. B.W. Lee, C. Quigg, and H. Thacker, Phys. Rev. Lett. 38, 883 (1977) and Phys. Rev. D16, 1519 (1977).
16. See B. Cabrera, this conference.
17. See W. Fairbank, this conference.
18. G. Kane and J. Leveille, U. of Michigan preprint UM HE 81-68 (Jan. 1982).
19. P. Lepage and S. Brodsky, Phys. Lett. 87B, 359 (1979).
20. E.g., D. Weingarten, Phys. Lett. 109B, 57 (1982); H. Hamber and G. Parisi, Phys. Rev. Lett. 47, 1792 (1982).
21. E.g., G. Bhanot and R. Dashen, Institute for Advanced Study preprint Print-82-0251 (Feb. 1982).
22. H. Georgi and H. D. Politzer, Phys. Rev. Lett. 40, 3 (1978).
23. R. Cahn, private communication.
24. G. Bodwin, S. Brodsky, and P. Lepage, Phys. Rev. Lett. 47, 1799 (1981); A. Mueller, Phys. Lett. 108B, 355 (1982); J. Collins, D. Soper, and G. Sterman, Phys. Lett. 109B, 388 (1982).
25. S. Gupta and H. Quinn, Phys. Rev. D25, 838 (1982).
26. J. Drees, Proceedings of the Bonn Conference on Lepton-Photon Physics (1981).
27. R. Feynman and R. Field, Nucl. Phys. B136, 1 (1978); B. Anderson et al., Z. Phys. C1, 105 (1979).
28. M. Shifman, A. Vainshtein, and V. Zakharov, Nucl. Phys. B147, 448 (1979).
29. E. Poggio, H. Quinn, and S. Weinberg, Phys. Rev. D13, 1958 (1976).
30. H. D. Politzer, Phys. Reports 14C, 129 (1974).
31. R. Hollebeek, Proceedings of the Bonn Conference on Lepton-Photon Physics (1981).

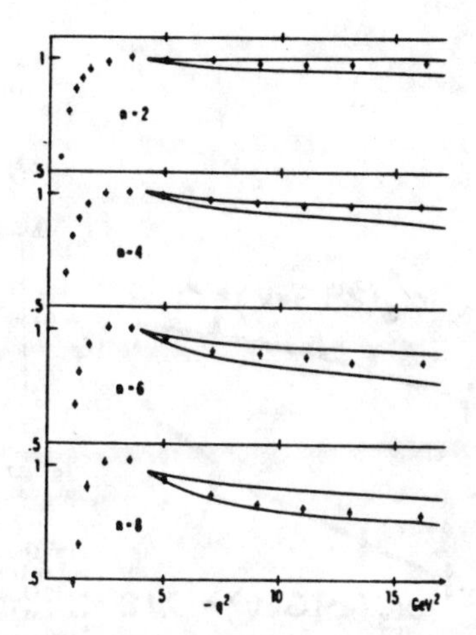

·ig. 9. Plots of $\int_0^1 \nu W_2^{\mathrm{proton}} x^{n-2}\,dx$ for $n = 2, 4, 6, 8$; normalized to one at $-q^2 = 4\ \mathrm{GeV}^2$. The solid lines are the maxi
minimal variation for $\bar{g}^2/4\pi|_{-q^2} = 0.1$.

ied predictions for the violations of scaling. In fig. 9 the predictions for the first few
are sketched for the $SU(3)_{\mathrm{physical}} \times SU(3)_{\mathrm{color}}$ model. The only free p
that $\bar{g}^2/4\pi = 0.1$ at $-q^2 = 4m_p^2$. Because the A^n are not calculable, th
zed to one at $4m_p^2$, and for each moment, there is a minimum and a max
n correspondin the smallest and largest eigenvalues of γ^n. The credib
because they represent a particular of the x v.
lly equivalent va es. Clearly, ov ange in
ically altered an $x' = x +$ he n
n $\bar{g}$ and 1/q^2. Fi
sing fix the

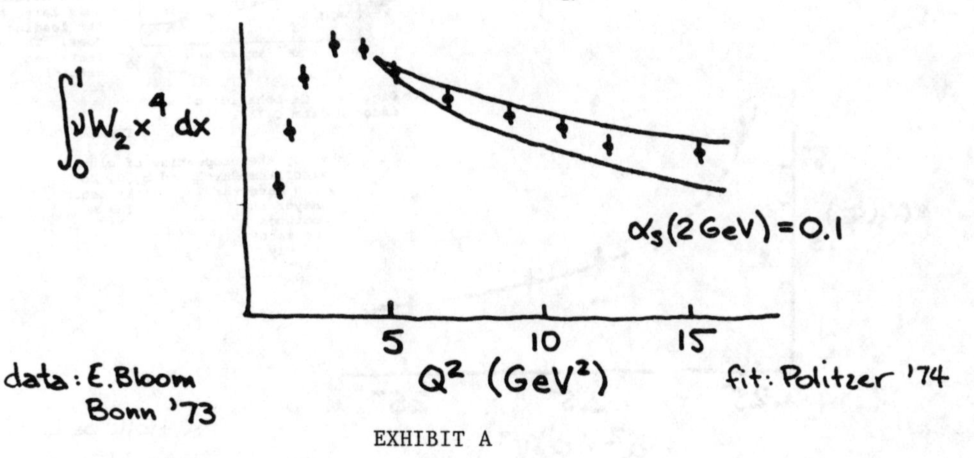

EXHIBIT A

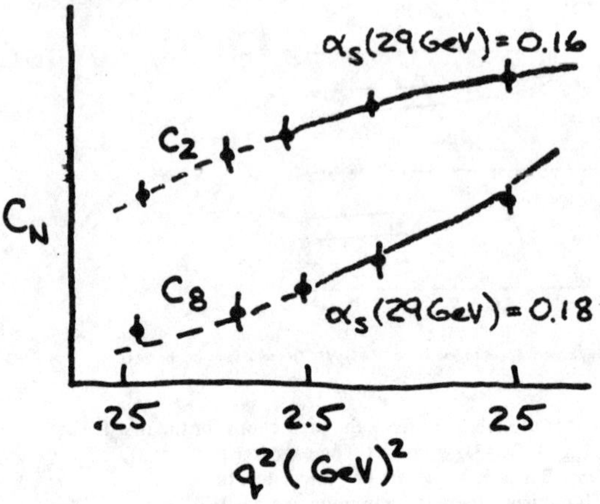

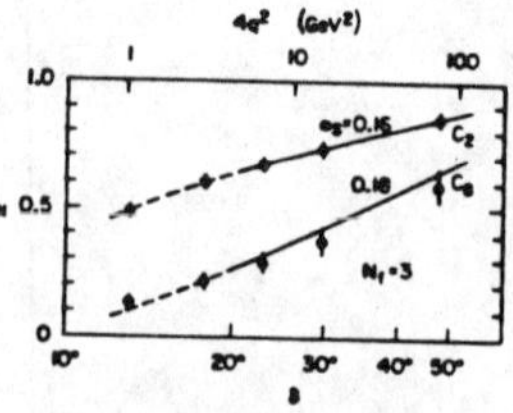

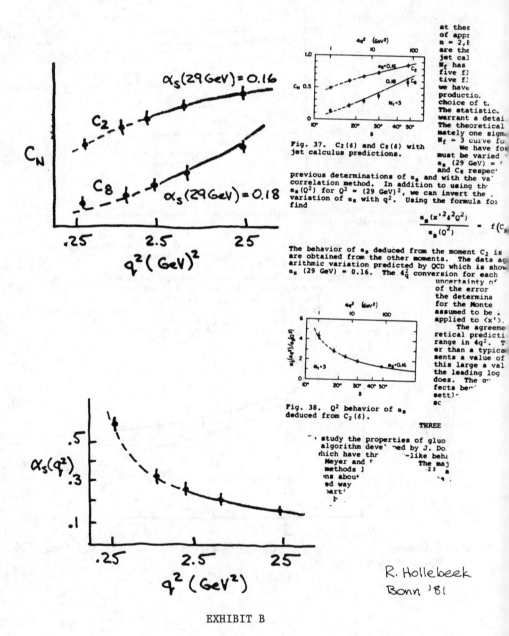

Fig. 37. $C_2(\delta)$ and $C_8(\delta)$ with jet calculus predictions.

at thes
of appr
n = 2,8
are the
jet cal
N_f has
five fl
tive fl
we have
productio.
choice of t.
The statistic.
warrant a detai.
The theoretical
mately one sigm.
N_f = 3 curve fo.
We have fo
must be varied
α_s (29 GeV) =
and C_8 respec

previous determinations of α_s and with the va
correlation method. In addition to using th
$\alpha_s(Q^2)$ for Q^2 = (29 GeV)^2, we can invert the
variation of α_s with q^2. Using the formula for
find

$$\frac{\alpha_s(x'^2\delta^2Q^2)}{\alpha_s(Q^2)} = f(C$$

The behavior of α_s deduced from the moment C_2 is
are obtained from the other moments. The data ag
arithmic variation predicted by QCD which is show
α_s (29 GeV) = 0.16. The 4_q^2 conversion for each
uncertainty of
of the error
the determina
for the Monte
assumed to be
applied to ⟨x'⟩.
The agreeme
retical predicti
range in $4q^2$. T
er than a typica
sents a value of
this large a val
the leading log
does. The o
fects be
settl
sc

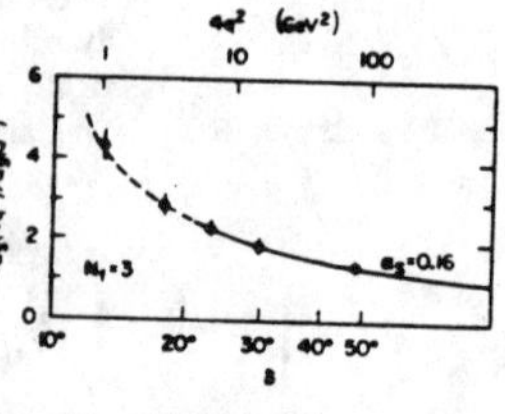

Fig. 38. Q^2 behavior of α_s deduced from $C_2(\delta)$.

THREE

study the properties of gluo
algorithm deve ed by J. Do
which have thr -like beha
Meyer and The maj
methods 23 a
ns about
ed way
part
h

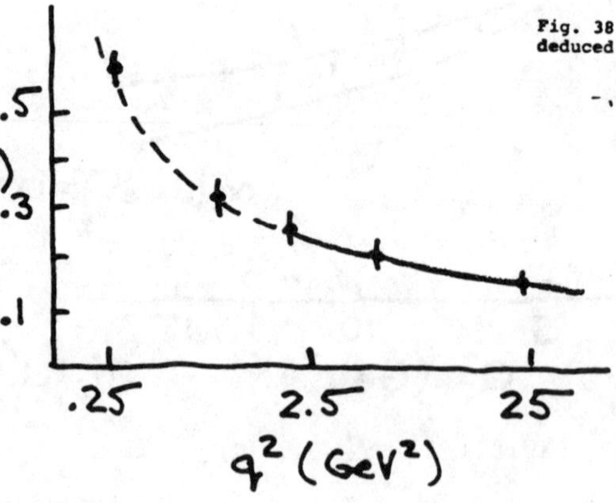

R. Hollebeek
Bonn '81

EXHIBIT B

CONFERENCE PROGRAM

Monday, May 24, 1982
- Welcoming Comments by Provost W. G. Holladay
- Free Quark Searches - (Chair - J. Orear, Cornell)
  - K. Lackner, Los Alamos (Quark Chemistry)
  - R. Bland, San Francisco State (New Results from the SF State Quark Search)
  - S. Freedman, Stanford (Free Quarks Search in High Energy e^+e^-)
  - R. Slansky, Los Alamos (Observable Fractional Charge in Broken QCD)
- Free Quark Searches, Cont'd. (Chair - R. Slansky, Los Alamos)
  - W. Fairbank, Stanford (Status of Results on Fractional Charge)
  - J. Orear, Cornell (Free Quarks in the Early Universe)
  - K. Ziock, Virginia (New Results from Quark Search at University of Virginia)*
  - W. Fairbank, Jr. (Optical Search for Quarked and Superheavy Atoms)*
- Theory (Chair - E. Berger, Argonne)
  - M. Dine, Inst. for Advanced Study (Axions: Visible and Invisible)
- D-Lifetimes (Chair - R. Majka, Yale)
  - R. C. Field, SLAC (Bubble Chamber Measurements)
  - S. Errede, Michigan; G. Bellini, Milano } Electronic and Emulsion Experiments
  - R. Majka, Yale (Summary/Critique)*

Tuesday, May 25, 1982
- Theory (Chair - E. Berger, Argonne)
  - L. Abbott, Brandeis (Composite Models of the Weak Interactions)
- e^+e^- Collisions (Chair - J. DeWire, Cornell)
  - P. Avery, Cornell (Results from CLEO)
  - P. M. Tuts, SUNY Stony Brook (Results from CUSB)
  - E. Hilger, Bonn (Results from PETRA)
  - W. Ford, Colorado (MAC/PEP Results)
  - W. D. Schlatter, SLAC (Mark II/PEP Results)
- Theory (Chair - E. Berger, Argonne)
  - J. Preskill, Harvard (Quark and Lepton Composites)*
- Neutrino and Beam Dump Experiments (Chair - H. Bingham, Berkeley)
  - T. Y. Ying, Ohio State (Beam Dump Experiments)
  - F. Merritt, U. of Chicago (Charm Production Measurements Using Prompt Muons)
  - E. Fisk, FNAL (Total Cross Sections and Structure Functions)
  - P. Igo-Kemenes, Columbia (Charged Current Total Cross Sections in Narrow Band Neutrino Experiments in Bubble Chambers)
  - H. Bingham, Berkeley (Anti-Neutrino Charged Current Total Cross Sections)

Wednesday, May 26, 1982
- Theory (Chair - E. Berger, Argonne)
  - M. K. Gaillard, Berkeley (Grand Unified Theories, etc.)
- Proton Decays (Chair - D. Sinclair, Michigan)
  - M. Marshak, Minnesota (US Experiments)
  - M. Rollier, Milano (European Experiments)

CONFERENCE PROGRAM (Cont'd.)

Monopole Result (Chair - J. Wikswo, Vanderbilt)
W. Fairbank for B. Cabrera, Stanford (From Flux Quantization to Magnetic Monopoles)*
Neutrino Oscillations, Neutrinoless Beta Decay (Chair - S. Csorna, Vanderbilt)
F. Boehm, Cal Tech
High Energy $\bar{p}p$ Collisions (Chair - F. Sannes, Rutgers)
D. Miller, Northwestern ($\bar{p}p$ at ISR)*
M. Yvert, LAPP Annecy } SPS Collider Experiments
H. Mulkens, Brussells, CERN }
Theory (Chair - E. Berger, Argonne)
H. David Politzer, Cal Tech (Puzzles, Problems, Prospects)

*Paper not prepared for publication in these proceedings.

Participants

L. Abbott, Brandeis University
M. S. Alam, Vanderbilt University
P. Avery, Cornell University
B. Barnett, Johns Hopkins University
G. Bellini, University of Milano
E. L. Berger, Argonne National Laboratory
H. H. Bingham, UC/Berkeley
R. Bland, San Francisco State University
F. Boehm, California Inst. of Technology
B. B. Brabson, Indiana University
T. T. Chou, University of Georgia
H. O. Cohn, Oak Ridge National Laboratory
H. Crater, UT/Space Institute
S. E. Csorna, Vanderbilt University
C. Darden, University of South Carolina
J. DeWire, Cornell University
M. Dine, Institute for Advanced Study
E. Eichten, Fermilab
K. Ellis, CERN
S. Errede, University of Michigan
F. L. Fabbri, Frascati
W. M. Fairbank, Stanford University
W. M. Fairbank, Jr., Colorado State University
W. Fickinger, Case Western Reserve University
R. C. Field, SLAC
H. E. Fisk, Fermilab
W. T. Ford, University of Colorado
S. Freedman, Stanford University/Argonne National Laboratory
M. K. Gaillard, UC/Berkeley
M. Goddard, Rutherford Laboratory
E. L. Hart, University of Tennessee
M. Hempstead, Harvard University
R. Hicks, Vanderbilt University
E. Hilger, University of Bonn
W. Holladay, Vanderbilt University
D. Hood, David Lipscomb College
P. Igo-Kemenes, Nevis Labs
A. Jawahery, Syracuse University
L. W. Jones, University of Michigan
J. A. Kadyk, Lawrence Berkeley Laboratory
K. Lackner, Los Alamos
R. L. Lander, UC/Davis
G. Levman, Louisiana State University
D. Lichtenberg, Indiana University
T. Y. Ling, Ohio State University
K. F. Liu, University of Kentucky
P. W. Lucas, Duke University

P. Mackenzie, Fermilab
R. Majka, Yale University
W. A. Mann, Tufts University
M. Marshak, University of Minnesota
T. Maruyama, SLAC
C. J. Maxwell, Rutherford/Appleton Labs
F. Merritt, University of Chicago
M. Mestayer, Vanderbilt University
D. Miller, Northwestern University
H. Mulkens, University of Brussels/CERN
F. A. Nezrick, DOE/Fermilab
J. Orear, Cornell University
R. S. Panvini, Vanderbilt University
H. D. Politzer, California Institute of Technology
J. Preskill, Harvard University
L. K. Rangan, Purdue University
K. Reibel, Ohio State University
M. Rollier, University of Milano
T. A. Romanowski, Ohio State
H. Sadrozinski, UC/Santa Cruz
F. Sannes, Rutgers University
R. Sard, University of Illinois
W. D. Schlatter, SLAC
H. Schneider, University of Karlsruhe
E. I. Shibata, Purdue University
D. Sinclair, University of Michigan
P. Skubic, University of Oklahoma
R. Slansky, Los Alamos
P. Steinberg, University of Maryland
S. Tether, M.I.T.
F. Turkot, Fermilab
P. M. Tuts, SUNY Stony Brook
J. P. Wikswo, Jr., Vanderbilt University
M. Yvert, LAPP, Annecy
K. Ziock, University of Virginia

AIP Conference Proceedings

		L.C. Number	ISBN
No.1	Feedback and Dynamic Control of Plasmas	70-141596	0-88318-100-2
No.2	Particles and Fields - 1971 (Rochester)	71-184662	0-88318-101-0
No.3	Thermal Expansion - 1971 (Corning)	72-76970	0-88318-102-9
No.4	Superconductivity in d-and f-Band Metals (Rochester, 1971)	74-18879	0-88318-103-7
No.5	Magnetism and Magnetic Materials - 1971 (2 parts) (Chicago)	59-2468	0-88318-104-5
No.6	Particle Physics (Irvine, 1971)	72-81239	0-88318-105-3
No.7	Exploring the History of Nuclear Physics	72-81883	0-88318-106-1
No.8	Experimental Meson Spectroscopy - 1972	72-88226	0-88318-107-X
No.9	Cyclotrons - 1972 (Vancouver)	72-92798	0-88318-108-8
No.10	Magnetism and Magnetic Materials - 1972	72-623469	0-88318-109-6
No.11	Transport Phenomena - 1973 (Brown University Conference)	73-80682	0-88318-110-X
No.12	Experiments on High Energy Particle Collisions - 1973 (Vanderbilt Conference)	73-81705	0-88318-111-8
No.13	π-π Scattering - 1973 (Tallahassee Conference)	73-81704	0-88318-112-6
No.14	Particles and Fields - 1973 (APS/DPF Berkeley)	73-91923	0-88318-113-4
No.15	High Energy Collisions - 1973 (Stony Brook)	73-92324	0-88318-114-2
No.16	Causality and Physical Theories (Wayne State University, 1973)	73-93420	0-88318-115-0
No.17	Thermal Expansion - 1973 (lake of the Ozarks)	73-94415	0-88318-116-9
No.18	Magnetism and Magnetic Materials - 1973 (2 parts) (Boston)	59-2468	0-88318-117-7
No.19	Physics and the Energy Problem - 1974 (APS Chicago)	73-94416	0-88318-118-5
No.20	Tetrahedrally Bonded Amorphous Semiconductors (Yorktown Heights, 1974)	74-80145	0-88318-119-3
No.21	Experimental Meson Spectroscopy - 1974 (Boston)	74-82628	0-88318-120-7
No.22	Neutrinos - 1974 (Philadelphia)	74-82413	0-88318-121-5
No.23	Particles and Fields - 1974 (APS/DPF Williamsburg)	74-27575	0-88318-122-3
No.24	Magnetism and Magnetic Materials - 1974 (20th Annual Conference, San Francisco)	75-2647	0-88318-123-1
No.25	Efficient Use of Energy (The APS Studies on the Technical Aspects of the More Efficient Use of Energy)	75-18227	0-88318-124-X

No.26	High-Energy Physics and Nuclear Structure - 1975 (Santa Fe and Los Alamos)	75-26411	0-88318-125-8
No.27	Topics in Statistical Mechanics and Biophysics: A Memorial to Julius L. Jackson (Wayne State University, 1975)	75-36309	0-88318-126-6
No.28	Physics and Our World: A Symposium in Honor of Victor F. Weisskopf (M.I.T., 1974)	76-7207	0-88318-127-4
No.29	Magnetism and Magnetic Materials - 1975 (21st Annual Conference, Philadelphia)	76-10931	0-88318-128-2
No.30	Particle Searches and Discoveries - 1976 (Vanderbilt Conference)	76-19949	0-88318-129-0
No.31	Structure and Excitations of Amorphous Solids (Williamsburg, VA., 1976)	76-22279	0-88318-130-4
No.32	Materials Technology - 1976 (APS New York Meeting)	76-27967	0-88318-131-2
No.33	Meson-Nuclear Physics - 1976 (Carnegie-Mellon Conference)	76-26811	0-88318-132-0
No.34	Magnetism and Magnetic Materials - 1976 (Joint MMM-Intermag Conference, Pittsburgh)	76-47106	0-88318-133-9
No.35	High Energy Physics with Polarized Beams and Targets (Argonne, 1976)	76-50181	0-88318-134-7
No.36	Momentum Wave Functions - 1976 (Indiana University)	77-82145	0-88318-135-5
No.37	Weak Interaction Physics - 1977 (Indiana University)	77-83344	0-88318-136-3
No.38	Workshop on New Directions in Mossbauer Spectroscopy (Argonne, 1977)	77-90635	0-88318-137-1
No.39	Physics Careers, Employment and Education (Penn State, 1977)	77-94053	0-88318-138-X
No.40	Electrical Transport and Optical Properties of Inhomogeneous Media (Ohio State University, 1977)	78-54319	0-88318-139-8
No.41	Nucleon-Nucleon Interactions - 1977 (Vancouver)	78-54249	0-88318-140-1
No.42	Higher Energy Polarized Proton Beams (Ann Arbor, 1977)	78-55682	0-88318-141-X
No.43	Particles and Fields - 1977 (APS/DPF, Argonne)	78-55683	0-88318-142-8
No.44	Future Trends in Superconductive Electronics (Charlottesville, 1978)	77-9240	0-88318-143-6
No.45	New Results in High Energy Physics - 1978 (Vanderbilt Conference)	78-67196	0-88318-144-4
No.46	Topics in Nonlinear Dynamics (La Jolla Institute)	78-057870	0-88318-145-2
No.47	Clustering Aspects of Nuclear Structure and Nuclear Reactions (Winnepeg, 1978)	78-64942	0-88318-146-0
No.48	Current Trends in the Theory of Fields (Tallahassee, 1978)	78-72948	0-88318-147-9
No.49	Cosmic Rays and Particle Physics - 1978 (Bartol Conference)	79-50489	0-88318-148-7

No.50	Laser-Solid Interactions and Laser Processing - 1978 (Boston)	79-51564	0-88318-149-5
No.51	High Energy Physics with Polarized Beams and Polarized Targets (Argonne, 1978)	79-64565	0-88318-150-9
No.52	Long-Distance Neutrino Detection - 1978 (C.L. Cowan Memorial Symposium)	79-52078	0-88318-151-7
No.53	Modulated Structures - 1979 (Kailua Kona, Hawaii)	79-53846	0-88318-152-5
No.54	Meson-Nuclear Physics - 1979 (Houston)	79-53978	0-88318-153-3
No.55	Quantum Chromodynamics (La Jolla, 1978)	79-54969	0-88318-154-1
No.56	Particle Acceleration Mechanisms in Astrophysics (La Jolla, 1979)	79-55844	0-88318-155-X
No. 57	Nonlinear Dynamics and the Beam-Beam Interaction (Brookhaven, 1979)	79-57341	0-88318-156-8
No. 58	Inhomogeneous Superconductors - 1979 (Berkeley Springs, W.V.)	79-57620	0-88318-157-6
No. 59	Particles and Fields - 1979 (APS/DPF Montreal)	80-66631	0-88318-158-4
No. 60	History of the ZGS (Argonne, 1979)	80-67694	0-88318-159-2
No. 61	Aspects of the Kinetics and Dynamics of Surface Reactions (La Jolla Institute, 1979)	80-68004	0-88318-160-6
No. 62	High Energy e^+e^- Interactions (Vanderbilt , 1980)	80-53377	0-88318-161-4
No. 63	Supernovae Spectra (La Jolla, 1980)	80-70019	0-88318-162-2
No. 64	Laboratory EXAFS Facilities - 1980 (Univ. of Washington)	80-70579	0-88318-163-0
No. 65	Optics in Four Dimensions - 1980 (ICO, Ensenada)	80-70771	0-88318-164-9
No. 66	Physics in the Automotive Industry - 1980 (APS/AAPT Topical Conference)	80-70987	0-88318-165-7
No. 67	Experimental Meson Spectroscopy - 1980 (Sixth International Conference , Brookhaven)	80-71123	0-88318-166-5
No. 68	High Energy Physics - 1980 (XX International Conference, Madison)	81-65032	0-88318-167-3
No. 69	Polarization Phenomena in Nuclear Physics - 1980 (Fifth International Symposium, Santa Fe)	81-65107	0-88318-168-1
No. 70	Chemistry and Physics of Coal Utilization - 1980 (APS, Morgantown)	81-65106	0-88318-169-X
No. 71	Group Theory and its Applications in Physics - 1980 (Latin American School of Physics, Mexico City)	81-66132	0-88318-170-3
No. 72	Weak Interactions as a Probe of Unification (Virginia Polytechnic Institute - 1980)	81-67184	0-88318-171-1
No. 73	Tetrahedrally Bonded Amorphous Semiconductors (Carefree, Arizona, 1981)	81-67419	0-88318-172-X
No.74	Perturbative Quantum Chromodynamics (Tallahassee, 1981)	81-70372	0-88318-173-8

No. 75	Low Energy X-ray Diagnostics-1981 (Monterey)	81-69841	0-88318-174-6
No. 76	Nonlinear Properties of Internal Waves (La Jolla Institute, 1981)	81-71062	0-88318-175-4
No. 77	Gamma Ray Transients and Related Astrophysical Phenomena (La Jolla Institute, 1981)	81-71543	0-88318-176-2
No. 78	Shock Waves in Condensed Matter - 1981 (Menlo Park)	82-70014	0-88318-177-0
No. 79	Pion Production and Absorption in Nuclei - 1981 (Indiana University Cyclotron Facility)	82-70678	0-88318-178-9
No. 80	Polarized Proton Ion Sources (Ann Arbor, 1981)	82-71025	0-88318-179-7
No. 81	Particles and Fields - 1981: Testing the Standard Model (APS/DPF, Santa Cruz)	82-71156	0-88318-180-0
No. 82	Interpretation of Climate and Photochemical Models, Ozone and Temperature Measurements (La Jolla Institute, 1981)	82-071345	0-88318-181-9
No. 83	The Galactic Center (Cal. Inst. of Tech., 1982)	82-071635	0-88318-182-7
No. 84	Physics in the Steel Industry (APS.AISI, Lehigh University, 1981)	82-072033	0-88318-183-5
No. 85	Proton-Antiproton Collider Physics - 1981 (Madison, Wisconsin)	82-072141	0-88318-184-3
No. 86	Momentum Wave Functions - 1982 (Adelaide, Australia)	82-072375	0-88318-185-1
No. 87	Physics of High Energy Particle Accelerators (Fermilab Summer School, 1981)	82-072421	0-88318-186-X
No. 88	Mathematical Methods in Hydrodynamics and Integrability in Dynamical Systems (La Jolla Institute, 1981)	82-072462	0-88318-187-8
No. 89	Neutron Scattering - 1981 (Argonne National Laboratory)	82-073094	0-88318-188-6
No. 90	Laser Techniques for Extreme Ultraviolt Spectroscopy (Boulder, 1982)	82-073205	0-88318-189-4
No. 91	Laser Acceleration of Particles (Los Alamos, 1982)	82-073361	0-88318-190-8
No. 92	The State of Particle Accelerators and High Energy Physics(Fermilab, 1981)	82-073861	0-88318-191-6
No. 93	Novel Results in Particle Physics (Vanderbilt, 1982)	82-73954	0-88318-192-4
No. 94	X-Ray and Atomic Inner-Shell Physics-1982 (International Conference, U. of Oregon)		0-88318-193-2